FOOD BIOTECHNOLOGY

Progress in Biotechnology

Progress in Biotechnology 17
FOOD
BIOTECHNOLOGY

Proceedings of an International Symposium organized by the Institute of Technical Biochemistry, Technical University of Lodz, Poland, under the auspices of the Committee of Biotechnology, Polish Academy of Sciences (PAS), Committee of Food Chemistry and Technology, PAS, Working Party on Applied Biocatalysis and Task Group on Public Perception of Biotechnology of the European Federation of Biotechnology, Biotechnology Section of the Polish Biochemical Society
Zakopane, Poland, May 9–12, 1999

Edited by

Stanislaw Bielecki
Technical University of Lodz, Institute of Technical Biochemistry,
Stefanowskiego 4/10, 90-924 Lodz, Poland

Johannes Tramper
Wageningen Agricultural University, Food and Bioprocess Engineering Group,
P.O. Box 8129, NL-6700 EV Wageningen, The Netherlands

Jacek Polak
Technical University of Lodz, Institute of Technical Biochemistry,
Stefanowskiego 4/10, 90-924 Lodz, Poland

2000

ELSEVIER
Amsterdam - Lausanne - New York - Oxford - Shannon - Singapore - Tokyo

ELSEVIER SCIENCE B.V.
Sara Burgerhartstraat 25
P.O. Box 211, 1000 AE Amsterdam, The Netherlands

First edition 2000

Library of Congress Cataloging in Publication Data
A catalog record from the Library of Congress has been applied for.

ISBN: 0-444-50519-9

∞ The paper used in this publication meets the requirements of ANSI/NISO Z39.48-1992
(Permanence of Paper).

Printed in The Netherlands.

Preface

Today it is expected from food biotechnologists that they satisfy many requirements related to health benefits, sensory properties and possible long-term effects associated with the consumption of food produced via modern biotechnology. This calls for an interdisciplinary approach to research, a necessity that can hardly be overemphasised, in view of the current public concern regarding the entire concept of biotechnology.

The aim of the International Symposium on Food Biotechnology held 9-12 May 1999 in Zakopane, Poland, was

1. to assess the impact of biotechnology on food production, and
2. to provide a meeting platform for scientists and engineers, both from academia and industry, involved in all aspects of food biotechnology, including the disciplines microbiology, biochemistry, molecular biology, genetic engineering, agro-biotechnology and food process engineering.

The symposium was organised by the Biotechnology Section of the Polish Biochemical Society and the Institute of Technical Biochemistry, Technical University of Lodz, under the auspices of the Working Party on Applied Biocatalysis, European Federation of Biotechnology, the Task Group on Public Perception of Biotechnology, and the Committee of Biotechnology and the Committee of Food Chemistry and Technology, Polish Academy of Sciences.

Over 120 participants with 86 contributions (oral or poster) attended this scientific event.

Delegates had the opportunity to hear lectures on genetically modified organisms, food processing and novel food products, measurement and quality control, and on legal and social aspects of food biotechnology. The papers included in these proceedings are categorised according to these topics.

During the symposium it became clear that much progress has been made in the last few years as result of the application of modern biotechnology throughout the whole food chain. However, because of lack of functionality in relation to consumer profits, questionable economics and difficult public acceptance, the question of better perspectives for modern biotechnology in that area is still open.

We wish to thank the DSM Food Specialities, The Netherlands, for a sponsorship, which covers the costs of publication of this book.

We hope that the symposium and this book, which contains most papers presented in Zakopane, will make a useful contribution to this key area, i.e. modern food biotechnology.

The Editors
Zakopane, Poland, May 1999

ACKNOWLEDGEMENTS

The Organizing Committee gratefully acknowledges the support of
the following sponsors:

DSM Food Specialities

State Committee for Scientific Research

ICN Biomedicals

Sugar Plant Ostrowy

Silesian Distillery

Brewery Okocim

Sigma-Aldrich

Technical University of Lodz

The Participants of an International Symposium "Food Biotechnology"

Contents

Session B FOOD PROCESSING AND FOOD PRODUCTS

Session C MEASUREMENT AND QUALITY CONTROL

Session D. LEGAL AND SOCIAL ASPECTS OF FOOD BIOTECHNOLOGY

KEYNOTE LECTURE

Food Biotechnology
S. Bielecki, J. Tramper and J. Polak (Editors)
© 2000 Elsevier Science B.V. All rights reserved.

Modern Biotechnology: Food for Thought

Johannes Tramper

Food and Bioprocess Engineering Group, Wageningen Agricultural University,
P.O. Box 8129, 6700 EV Wageningen, The Netherlands

Hans.Tramper@algemeen.pk.wau.nl

Keywords: BST, Chymosin, Soy, Potato Starch, Lactoferrin

I don't want the Cobra event to be seen as anti-biotechnology or anti-science, since it isn't. In the introduction I compare genetic engineering to metallurgy – it can be used to make plowshares or swords. The difference is human intent. *

1. THE BIOTECH CENTURY

Biotechnology is older than written history, dating back as far as 4000 BC when malting and fermentation were practiced in Mesopotamia. Despite of this long history, in a cover story of a 1998 issue of Business Week magazine, the 21st century was nominated as 'The Biotech Century', with biology replacing physics as the dominant discipline. A major switch of leading chemical companies to the life sciences is already occurring. Now, at the end of the 20th century, the discussion about the introduction of transgenic animals and plants, products of modern biotechnology, is fierce, especially when food is involved. In the Daily Telegraph of 10 June 1998, Prince Charles for instance in the article 'Seeds of Disaster' takes a strong position against genetically modified crops. In the mean time he has opened his own web site to discuss these matters.

In contrast to this are expectations such as expressed for instance by Rifkin, president of The Foundation of Economic Trends in Washington, D.C., in his article 'Will Genes Remake the World?' in Genetic Engineering News of 1 April 1998: *An increasing amount of food and fiber will likely be grown indoors in tissue culture in giant bacteria baths, partially eliminating the farmer and the soil for the first time in history.* Animal cloning probably will become commonplace, with 'replication' increasingly replacing 'reproduction', so the farmer

---

*Interview with Preston, writer of 'The Cobra Event', in Genetic Engineering News, March 1, 199

4

on the other hand, may be replacing "indoor cell factories". Monsanto, for instance, recently announced the foundation of Integrated Protein Technologies, a unit formed to produce transgenic pharmaceutical proteins, vaccines and industrial enzymes, initially focussing on plants. According to this company, it takes about two years to produce clinical material and three years for commercial quantities of a protein using a corn system (Genetic Engineering News, February 15, 1999).

The discussion is much less heated if it concerns non-food applications. In particular when a life-saving drug is the target, the voices against are much less loud. The question remains whether efforts should be mostly directed to transgenesis of animals or plants, or to the genetic modification of microbial, plant and animal cells. Functionality, economics and acceptance by society, are obviously the decisive factors. In this paper a non-comprehensive, personal (Dutch) view on these matters is given, using the examples given in the keywords.

2. RECOMBINANT-DNA TECHNOLOGY

As already said, biotechnology is older than written history. However, what has been labeled as modern biotechnology finds its origin much and much later, that is in the second half of this century or, to be exact, in 1973. In that year Stanley Cohen, Annie Chang and Herbert Boyer from Stanford University and the University of California did the first successful recombinant-DNA experiments. They introduced a gene for resistance against kanamycine in a plasmid with resistance against tetracycline. The bacterial strain, *E. coli*, in which copies of that plasmid were added to the authentic genetic material, showed resistance against two antibiotics, i.e. kanamycine and tetracycline. Realizing the enormous impact their finding could have, they first introduced a voluntary moratorium to discuss the potential danger of this new technology before moving on with further experiments.

The first commercial application of this technology followed a little less than a decade, in 1982. The Eli Lilly Company (Indianapolis) then introduced insulin produced by a genetically modified bacterial strain, i.e. also *E. coli*. As a result of that, by way of speaking an unlimited amount of insulin became available for an economic prize. And, in contrast to the old product, without allergic side-reactions. It immediately also made the complexity of the issue clear. A German company developed a similar commercial process, but under pressure of the "Grünen" the company did not get permission from the government to produce. The German diabetics, however, insisted quite rightly on having the superior new product, resulting in the hypocritical situation that production in Germany was forbidden, while the similarly produced product was imported and marketed.

Again a little less than a decade later, at the end of the eighties and the beginning of the nineties, genetically modified, so-called transgenic plants and animals followed. The tomato Flavr Savr and the Dutch bull Herman are the front runners. In Flavr Savr the gene coding for polygalacturonase has been blocked. During ripening of normal tomatoes this enzyme is expressed and degrades pectin, thus softening the fruit and accelerating the rotting process. The latter processes are considerably delayed in Flavr Savr, so that picking can be done when the tomato is fully ripe while the keeping quality is maintained or even better than that of tomatoes prematurely picked. The second example, Herman, is a bull with an extra gene coding for human lactoferrin. The last ten years he has been a topic of fierce debate in the Dutch papers (see below).

In the mean time modern biotechnology has a spin off in the form of a considerable number of realized applications and even more in the pipeline. Biotechnological companies and institutes have introduced new medicines, vaccines, diagnostic tests, medical treatments, environmental-friendly products and food and feed. One of the latest developments is the cloning of adult mammals. The examples given as keywords are worked out in somewhat more detail below.

2.1. rBST

The magazine Genetic Engineering News (GEN) contains a column called Point of View. In the issue of 15 January 1998 this column has the title 'Public Education Still Needed on Biotech' and concerns the opinion of Isaac Rabino, professor in biological and health sciences at State University of New York. What he writes among others is:

The complexity of biotechnology issues can be seen in the production of genetically engineered bovine somatotropin (BST), which increases milk production in cows. Use of BST was opposed by consumer groups, who feared that the mastitis or inflammation of the udder, caused by increased production would result in wider use of antibiotics, which could find its way into the milk supply. ... For these reasons Canada and the European Union put a moratorium on the use of BST.

This ban has been enforced in the EU since 1988.

BST has been an issue of controversy again in the daily papers of the last years, at least in the Netherlands. About three years ago the papers started to express the thought that the importation of meat coming from cows that have been treated with BST could not much longer be hold up. An international committee of scientists, namely, concluded in 1995 that this meat is no risk for human health when the BST is given under strictly controlled conditions. In a new GATT-agreement, signed by both USA and EU also in 1995, it is regulated that hormone-products and genetically modified products can only be prohibited on scientific grounds. To get her rights, the US government approached the World Trade Organization (WTO comes forth from GATT) the end of 1995 with the request to lift the ban. The EU, on the contrary, is trying anything in her power to prevent the BST-meat from coming on the European market. Fast development of the skeleton, accompanied by pain, tumor formation, reduced fertility, increased stress and aggressiveness, all are used as arguments.

Two years later the meat was still not on the market and one could read in the papers then that the fight no longer concerned the meat alone anymore. It had escalated into fight over a possible ban of animal tallow and gelatin used in pharmaceuticals, and, as one might have guessed, BSE and not BST was the issue any longer. In the American papers this was entitled as the 'mad bureaucrats disease' of Brussels. Now the fight in particular concerns six other (sex) hormones, forbidden in the EU, but quite generally used in the USA as growth stimulators. Although the issue is not settled at the moment of writing this, it seems close to an end. Either 'normal' meat will be labeled as hormone free, or the US guarantees that the to-the-EU-exported meat is hormone free. Anyway, we should prevent a hypocritical situation such as the insulin case was in Germany.

Two years ago it could be read in the Dutch papers that the Americans did not worry too much about products coming from BST cows. Nowadays, milk products on which it is clearly stated that they are guaranteed free of recombinant BST can be found on the shelves in the supermarkets. This is clearly in line with the increasing concern of the American society for the products of modern biotechnology as expressed by a law suit filed May 1998 in Washington (source: Greg Aharonian, Internet Patent News Service, May 28, 1998):

A coalition of scientists, public interest organizations and religious groups filed suit against the FDA seeking to have 36 genetically engineered foods taken off the market and asking the FDA be forced to comprehensively test and label such products.

2.2. rChymosin

More than 7 ages before Christ, Homer, poet of the Iliad and Odyssey, the oldest preserved writings of Greek literature, described (without knowing) a simple but very interesting biotechnological experiment. What he wrote is the following. Take a small fig twig and squeeze it. Then stir the squeezed part through milk and what you see is the formation of solid material in the fluid. The fluid can easily be decanted. The remaining solid mass tastes well and can be kept for a longer period than milk. What he described obviously was the making of cheese. What Homer did not know and could not know at that time is that from the squeezed fig twig juice was leaking into the milk. In this juice the enzyme ficin is present, which catalyzes the hydrolysis of the casein in the milk into paracasein and a protein soluble in the fluid. The paracasein micelles agglomerate and a gel is formed.

Also stemming from ancient times is another cheese story. This ancient story is that if a young calf is slaughtered and the rennet-stomach is taken out and filled with milk, a similar phenomenon occurs as with the fig twig: a gel is formed in the fluid. From the stomach wall hydrolytic enzymes, predominantly chymosin, leak into the milk and catalyze the same reaction. For the observers in those early days a magic but usable happening; seen from our perspective one of the first biotechnological applications.

A little understanding of what happened on molecular level originated in the nineteenth century. In this century also a first company was founded that commercialized a standard preparation for the cheese manufacturers. The founder was Christian Hansen who started to buy the rennet-stomach from slaughtered young calves and extract these with salt water. The extract, the so-called rennet-ferment or rennin, is one of the first standardized, industrial products for application in a biotechnological process, i.e. making cheese. Till today the company Christian Hansen is still producing rennin in quite the same manner.

Per ten thousands liters of milk about one liter of rennin is used in cheese manufacturing. That does not look very much, but in the Netherlands alone already about 700.000 tons of cheese are produced yearly, meaning utilization of roughly 1 millions liters of rennin per year. If you realize that, you can imagine that the demand world wide for rennet-stomachs from young calves is very, very high. Therefore rennin always has been scarce and expensive and alternatives have been searched for by industry for a long time already. For instance microbial rennins have been marketed, but without much success due to inferior quality of the cheeses.

In the beginning of the eighties a few companies started to do experiments in which DNA coding for calf chymosin was transformed to microbial strains. This created the possibility to produce authentic calf chymosin by fermentation and this procedure would proof to satisfy a market request, that is

- a product of a constant high quality
- constant availability at a stable price
- cheaper than rennin (rennet-ferment)

About 10 years ago the Dutch company Gist-brocades (presently part of DSM) was the first to market a product with such qualifications and Switzerland was the first country to approve acceptance. Before marketing, first a very extensive testing took place to show absolute safety and superior quality. In the mean time the product is accepted and used in many countries all over the world, while several other companies have come on the market as well with a calf chymosin produced by recombinant microorganisms. France was one of the countries delaying approval for a long time, but under pressure of the BSE issue, it allowed use in 1998. The new source also opened new markets. Cheese produced by recombinant chymosin namely is allowed also for people who eat vegetarian, kosher or halal.

Ironically enough, one of the countries where it is not used yet is the Netherlands. Though late for a country where it was first produced and with such a huge cheese production, acceptance was approved in 1992. Nevertheless, cheese manufacturers in the Netherlands are still not using it in fear that the German people will not buy Dutch cheese any longer. Germany is the biggest market for Dutch cheese and there always has been a strong lobby against products of modern biotechnology, although use of recombinant chymosin has been approved in 1997. Making things even more complex is the fact that much of the cheese imported to the Netherlands is produced by recombinant chymosin.

3. TRANSGENIC PLANTS

Plants are genetically modified for the following three application-oriented reasons:

1. improving output traits, that is to say obtaining a qualitatively better product such as the above-mentioned Flavr Savr. Goals are improving taste, keeping qualities and/or nutritional value, and prevention and healing of diseases and ailments. The science in this field is still in its infancy, but it is just in this field where expectations for the long run are highest. The DuPont Company is generally viewed as the leader in this field. Also a company like Proctor & Gamble is actively exploring this field, for instance with their fat substitute Olestra.

2. improving input traits with the aim to cultivate the plant more easily and economically. Disease and pest resistance, protection against low temperatures, drought and/or frost, immunity against herbicides, these are the properties one aims to give the plant. Monsanto appears to have booked the first big success with this. In 1996 this company has marketed genetically modified soy seed that grows into a soy plant resisting the much-used herbicide Roundup; using this seed, it is claimed that the farmer can suffice with less herbicide. Other big players in this field are Dow Chemical, Novartis, AgrEvo and Zeneca. In the mean time on a significant part of the cultivable land in the USA and Canada such transgenic crops (soy, maize, cotton, etc.) are grown.

3. obtaining plants that produce pharmaceuticals and other high-value compounds. Although plants already since ages yield many ingredients for the pharmaceutical industry - a quart of our medicines contain compounds of plant origin - this field for transgenic plants is still virtually unexplored, certainly in comparison to transgenic animals. First experiments in this field are however promising, moreover as diseases which can be carried over to human form much less a problem than in the case of animals.

3.1. Amylopectin Potato Starch

Recombinant soy was rather abruptly and aggressively (for Dutch standards) marketed by Monsanto in Europe in November 1997, raising a storm of protests. In contrast, the Dutch starch company Avebe introduced step by step over a number of years a transgenic potato without receiving much opposition. In the beginning of the ninetieth this potato was developed in collaboration with the Laboratory for Plant Breeding of the Wageningen Agricultural University. The aim was to obtain an amylose-free potato. In normal potatoes about 80% of the starch consists of amylopectin, branched chains of glucose molecules, while 20% is amylose, merely a linear chain of glucose units. To get a good starch, amylose should be removed by a rather complex, environmental unfriendly, expensive separation process. Therefore alternatives were badly needed.

By means of genetic modification anti-sense DNA of the GBSS-gene was inserted in potato DNA; the GBSS-gene is responsible for the amylose production. After translation the anti-sense RNA binds to GBSS-RNA so that it can not be transcribed anymore, preventing thus amylose production. The pertinent transgenic potato is indeed largely free of amylose. The processing is facilitated to a large extent, the energy consumption reduces by about 60%, pollution halves, while yield increases by 30%; so all very favorable for the transgenic potato.

Starting in contained lab, followed by green house and small blocks of land, commercial production started a couple of years ago. In 1998 the transgenic potato was grown on about 1500 hectares. In that year Avebe also applied in Brussels to get permission to bring this potato on the whole EU market. Even though extensive safety tests had been executed, the EU did not grant permission. Probably as result of all the commotion around transgenic food and fear for accelerated resistance of bacteria against antibiotics, it seems that Brussels has sharpened the rules. For selection reasons, in addition to the anti-sense DNA also a gene construct giving resistance against the antibiotic kanamycin had been inserted, which is quite commonly done. Some researchers now fear that such resistance may be transferred to bacteria, for instance in the stomach of somebody who has eaten the transgenic product. Although there is no proof that this can happen, in the journal Nature a strong position against the use of antibiotic resistance in plants was taken in 1998. Yet it is not resistance against kanamycin what troubles the EU. This antibiotic is not used very much anymore in medical practice and for that reason widely used and accepted in biotechnology. Accidentally, however, another piece of DNA has been inserted as well in the potato, causing resistance against another antibiotic, i.e. amikacin, a very potent antibiotic which is sparsely used by physicians to keep it as a last weapon against bacteria which have become resistant against other, more conventional antibiotics. Even though Avebe clearly states that the potatoes are not intended for human consumption, but for producing starch to be used in the textile and paper industry, the EU requires more proof that this amikacin resistance can not be transferred. For 1999 cultivation has been cancelled in the Netherlands too and Avebe is presently considering whether or not it will continue the cultivation of the amylopectin-potato and where. Clear is that there is a need for only very well defined and controlled changes in DNA. For microorganisms this appears to be possible now (see paper of Groot in this book).

INTERMEZZO

At the end of the millenium, an interesting question to address is whether in the next century transgenic plants (and animals, see below) will largely replace microbial fermentation and cell cultivation. An interesting neck and neck race at the turn of the century is the phytase case (see above The Biotech Century).

Supplementation of the diet with selected enzyme activities may promote a decrease in the overall pollutive effect of animal excreta. This is particularly true in the case of dietary phosphorus, a large proportion of which remains unassimilated by mono-gastrics (main source of this paragraph: Walsh, G.A., Power, R.F. and Headon, D.R., Enzymes in the animal-feed industry, TIBTECH **11** (1993) 424-430). In the region of 60-65% of the phosphorous present in cereal grains exists as phytic acid (myo-inositol-hexaphosphate) which, accordingly, represents the major storage form of phosphate in plants. However, in this form, the phosphate remains largely unavailable to mono-gastrics as these species are devoid of sufficient, suitable, endogenous phosphatase activity that is capable of liberating the phosphate groups from the phytate core structure. The animals' inability to degrade phytic acid has a number of important nutritional and environmental consequences. Phytic acid is considered anti-nutritional in that it chemically complexes a number of important minerals such as iron and zinc, preventing their assimilation by the animal. The lack of available phosphorus also forces feed compounders to include a source of inorganic phosphate (such as dicalcium phosphate) in the feed, with the result that a large proportion of total phosphate is excreted. It has been estimated that in the USA alone, 100 million tons of animal manure is produced annually, representing the liberation of somewhere in the region of 1 million tons of phosphorus into the environment each year. The potential pollutive effect of this in areas of intensive pig production is obvious. Many countries are enacting tough, new anti-pollution laws in an attempt to combat the adverse effect of animal waste on the environment.

Several microbial species (in particular fungi) produce phytases. The incorporation of suitable, microbial-derived phytases in the diet can confer the ability to digest phytic acid on the recipient animals. This would have a threefold beneficial effect: the anti-nutritional properties of phytic acid would be destroyed; a lesser requirement for feed supplementation with inorganic phosphorous would exist; and reduced phosphate levels would be present in the faeces. Several trials have confirmed that the inclusion of phytase in animal feed promotes at least some of these effects. However, the enzyme is not yet used in many countries. This may be explained, in part, by the fact that most microbial species only produce low levels of phytase activity that, obviously, has an effect on the cost of the finished product. It seems likely that widespread utilization of phytase within the industry will only be made possible by the production of this enzyme from recombinant sources, and at least two major enzyme companies are marketing such an enzyme for a number of years now.

Within foreseeable time plants will also produce enzymes on a commercial scale. The Dutch company Plantzyme succeeded in producing the feed enzyme phytase in rape seed. Using modern biotechnology the phytase gene has been introduced into the genetic material of rapeseed. In the rapeseed plant the enzyme is produced in the seeds. The advantage of that is that it can be easily harvested, namely in the same manner as regular rapeseed. Besides the enzyme is packed in the seed, such that it can be stored for several years. The seeds can be added directly to the feed, without having to isolate the enzyme first. Feed for chickens and pigs mainly consists of wheat and barley and it already also contains a small amount of rapeseed. The research is in the final phase before the product goes into the admittance and registration procedure. More information can be found in the paper of Zyla et al. (this book).

4. TRANSGENIC ANIMALS

Animals are genetically modified for the following three application-oriented reasons:
1. to make animals suitable for production of high-value (human) proteins, especially pharmaceuticals. Tens of medicines of transgenic animals are meanwhile in the various stages of clinical tests, among which lactoferrin isolated from the milk of the daughters of Herman. Depending on the application one has to think of herd sizes ranging from tens, such for instance for Factor IX from transgenic sheep or pig milk, to several thousands in case of α-1-anti-trypsin from sheep or goat milk.
2. to make animals suitable as organ donor for xenotransplantations. Xenotransplantation is ethically acceptable so writes a committee of the Dutch Health Council beginning 1998 to Mrs. Borst, minister of health at that time in the Netherlands. This is in line with most other advises in the rest of the world. However xenotransplantation is for the present not allowed as the dangers and risks are too unknown or too little identified yet. Nevertheless the multi-national Novartis is ready since 1996 to start experiments with organs of pigs for use in humans, this in cooperation with the British high-tech company Imutran. It concerns transgenic-pig organs that are supposed to give less immunogenic reactions.
3. to obtain better domestic animals and fishes. Disease and pest resistance, pigs that can metabolize cellulose, pigs with leaner meat, sheep that can utilize cystein-poor feed, frost-resistant salmons (with anti-freeze genes from cold-resistant fish from the Pool seas), gigantic salmons, are examples of topics of ongoing studies. The first transgenic salmon has recently come on the Chinese food market (Twardowsky, personal communication).

4.1. Human lactoferrin (hLF)

Human lactoferrin is an iron-binding protein occurring in mother milk. It plays a role in inflammation and immune reactions. The purified protein has in its iron-free form an inhibiting effect on the growth of a wide variety of bacteria, at least in the test tube; the explanation is that it strongly binds the iron the bacteria need for their growth. Also binding of hLF to the surface of several bacteria with loss of viability has been observed and is ascribed to damage of the cell membrane. Potential applications of hLF are prevention or curing intestinal infections by adding it to anti-diarrhea preparations or to special diets as neutraceutical for premature-born babies and patients with a weakened immune system, among others HIV and cancer patients. Other secondary applications mentioned are 'natural' antibiotic, i.e. as a preservative in among others cremes for wound healing and in products for personal care and contact lens fluids.

Biomedical application requires production of large quantities of hLF. One possibility is the use of transgenic animals. Pharming is presently an independent Dutch company, founded ten years ago as the European branch of the American Gene Pharming, with just that aim and more generally production of high-value pharmaceutical compounds using transgenic animals. To obtain transgenic cows the hLF gene was transferred by micro-injection to an *in vitro* fertilized egg cell. After cultivation in a test tube to a multi-cellular embryonic stage, the transgenic embryos were transferred to carry-mothers. From the 21 pregnancies 19 calves were born. In 2 cases the injected DNA appeared to be inserted in the DNA of the animal. In one case the gene was not found in all tissue types. For the other animal, a bull named "Herman", it was found that 3 copies of the hLF gene are present in all investigated tissue, including sperm. The last 10 years Herman has been the subject of fierce discussions in the Dutch papers.

These discussions first of all triggered acceleration in legislation and in 1992 an advanced law concerning GMO's was implemented (see below). Meanwhile Herman had gotten offspring and the discussion whether or not the daughters of Herman should be allowed to have calves started. Prevention of mastitis, expected to be the result of the hLF in the milk, was not considered a weighty enough reason. In 1994 a special committee investigated the other questions requested to address by law (see below). In their report (important source of information for this paragraph) the committee concluded that as result of still existing uncertainties it was in this stage impossible to come to a solid comparison of the different alternatives. *Production of biomedical proteins is, seen the complexity of some of these proteins, not in all cases possible in microorganisms. Next to production by cell lines and purification of the protein from human material, transgenic animals seem at this moment for (large-scale) production of complex biologically active biomedical proteins technically suitable systems. The lactation system of cows can offer in this respect very good possibilities.*

Although acceptable alternatives were thus not excluded, further experimentation was approved and the daughters of Herman appeared to have the hLF in their milk. Presently clinical studies are under way to investigate the biological activity and safety of hLF. Concerning Herman it was not the end of the story, although according to the contract he should be slaughtered at the end of the experiment with him. After being the subject of national discussion for so many years, it was decided to castrate him, let him die in a natural way and then preserve the body for the museum. Presently, that is May 1999, Herman is still alive and can be visited on the farm.

5. LEGISLATION

Concerning legislation for processes and products of modern biotechnology the Netherlands is rather progressive in comparison to most other countries. Both with respect to processes and products laws are far advanced and clear:

- The laws for making of and working with genetically modified microorganisms and plant and animal cells are fixed, clear and unambiguous.
- For genetically modified, i.e. transgenic plants there is a step-by-step approach: first show in the lab that risks are acceptable, then proceed in the glass house and, finally, execute contained field tests. There is, however, no societal consensus yet on what acceptable risks are.
- For genetically modified, i.e. transgenic animals there is a "No, provided that" law, that is to say:
  only if the aim is weighty enough
  only if there are no unacceptable consequences for human health and well-being
  only if there are no unacceptable consequences for the animals

> only if there are no unacceptable consequences for the environment
>
> only if there are no acceptable alternatives

Like for transgenic plants, there is not yet a societal consensus over the meaning of weighty enough, unacceptable consequences and acceptable alternatives. Ongoing studies and public debates are still needed to further concretize these terms.

Also product legislation is rather progressive: one committee investigates the safety of a product, another determines what has to go on the label. The 'Informal Biotechnology Talks' with representatives of all interest groups, has set up guidelines for labeling of products where modern biotechnology in one way or another has been involved. These guidelines are widely accepted and implemented.

6. CONCLUSIONS

There is no doubt that recombinant microorganisms and transgenic plants and animals are here now to stay. The most important issue at the moment is to increase public acceptance. Education of and continuous communication with the public is essential to change the perception that people have of modern biotechnology. Badly needed to accomplish this is to come up with positive examples, in particular with respect to transgenic plants that are to be used in the food production; positive with respect to consumers, environment and third-world countries. Educating biotechnology students in an interdisciplinary fashion (integration of social and technical sciences) will certainly facilitate this in the future.

ACKNOWLEDGEMENTS

The author likes to thank Prof. Stan Bielecki, chairman of the symposium, and the other members of the organizing committee. They all greatly facilitated the writing of this paper.

GMO IN FOOD BIOTECHNOLOGY

Food Biotechnology
S. Bielecki, J. Tramper and J. Polak (Editors)
© 2000 Elsevier Science B.V. All rights reserved.

New properties of transgenic plants

Niemirowicz-Szczytt K.

Department of Plant Genetics, Breeding and Biotechnology, Warsaw Agricultural University, Nowoursynowska 166, 02-787 Warsaw, Poland

Genetically modified (GM) food has not been produced so far in Poland. However there are new transgenic plants available for experiments and works are in progress to design respectable regulations in this area. The new legislation is indispensable in order to answer the questions of potential consumers such as: could GM crops directly damage health, lead to antibiotic resistant bacteria, increase allergies in humans, create uncontrollable "super weeds", damage the environment, reduce biodiversity and lead to monopolies for large biotechnology companies? Clear and efficient structures and procedures are expected to ensure the supervision of GM products.

In this situation it is important to realise what the characteristics of transgenic plants are and how they differ from naturally modified organisms.

The greatest number of transgenic crops has been obtained in US. The American Animal and Plant Health Inspection Service (APHIS) has approved several crops with new traits (Table 1 and 2). Some of them are still waiting to be accepted.

Table 1
The list of transgenic crops for fresh consumption approved by APHIS (1998)

CROP	TRAIT (number of lines)	INSTITUTION
Chicory	Male sterile (1)	Bejo Zaden BV
Papaya	VR-PRSV (2)	Cornell Univ.
Squash	VR-CMV+WMV2+ZYMV (1)	Asgrow
Tomato	Altered fruit ripening (1)	Calgene, Agritope Monsanto, DNA Plant T.
	Fruit PG (3)	Zeneca+Petoseed
	Lepidopteran resistance (1)	Monsanto

Abbreviations: VR-virus resistance; PG- polygalacturonase

Table 2
The list of transgenic crops for processing approved by APHIS (1998)

CROP	TRAIT (number of lines)	INSTITUTION
Canola	Phosphinothricin tolerance (1)	AgrEvo
(Rapeseed)	Oil profile altered (2)	Calgene
Corn	Glyphosate (1)	Monsanto
		Dekalb G.
	Eur. Corn Borer resistance (1)	Dekalb G.
	Eur. Corn Borer resistance + Glyphosate	Monsanto
	(1)	AgrEvo
	Phosphinothricin tolerance (1)	AgrEvo
	Phosphinothricin t.+ Eur. Corn Borer r. (1)	Plant G.
	Male sterile (1)	
Potato	Colorado potato beetle resistant (7)	Monsanto
Soybean	Phosphinothricin t. (5)	AgrEvo
	Glyphosate (1)	Monsanto
	Oil profile altered (1)	Du Pont
Sugar beet	Glyphosate (1)	Novartis S.
	Phosphinothricin t. (1)	AgrEvo
Cotton	Sulfonylurea t.	Du Pont
	Glyphosate (2)	Monsanto
	Lepidopteran r. (3)	Monsanto
	Bromoxynil t. (1)	Calgene
	Bromoxynil t.+ Lepidopteran r. (2)	Calgene

Abbreviations: r.- resistance; t.- tolerance

The commercialised transgenic crops (grown in temperate climate areas) of which the unprocessed fruits or leaves can be consumed include tomato, chicory and squash. The following properties were modified: virus resistance (TMV, CMV), altered fruit ripening, decreased fruit polygalacturonase level and glyphosate tolerance in tomato, male sterility in chicory and virus resistance (CMV) in squash.

Other transgenic plants that require processing before consumption comprise corn, rapeseed (canola), soybean, potato and sugar beet. In this group the new traits are as follow: Lepidopteran insect resistance, glyphosate and phosphinothricin tolerance in corn, oil profile altered and glyphosate tolerance in rapeseed, oil profile altered, glyphosate and phosphinothricin tolerance in soybean, Colorado potato beetle resistance in potato and glyphosate tolerance in sugar beet. The range of new properties is rather limited. Thus it is obvious that new GM crops are difficult to obtain and commercialise.

A lot of research on transgenic plants is being done and there are many proposals for developing new traits. Transgenic plants offer the potential to be one of the most economical systems for large-scale production of proteins and other products for industrial, pharmaceutical, veterinary and agricultural use.

At the present level of knowledge it is possible to distinguish four basic areas of interest related to food biotechnology:
-   biosafety of transgenic food;
-   quality of transgenic products;
-   transgenic plants as a source of new proteins;
-   edible vaccines.

With new transgenic plants entering production there is a lot of discussion on the safety assessment of new products. Potential users fear that modified food may be harmful to the consumers' health (by introducing new types of excessive amounts of allergens or residues). Scientific experiments conducted so far seem to confirm that transgenic food is not produced yet on such a large scale as it is advertised and proves to have no negative effect on human health. For instance, during the sugar manufacturing from transgenic sugar beets nucleic acid and proteins were eliminated (Klein et al. 1998). This means that sugar made from transgenic plants is free of transgenic residues.

Transgenic maize used as a diet component did not show any adverse effect on the survival or body weight of broiler chickens. Broilers raised on diets prepared from transgenic maize exhibited significantly better feed conversion ratios and improved yield of breast muscle (Brake and Vlachos 1998).

Attempts have been made to monitor the share of glyphosate-tolerant soybeans imported from North America to Japan. It was estimated that transgenic soybean forms approximately 1.1% of the commercial soybeans. This percentage is lower than that announced officially (Shirai et al. 1998).

Another possible method of monitoring food safety consists in establishing databases for assessing the potential allergenicity of proteins used in transgenic food (Gendel 1998). The National Centre for Food Safety and Technology (US) is known to have constructed non-redundant allergen sequence databases that contain all currently available sequence variants for food and non-food allergens and a separate database of wheat gluten protein sequences. The information will be available on-line.

The quality of food (juice, pulp and puree) obtained from transgenic tomato fruits has been tested (Porretta and Poli 1997; Porretta et al. 1998; Errington at al. 1998). Transgenic tomato fruits with reduced polygalacturonase (PG) show better properties (improved viscosity, colour and other sensory attributes) as compared to the control.

It seems that there is a great chance to improve the quality of wheat flour by accummulating high-molecular-weight glutenin subunits (HMW-GS). A number of transgenic wheat lines have already been developed in which the accumulation of the introduced gene product is additive to that of the endogenous HMW-GS (Blechl et al. 1998).

Commercial crops, such as maize, clover, bean and potato can be used as a source of new recombinant proteins. For instance, it was possible to obtain stable expression and accumulation of avidin (chicken egg white) and GUS (beta-glucuronidase) in transgenic maize kernels. The accumulation levels were 5.7% and 0.7% of extractable protein respectively. Biochemical properties of purified avidin and GUS were similar to those of the corresponding native proteins (Kusnhadi et al. 1998). Proteins in maize and other crops can also be improved by the introduction of genes encoding proteins with high methionine content (Tabe and Higgins 1998).

Some experiments confirm that it is possible to improve forage quality by the introduction of modified genes encoding storage proteins into some fodder crops. A modified gene encoding delta-zein was introduced into white clover (Sharma et al. 1998). All the transgenic plants accumulated delta-zein in their leaves. The genes were relatively stable and the accumulation of delta-zein increased with the age of leaves: from 0.3% (in the youngest leaves) to 1.3% (in the oldest leaves) of total water-soluble protein.

Preliminary experiments confirm that antigens produced in transgenic plants, such as potato, banana and lettuce could provide an inexpensive source of edible vaccines and antibodies. The first commercial applications will probably be made in animals. There are two possible ways of using this type of vaccine - either by the consumption of fruits or leaves or by purification of recombinant proteins from plant crude extracts. Transgenic potato with the synthetic gene encoding for the *Escherichia coli* heat-labile enterotoxin B subunit (LT-B) was tested on mice and human volunteers as a source of immunisation against diarrhoea (Tacket et al. 1998, Mason et al. 1998). The results show that a vaccine antigen delivered by an edible transgenic potato was processed by the human immune system. Potato vaccine in higher doses gave similar effects as bacterial vaccine.

The idea of using plants as a biofactory for recombinant antibodies and other proteins provides a great temptation for researchers. At the moment, however, it is difficult to predict the real advantages and disadvantages of this solution. So far no crop containing required protein has been commercialised.

REFERENCES

1. A.E. Blechl, H.Q. Le, O.D. Anderson, K. Muntz,Journal-of-Plant-Physiology, 152 (1998) 703-707.
2. J. Brake and D. Vlachos, Poultry-Science, 77 (1998), 648-653.
3. N. Errington, G.A. Tucker, J.R. Mitchell, Journal of the Science of Food and Agriculture, 76 (1998) 515-519.
4. S.M. Gendel, Advances-in-Food-and-Nutrition-Research,42 (1998) 63-92.
5. J. Klein, J. Altenbuchner, R. Mattes, Journal of Biotechnology, 60 (1998) 145-153.
6. A.R. Kusnadi, R.L. Hood, D.R. Wichter, J.A. Howard and Z.L. Nikolov, Biotechnology-Progress, 14 (1998) 149-155.
7. H.S. Mason, T.A. Haq, J.D. Clements, C.J. Arntzen, Vaccine, 16 (1998) 1336-1343.
8. S. Porretta and G. Poli, International-Journal-of-Food-Science-and-Technology, 32 (1997) 527-534.
9. S. Porretta, G. Poli, E. Minuti, Food-Chemistry, 62 (1998) 283-290.
10. S.B. Sharma, K.R. Hancock, P.M. Ealing, D.W.R. White, Molecular-Breeding, 4 (1998), 435-448.
11. N. Shirai, K. Momma, S. Ozawa, W. Hashimoto, M. Kito, S. Utsumi, K. Murata, Bioscience,-Biotechnology-and-Biochemistry, 62 (1998) 1461-1464.
12. L.Tabe, T. J.W. Higgins, Trends-in-Plant-Science, 3(1998) 282-286.
13. C.O. Tacket, H.S. Mason, G. Losonsky, J.D. Clements, M.M. Levine, C.J. Arntzen, Nature-Medicine, 4 (1998) 607-609.

Food Biotechnology
S. Bielecki, J. Tramper and J. Polak (Editors)
© 2000 Elsevier Science B.V. All rights reserved.

Modulation of carbohydrate metabolism in transgenic potato through genetic engineering and analysis of rabbits fed on wild type and transgenic potato tubers

A. Kulma, G. Wilczyński, M. Milcarz, A. Prescha and J. Szopa

Institute of Biochemistry, University of Wrocław,
Przybyszewskiego 63/77, 51-148 Wrocław, Poland
E-mail: szopa@angband.microb.uni.wroc.pl

Previously we described generation of two types of transgenic potato plants. In one of them the 14-3-3 protein derived from *Cucurbita pepo* was overexpressed, in the other one respective endogenous protein was repressed. Detailed analysis of those plants suggested that the analysed 14-3-3 isoform controlled plant senescence.

In this study the carbohydrate contents, adenine nucleotide level and catecholamine contents in tubers and leaves of transgenic plants grown in a greenhouse and in a field were compared. Overexpressing of 14-3-3 protein led to an increase in catecholamine and soluble sugars contents in leaves and a reduction in tubers size and starch content. The repression of 14-3-3 synthesis led to opposite effect, namely to a decrease in catecholamine and soluble sugars content in leaves and to an increase in tubers size and in starch content. It is proposed that 14-3-3 protein affects carbohydrate metabolism in potato via regulation of catecholamine synthesis. The transgenic potato tubers differing in soluble sugars to starch ratio and in carbohydrate content in tubers were used for rabbit feeding. The increase in body weight correlated with soluble sugars content in potato tubers. There were only slight changes in leucocyte and erythrocyte numbers measured in peripheral blood and also almost no change in hemoglobin content and erythrocyte volume.

1. INTRODUCTION

The development of plant transformation techniques in the last decade has positively affected plant biotechnology and subsequently food production. Almost all agriculturally important plants can be transformed and thus modified. Potato is among those plants, which are most extensively engineered and manipulation concerned mainly carbohydrate metabolism. Besides carbohydrates are the compounds accumulated in storage organs and are the main components of our diet. In addition the carbohydrates provide a carbon skeleton for the synthesis of amino acids, nucleotides and other organic compounds.

So far the manipulations of Calvin cycle enzymes, enzymes involved in starch biosynthesis in photosynthetic tissues and storage organs, enzymes of the major route for chaneling photoassimilates into the cytosol and enzymes active in translocation of photoassimilates from source to sink tissues in transgenic potato plants have been reported [1,2]. The results of these experiments show that one can change plant metabolism through genetic engineering.

The question remains open whether such a change can be specifically directed to altering the quantity of carbohydrates in a desired way.

The response of plant carbohydrate metabolism to genetic engineering of specific enzymes varies from predictable to unexpected. This is due to the existence of alternative metabolic routes within the same tissue.

In the case of a complex metabolic pathway such as carbohydrate synthesis, it is often impossible to manipulate metabolism in a desired way by changing a single enzyme. There are usually alternative pathways by which the action of missing enzyme is compensated. An alternative way of changing metabolic pathway is to affect certain metabolic step indirectly by adding or removing protein influencing the activity the target enzymes. Potential candidate for such a manipulation is the 14-3-3 protein functioning as an adaptor protein for several enzymes such as sucrose-P-synthase and nitrate reductase.

The first 14-3-3 protein has been initially isolated from bovine brain as an abundant, acidic, brain-specific polypeptide. Since this discovery several different functions have been proposed for this class proteins. They are highly conserved and are found in a broad range of organisms including mammals, insects, yeast and plants [3,4].

Many recent findings point out to the participation of these proteins in cell cycle control and gene expression. Members of the 14-3-3 family activate neurotransmitter synthesis, activate ADP-ribosylation of proteins, regulate protein kinase C and nitrate reductase, display a phospholipase A_2 activity and associate with the product of proto-oncogenes, oncogenes and the cdc 25 gene [5]. The broad spectrum of activities that are affected by 14-3-3 proteins suggests that there are several isoforms and each isoform may have its own unique function.

Previously the cDNA from *Cucurbita pepo var. patissonina* has been isolated and sequenced. The nucleotide and deduced amino acid sequences show very high similarity to known 14-3-3 protein sequences from other sources. We have also characterised the transgenic potato plants overexpressing the *Cucurbita* 14-3-3 cDNA [6-7].

Recently six cDNA sequences encoding potato 14-3-3 isoforms were described [8]. The homology between the sequences ranges from 74% to 87%. Western blot analysis revealed in leaf extracts five protein bands of a molecular mass ranging from 32.5 kDa to 26.4 kDa. We also reported that the quantity of particular 14-3-3 isoform present in leaf depends upon age of the tissue. In order to study the importance of 14-3-3 for regulation of plant senescence we repressed the 14-3-3 protein synthesis in potato by expressing the antisense RNA. Analysis of the transgenic plants revealed that the antisense plants lost chlorophyll faster during their growth than the control plants showing accelerated senescence.

In this study we compared carbohydrate contents, adenine nucleotide level and norepinephrine contents in tubers and leaves of transgenic plants grown in greenhouse and in a field. We have found that overexpression of 14-3-3 protein induced an increase in catecholamine contents and soluble sugars in leaves and a reduction in tubers size and starch content. The repression of 14-3-3 synthesis led to an opposite effect, a decrease in catecholamine contents and soluble sugars in leaves and an increase in tubers size and starch content. It is proposed that 14-3-3 protein affects the carbohydrate metabolism in potato *via* regulation of catecholamine synthesis.

The transgenic potato plants which differed in ratio of soluble sugars to starch and in carbohydrate content in tubers were used for rabbit feeding. After eight week raw tubers feeding, the rabbits were bled and the peripheral blood as well as organs were analyzed.

We have noticed an increase in body weight in animals fed on transgenic tubers as compared to control rabbits. There was only slight changes in leucocyte and erythrocyte numbers counts measured in peripheral blood. Almost no changes in hemoglobin content and erythrocyte volume was detected.

2. MATERIALS AND METHODS

2.1. Plant material and bacterial strains

Potato plants (*Solanum tuberosum* L. cv. Desiree) were obtained from „Saatzucht Fritz Lange KG" (Bad Schwartau, FRG). Plants in tissue culture were grown under a 16 h light – 8 h dark regime on MS medium [9] containing 0.8% sucrose. Plants in the greenhouse were cultivated in soil under 16 h light (22°C) - 8 h dark (15°C) regime. Plants were grown in individual pots and were watered daily. Tubers were harvested 3 months after transfer of the tissue culture plants to the greenhouse. Field trials were performed in the vicinity of Wrocław, Poland between April and September of 1996, 1997 and 1998.

Escherichia coli strain *DH 5a* (Bethesda Research Laboratories, Gaithersburg, USA) was cultivated using standard techniques [10]. *Agrobacterium tumefaciens* strain *C58C1* containing plasmid pGV2260 [11] was cultivated in YEB medium [16].

2.2. Recombinant DNA techniques

DNA manipulations were done as described in Sambrook et al. [10]. DNA restriction and modification enzymes were obtained from Boehringer Mannheim (Germany) and New England Biolabs (Beverly, MA). *E coli* strains *DH5a* and *XL1-Blue* were used for bacterial work.

2.3. Construction of sense and antisense 14-3-3 gene

The 1.2 kb *SmaI-Asp718* fragment of plasmid A215 [7] encoding a 14-3-3 protein from *Cucurbita pepo* was ligated in the sense orientation into the *SmaI* site of the plant transformation vector BinAR.

The 1.2 kb *XbaI/Asp718* fragment of plasmid RA215 (EMBL/GenBank database acc no. X87370) encoding a 14-3-3 protein from potato was ligated in the reverse orientation into the *Asp718/XbaI* site of the vector BinAR.

The vectors were introduced into the *Agrobacterium tumefaciens* strain *C58C1:pGV2260* as described before [12] and the integrity of the plasmid was verified by restriction enzyme analysis.

2.4. Transformation of potato

Young leaves of wild-type potato *Solanum tuberosum* L. cv. Desiree were used for transformation with *A. tumefaciens* by immersing the leaf explants in the bacterial suspension. *A. tumefaciens*-inoculated leaf explants were subsequently transferred to callus induction and shoot regeneration medium [12].

2.5. Screening of transgenic plants.

Transgenic potato plants were screened using Northern and Western blot analysis with a 14-3-3 specific cDNA fragment as a probe and antiserum against recombinant 14-3-3 protein, respectively, as described previously [7,13].

2.6. SDS PAGE and protein immunoblotting

Solubilized protein extracts were run in 12% SDS-polyacrylamide gels and blotted electrophoretically onto nitrocellulose membranes (Schleicher and Schuell). Following transfer, the membrane was sequentially incubated with blocking buffer (5% dry milk) and antibody directed against the 14-3-3 recombinant protein (1:2000 dilution). Formation and detection of immune complexes was performed as previously described [14]. Alkaline phosphatase-conjugated goat anti-rabbit IgG served as a second antibody and was used at a dilution of 1:1500.

2.7. Protein extraction

The tissues were powdered in liquid nitrogen and extracted with 50 mM HEPES-NaOH buffer, pH 7.4, containing 5 mM $MgCl_2$, 1 mM EDTA, 1 mM EGTA, 0.5 mM PMSF, 10% glycerol, 0.1% Triton X-100 and 0.2% 2-mercaptoethanol (buffer E) [15]. After a 20 min centrifugation at 13000 rpm (HERAEUS minifuge) and 4°C, the supernatant was used immediately or frozen in liquid nitrogen and stored at -70°C until use.

2.8. Tissue extraction for adenine nucleotides and norepinephrine (protein free extract) assay

The frozen plant tissue was powdered in liquid nitrogen and extracted with 10% TCA. The TCA extract was processed basically as described by Steiner [16]. TCA supernatant was extracted six times with 10 volumes of ethyl ether. The extracted phase was evaporated in a vacuum.

2.9. Determination of starch and soluble sugars

The potato tuber slices and leaf discs were extracted with 80% ethanol-50 mmol/L HEPES-KOH, pH 7.4 at 80°C. The supernatant was used for enzymatic analysis of glucose, fructose and sucrose [17]. For starch measurement extracted plant material was homogenized in 0.2 mol/L KOH and following incubation at 95°C adjusted to pH 5.5 with 1 mol/L acetic acid. Starch was hydrolysed with amyloglucosidase and the released glucose determined enzymatically.

2.10. Catecholamine assay

HPLC assay system (Bio-Rad) has been used which measures all three catecholamines (dopamine, norepinephrine and epinephrine). The method consists of two purification steps and electrochemical detection with the use of Merck-Hitachi HPLC Instrument, model D-7000. In the first step, protein free extract was chromatographed on cation exchange column and finally on analytical reversed phase column type JEC, both columns are included into Bio-Rad kit.

2.11. Animal feeding

Two weeks old rabbits were fed on wild type or transgenic potato tubers obtained from field trials. A half of rabbits daily dose was replaced by potato tubers in the first three weeks it was 150 g per day and then 225 g fresh weight tubers per day up to the end of experiment.

3. RESULTS AND DISCUSSION

The members of the 14-3-3 protein family display several activities, e.g. tyrosine and tryptophan hydroxylase activation, regulation of protein kinase C and endonuclease, phospholipase A_2 activity and binding to the DNA G-box via 67 kDa protein. The broad spectrum of activities that are affected by 14-3-3 proteins suggests that there are several isoforms and each isoform may have its own unique function.

3.1. The 14-3-3 protein content in control plants

When the protein extract of tissue culture or greenhouse growing potato was resolved on SDS PAGE and probed with antiserum against recombinant protein derived from truncated *Cucurbita pepo* cDNA in which highly variable 3'-end was removed [7], five distinct protein fraction with molecular mass ranging from 32.5 kDa to 26.4 kDa were recognized (Fig.1). The gradient of 14-3-3 protein isoforms is clearly visible along the plant stem and we called them now P14-3-3a (32.5 kDa), P14-3-3b (30.4 kDa), P14-3-3c (28.7 kDa), P14-3-3d (27.2 kDa) and P14-3-3e (26.4kDa).

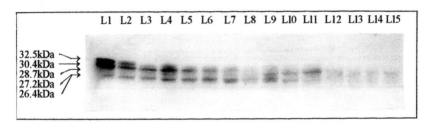

Figure 1. Western blot analysis of 14-3-3 isoforms level in the youngest (L1, 1st leaf from the top) and subsequent leaves from control potato plant. Tissue extracted with buffer E, 40 μg of proteins were applied to each slot. On the right molecular mass is marked.

It was interesting that the level of individual fractions varied depending upon the plant sector examined. It is seen that in the youngest leaf (1st leaf from the top) two protein bands P14-3-3a and P14-3-3b are recognized by the polyclonal antibody against truncated 14-3-3 protein. In the subsequent leaves the expression of both isoforms dramatically decrease and completely disappears by third leaf in case of P14-3-3a isoform and by eigth leaf in case of isoform P14-3-3b. Parallely the isoform P14-3-3c, P14-3-3d and P14-3-3e increases and reach the expression maximum in 9th-11th, 5th-7th and 8th-10th leaves from the top, respectively. In the oldest leaves starting from 12th leaf (from the top) the amount of all 14-3-3 isoforms markedly decreases.

In order to investigate physiological function of these proteins in more details, transgenic potato plants overexpressing 14-3-3 protein called J2 from *Cucurbita pepo* and with repression (J4) of this protein (P 14-3-3c isoform) were obtained.

24

3.2. Sense and antisense constructs and cDNA overexpression and repression

To analyze the function of the cDNA encoding *Cucurbita pepo* and potato P14-3-3c isoform, they were cloned in the sense or antisense orientation into the marked restriction sites of the BinAR vector (Fig.2)

Potato leaf explants were transformed with *Agrobacterium tumefaciens* containing the BinAR-14-3-3 sense or antisense construct. In order to screen and select transgenic potato, leaf samples (8th to10th leaves from the top) were taken from 2 month old transgenic and control plants and analysed for 14-3-3 mRNA content as described [7-8]. Finally, transgenic lines J2.52, J2.11 and J2.28 overexpressing 14-3-3 protein and J4.3, J4.12 and J4.55 with repressed P14-3-3c isoform were selected and further analysed.

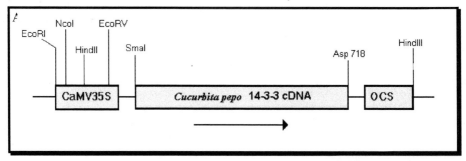

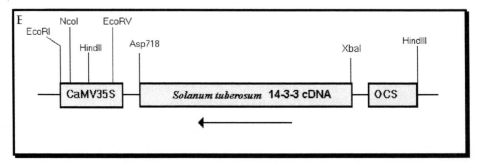

Figure 2. The structure of the chimeric 14-3-3 gene in sense (A) and antisense (B) orientation used for transformation of potato plants (for details see Materials and Methods).

3.3. Analysis of 14-3-3 protein in transgenic plant

The gradient of the 14-3-3 protein isoforms along the stem could be very well seen in the Fig.3 where we compared the immunoblotting of extracts from leaves (3rd, 5th, 9th and 11th from the plant top) of control and transgenic plants with overexpression (J2) and repression (J4) of 14-3-3 protein synthesis probed with antisera anti-14-3-3.

Overexpressed 14-3-3 migrates in SDS PAGE the same as P14-3-3c isoform which was reduced in antisense plants.

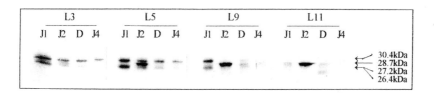

Figure 3. Immunoblotting of protein extracts from 3rd (L3), 5th (L5), 9th (L9) and 11th (L11) leaf from the top of control (D) and transgenic plants with overexpression (J2) and repression (J4) of 14-3-3 protein synthesis probed with antisera anti-14-3-3. Included is also transgenic potato with repression of ADP-ribosylation factor synthesis (J1). Leaf extracted with buffer E, 40 μg of proteins applied onto each slot. On the right, molecular mass is marked.

We also included transgenic plant (J1) with the repression of ARF (ADP-ribosylation factor) protein synthesis [18] which shows an increase in P14-3-3a isoform content as a consequence of ARF protein content decrease [13]. It is interesting to note that leaves of the upper half of ARF antisense plant were smaller and narrower than in controls and strongly resembled the very young (sink) leaves appearing at the top of stem of control plants [13]. This phenotypic observation strongly support the finding that isoform P14-3-3a is primarily synthesized in young leaves and its expression correlates with leaf shape.

3.4. Phenotype of transgenic plants

14-3-3 sense and antisense plants maintained in tissue culture were visually indistinguishable from non-transformed control plants. When grown in the greenhouse, again there was no dramatic change in the phenotype of the aerial parts of the transgenic plants. However plants overexpressing 14-3-3 were flowering earlier than control and the leaves of transgenic potato plants contain more chlorophyll and lose it slower than control plants (Table 1).

Reduction of 14-3-3 content in antisense plants showed opposite effect. The leaves lost chlorophyll faster during their growth than the control plants. It should be pointed out that a significant change in the phenotype of the tubers formed was observed (Table 2).

The sense transgenic plants grown under standard greenhouse conditions showed an increase in tuber number and a decrease in tuber size. The antisense plants showed again opposite characteristic. A decrease in tuber number and increase in tuber size was a characterisitc feature of these plants.

3.5. The 14-3-3 protein and carbohydrate synthesis

There was significant alteration in carbohydrate content of leaves and tubers from plant grown in greenhouse with different levels of 14-3-3 protein. The results presented in Table 3 show that increased levels of 14-3-3 protein significantly affect the ratio of soluble sugars to starch in leaves what resulted from an increase in glucose content and decrease in starch synthesis.

Table 1
Total protein and chlorophyll content in the leaves (from plant top) from control (D) and transgenic potato plants overexpressing (J2)and with repressed (J4)14-3-3 protein grown for 8 weeks in the greenhouse

Leaf number from top		2	4	6	8	10	12	14	16
Total protein	D	100	99.3 ±6.7	82.5 ±6.1	68.4 ±4.3	61.0 ±4.3	39.5 ±3.5	38.9 ±3.5	22.1 ±4.5
	J2	88.5 ±5.8	97.9 ±7.0	85.9 ±7.0	65.0 ±6.0	63.0 ±5.0	39.6 ±4.5	38.0 ±5.0	23.5 ±4.0
	J4	98.5 ±5.8	97.9 ±6.8	81.9 ±6.5	65.0 ±5.1	58.0 ±5.1	37.6 ±4.7	38.9 ±5.0	20.5 ±4.1
Chlorophyll*	D	100	--	98.0 ±7.1	--	87.0 ±5.5	--	77.0 ±6.1	--
	J2	125.9 ±6.1	--	121.4 ±8.4	--	117.0 ±7.5	--	105.1 ±8.1	--
	J4	94.0 ±7.5	--	71.0 ±6.3	--	53.0 ±6.0	--	38.0 ±4.3	--

Values for total protein and chlorophyll are given as percentage of the control. At least two determination from three plants with similar size were performed for each value. The absolute values for total protein and chlorophyll for the control 2nd leaf were 14.9 mg/g tissue and 937 ng/g tissue, respectively.
* determined by A_{667} measurements of acetone (90%) extract.

Table 2
The effect of 14-3-3 manipulation on tuber development. Transgenes with overexpression (J2) and repression (J4) of 14-3-3 protein synthesis and repression of ADP-ribosylation factor were analysed and compared to the control (D)

	Fresh weight per plant (g)	Tuber number per plant	Tuber density (g/cm^3)	Starch (µmole/g FW)	Mean fresh weight per tuber (g)
D	64.9±6	7.3±0.6	1.080±0.004	557±45	8.9
J2	66.1±7	10.5±0.7	1.090±0.004	372±45	6.3
J4	82.1±7	4.7±0.5	1.081±0.004	679±50	17.5
J1	57.4±5	4.9±0.3	1.101±0.005	653±50	11.7

Data were obtained from three to four independent transgenic lines selected from each transgene (J2, J4, J1) and each line was represented by at least six plants

3.5. The 14-3-3 protein and carbohydrate synthesis

There was significant alteration in carbohydrate content of leaves and tubers from plant grown in greenhouse with different levels of 14-3-3 protein. The results presented in Table 3 show that increased levels of 14-3-3 protein significantly affect the ratio of soluble sugars to starch in leaves what resulted from an increase in glucose content and decrease in starch synthesis.

Table 3
Carbohydrate content in leaves of control (D) and transgenic potato plants overexpressing *Cucurbita pepo* 14-3-3 protein (J2) and with repression of either potato 14-3-3 protein (J4) or ADP-ribosylation factor (J1) in μmole /g FW

	Glucose	Fructose	Sucrose	Starch	Soluble sugars per starch	Sucrose per starch
D	2.24±0.4	1.27±0.3	3.31±0.5	4.25±0.5	1.6	0.78
J2	3.36±0.4	1.20±0.3	2.39±0.3	2.60±0.3	2.7	0.92
J4	1.37±0.3	1.20±0.3	2.40±0.3	3.79±0.4	1.3	0.63
J1	2.04±0.3	1.10±0.2	2.83±0.4	4.86±0.5	1.2	0.58

Leaves were harvested after 8h illumination; data were obtained from three to four independent transgenic lines selected from each transgene (J2, J4, J1) and four to six plants represented each line.

The reduction of 14-3-3 protein level in antisense transgenic plants showed opposite effect. In tubers a decrease in starch content of sense transgenic plants and an increase in starch of antisense plants was also detected (Table 3).

In ARF-antisense plants where overproduction of 14-3-3 protein occurs, the ratio of soluble sugars over starch in leaves is similar to 14-3-3 antisense plants. J1 plants accumulated more starch in tuber and produced low tuber number as 14-3-3 antisense did but fresh weight of tuber per plant was lower when compared to the control.

3.6. Field trials of potato transgenic plants

In order to prove the results obtained from transgenic plants grown in standard greenhouse condition the field trials experiment was performed and the data are presented in Table 4.

In three years experiments (1996, 1997 and 1998) the significant changes in several measured parameters was found. The reduction in 14-3-3 protein level resulted in an increase of tuber starch content and fresh weight per plant and of mean fresh weight per tuber and in a decrease of tuber number per plant. Opposite effect for 14-3-3 overexpressing plants was detected. Thus field trials experiment confirm the results obtained for transgenic plants grown in greenhouse condition and this also concerns ARF-antisense plants.

Table 4
Field trials of potato wild type (D) and transgenes with overexpression (J2) and repression (J4) of 14-3-3 protein and with repression of ADP-ribosylation factor(J1)

	Tuber fresh weight(g) per plant			Tuber starch content, % FW			Tuber number per plant			Mean fresh weight(g) per tuber		
	Year of experiment			Year of experiment			Year of experiment			Year of experiment		
	1996	1997	1998	1996	1997	1998	1996	1997	1998	1996	1997	1998
D	635 ±30	539 ±30	646 ±35	13.6 ±0.3	10.4 ±0.2	10.8 ±0.2	11.2 ±0.4	10.4 ±0.4	8.3 ±0.3	56.7	51.8	78.0
J2	700 ±30	476 ±25	623 ±25	12.1 ±0.2	9.5 ±0.1	10.0 ±0.1	13.8 ±0.4	11.4 ±0.3	7.8 ±0.3	50.7	41.8	80.0
J4	763 ±40	578 ±25	789 ±30	14.1 ±0.3	10.6 ±0.2	10.9 ±0.2	6.8 ±0.3	8.3 ±0.3	8.7 ±0.3	112.2	69.6	84.0
J1	540 ±35	534 ±35	522 ±30	14.2 ±0.3	12.8 ±0.2	10.0 ±0.2	9.3 ±0.3	9.2 ±0.3	5.2 ±0.2	68.8	58.0	100.0

For control and transgene 75 plants were grown

3.7. Analysis of catecholamines in potato leaves

The striking change with respect to carbohydrate content in transgenic plants arise the question of how 14-3-3 protein involves carbohydrate metabolism. A possible mechanism through catecholamine regulation was considered. It is known that catecholamines in animal cells regulates glycogen metabolism. Epinephrine and norepinephrine markedly stimulate glycogen breakdown and their action is mediated by cyclic AMP. Further, it is well established that 14-3-3 protein regulates tyrosine hydroxylase activity, the rate limiting enzyme in catecholamine synthesis pathway. With the use of monoclonal antibody anti-tyrosine hydroxylase (Sigma) the respective protein band in plant material was detected and the same band was recognized by anti-phosphothreonine antibody suggesting that tyrosine hydroxylase in phosphorylated state in potato leaves is present [19].

The content of tyrosine hydroxylase does not change when subsequent leaves were analysed which might suggests that the protein is constitutively expressed in potato plant. It should be also pointed out that there was no change in tyrosine hydroxylase content in leaves from control and transgenic plants when analysed by immunodetection (not shown).

Thus we analysed catecholamines content in potato control and transgenic plants. The TLC chromatography of TCA extracts of leaves reveals the presence of bands corresponding to catecholamine standards and the amount of particular catecholamine depends upon the content of 14-3-3 protein [19]. It should be pointed out that epinephrine was not detected however, methylepinephrine was found instead.

It was also proved that culturing of plants on the medium containing radioactive dopamine resulted in its conversion to other catecholamines [19]. The conversion of dopamine to norepinephrine and epinephrine was not dependent on light suggesting, that it is an active endogenous process [19].

The HPLC examination of catecholamines in TCA extracts of leaves from control and transgenic plants was done with the use of Bio-Rad kit commonly used in Clinics for urine examination. The examinations were repeated several times and based on the retention time (RT), dopamine (RT 9.66) and norepinephrine (RT 3.85) can be identified. Epinephrine (RT 4.41) however, was not found but similarly to TLC data methylepinephrine (RT 4.72) was detected. Dihydroxybynzylamine was used as an internal standard and gave retention time around 6.20.

Identical HPLC profile was obtained by Takimoto et al. [20] when extract from *Lemna paucicostata* was analysed. The authors also reported an important effect of norepinephrine on induction of plant flowering. It should be noted that transgenic potato overexpressing 14-3-3 protein flowered earlier than control plants. The content of catecholamines in leaves and tubers in control and transgenic plants measured by HPLC method is presented in Table 5.

Table 5
Norepinephrine content of leaves and tubers from control (D) and transgenic potato plants with overexpression (J2) and repression (J4) of the 14-3-3 protein and repression of the ADP-ribosylation factor(J1)

	Leaves			Tubers		
	Norepinephrine nmol /g FW^{-1}	Soluble sugars per starch	Dry weight %	Norepinephrine pmol /g FW^{-1}	Soluble sugars per starch x 10^{-2}	Dry weight %
D	77.4±8.7	1.6	10.9±0.9	57.6±7.2	4.4	18.90±0.02
J2	95.2±9.8	2.7	8.7±0.08	169.0±15.0	5.6	15.60±0.02
J4	46.7±6.5	1.3	8.4±0.08	5.6±0,6	3.8	17.95±0.02
J1	34.5±4.3	1.2	10.1±0.8	24.0±4.0	3.6	17.90±0.08

Data were obtained from three to four independent transgenic lines selected from each transgene (J2, J4, J1) and each transgenic line and wild type was represented by four to six plants. Leaves were harvested after 8h illumination

The obtained data correspond to those from TLC chromatography. Plants overexpressing 14-3-3 protein produces more catecholamines than control plants. Reduction in 14-3-3 protein level in transgenic plants resulted in a decrease in catecholamine content of leaves. Concomitant with the changes of norepinephrine contents in leaves and tubers ,the ratio of soluble sugars over starch also varies. Less catecholamines, lower ratio of soluble sugars to starch can be observed.

Interestingly, ARF-antisense transgenic plants (J1) contained as low amounts of catecholamine as 14-3-3 antisense plants but, in this plant overproduction of P14-3-3a isoform was detected. Thus it can be speculated that P14-3-3a isoform is not involved in the regulation of catecholamine synthesis and in this transgenic plants another mechanism which reduce catecholamine contents is operating

3.8. Rabbit feeding experiment
Two weeks old rabbits were fed on wild type and transgenic potato tubers obtained from field trials. The animals were weighed weekly (Table 6).

Table 6
Analysis of body weight and basic parameters of peripheral blood of rabbits fed on wild type and transgenic potato tubers

	Body weight			Peripheral blood analysis	
	4 weeks	8 weeks	9 weeks	Protein (g/l)	Glucose(mg%)
D	510±40	770±40	740±40	56.7±6	120±20
J2	620±40	850±40	860±60	64.0±2	97±7
J4	520±30	680±60	620±60	65.1±2	110±30

The highest increase in body weight of rabbits fed on tubers from J2 transgenic plant was observed. Rabbits fed on tubers from transgene J4 showed the lowest increase in body weight. After nine weeks feeding, rabbits were bleeded and peripheral blood blood analysed for protein and sugar content (Table 6) and the results of analysis of few other parameters are presented in Table 7.

Table 7
Analysis of blood morphology parameters of peripheral blood of rabbits fed on wild type and transgenic potato tubers

	Leukocytes $\times 10^3 \mu l^{-1}$	Erythrocytes $\times 10^6 \mu l^{-1}$	Hemoglobin g%	Hematocrit %	Volume of erythrocyte μm^3
D	5.6±0.1	5.6±0.3	11.3±0.3	32.7±0.6	58.3±2.1
J2	3.9±1.6	5.5±0.3	11.9±0.5	33.1±1.7	60.0±0.2
J4	5.7±0.1	4.6±0.5	10.0±0.7	29.8±0.9	65.2±4.4

All measured parameters were quite close to the normal values. There was only slight changes in leukocyte and erythrocyte number counts measured in peripheral blood and also almost no changes in hemoglobin content and erythrocyte volume was detected.

4. CONCLUSIONS

Manipulation of 14-3-3 protein contents in potato plant resulted in significant changes in tuber size and yield. Clearly, overexpression of 14-3-3 protein (P14-3-3c homolog) in potato plants resulted in a decrease of tuber size and starch content. Opposite effect in 14-3-3 antisense transgenic plants (P14-3-3c isoform repressed) was detected. In these plants an increase in tuber size, yield and starch content was detected (Fig. 4).

Somehow unexpected results were obtained from analysis of ARF-antisense plants which contain an increased quantity of 14-3-3 protein. Basically phenotypical and biochemical analysis of these plants strongly resembled rather those plants with 14-3-3 reduction than those with protein overexpression. However, it should be noted that in ARF-antisense plants activation concerns P14-3-3a isoform. This suggest that at least P14-3-3c that was repressed or overexpressed and P14-3-3a isoforms play different function in potato. The mechanism by which 14-3-3 protein regulates carbohydrate metabolism in potato possibly involves regulation of catecholamine synthesis presumably by affecting tyrosine hydroxylase activity. As in animal cells, catecholamines may activate cAMP synthesis which in turn activates starch mobilization, in fact an increase in adenine monophosphate nucleotide level concomitant with a decrease in other adenine nucleotides content was detected (compare Table 5).

Table 8
Adenine nucleotide level (nmole/gFW) in leaves from control(D) and transgenic potato plants with overexpression (J2) and repression (J4) of the 14-3-3 protein and repression of the ADP-ribosylation factor (J1)

	AMP and cAMP	ADP	ATP
D	12.3±3.0	1012.0±90	1186.7±100
J2	13.8±3.0	864.0±60	1024.0±60
J4	18.6±3.0	807.0±124	1005.0±92
J1	26.7±2.9	939.7±80	553.1±70

The measurement were performed exactly as described by Gilmore and Bjorkman[21] using HPLC (BioRad HRLC system) and Vydac 303-NT 405 nucleotide analysis column. Data were obtained from four independent transgenic lines selected from each transgene (J2, J4, J1) and four to six plants represented each line.

Biochemical evidence also indicates that 14-3-3 protein can regulate senescence in plants. Treatments of control plants with alprenolol, the catecholamine antagonist resulted in appearing of senescence symptoms [19]. The examination of transgenic plants grown in a dark strongly support the view that catecholamines play critical role in senescence (Fig. 4). In transgenes overexpressed 14-3-3 protein and thus containing higher level of catecholamines senescence is markedly delayed. Reduction of catecholamine contents resulted in earlier appearing of senescence syndrome.

Obtained transgenic potato plants differed in ratio of soluble sugars to starch and in carbohydrate and protein content in tubers were used for rabbit feeding, during experiment the animals were weighed weekly. After eight weeks of feeding with raw tubers, the animals were bled and peripheral blood and organs were analyzed.

The increase in body weight concomitant with soluble sugars content in potato tubers can be easily noticed. Transgenic plant where repression of P14-3-3c isoform was induced displayed exactly opposite effect on rabbit feeding. There were only slight changes in leukocytes and erythrocytes numbers measured in peripheral blood and also almost no changes in hemoglobin content and erythrocyte volume was detected.

Thus, there is no side negative effect on rabbit fed on transgenic potato and increase in animal body weight seems to be a function of soluble sugars content in food.

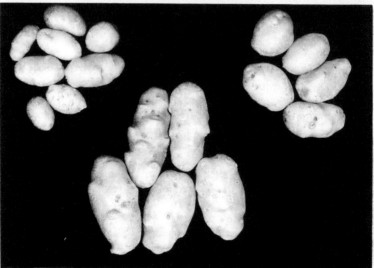

Figure 4. Two months old potato plants were transferred to the dark and grown for two weeks. On the right transgenic plant overexpressed 14-3-3 protein, control plant at the center and with repression of P14-3-3c on the left (upper panel)
Repression of the 14-3-3 protein gene leads to changes in tuber size (lower panel). The largest tuber of five control plants (left), five tubers of different transgenes with 14-3-3 repression (middle) and eight tubers of transgenic lines with overexpression (right) are presented.

ACKNOWLEDGEMENT

The authors wish to thank Professor H. Słowiński and Dr. M. Pytlarz-Kozicka for help in performing of the field trials and B. Dudek for greenhouse work and Professor A.F. Sikorski for linguistic advice. This work is supported by grant No 6P04A04310 from KBN.

REFERENCES

1. T. Rees, TIBTECH, 13 (1995) 375.
2. O.J.M. Goddijn and J. Pen, TIBTECH, 13 (1995) 379.
3. A. Aitken, D.B. Collinge, B.P.H. van Heuseden, T. Isobe, P.H. Roseboom, G. Rosenfeld and J. Soll, TIBS 17 (1992) 498.
4. G. Wilczyński, A. Kulma., E. Markiewicz and J. Szopa, Cell.Mol.Biol.Lett. 2 (1997) 239.
5. A. Aitken, TIBS 20 (1995) 95.
6. J. Szopa, Acta Biochim. Polon. 42 (1995) 183.
7. P.D. Burbelo and A. Hall, Curr. Biol. 5 (1995) 95.
8. E Markiewicz., G. Wilczyński, R. Rzepecki, A. Kulma and J. Szopa, Cell.Mol.Biol.Lett. 1 (1996) 391.
9. G. Wilczyński, A. Kulma and J. Szopa, J. Plant Physiol., 153 (1998) 118.
10. T. Murashige. and F. Skoog, Physiol. Plant. 51 (1962) 493.
11. J. Sambrook, E.F.Fritsch, and T. Maniatis, Molecular Cloning: A Laboratory Manual, 2nd ed. (Cold Spring Harbor, NY: Cold Spring Harbor Laboratory Press), 1989.
12. R Deblaere., B. Bytebier. H. de Greve, F. Deboeck, J. Schell, M. Van Montagu and J. Leemans, Nucl. Acids Res. 13 (1985) 4777.
13. M Rocha-Sosa, U. Sonnewald, W. Frommer, M. Stratmann, J. Schell and L. Willmitzer, EMBO J. 8 (1989) 23.
14. G. Wilczyński, A. Kulma., A.F. Sikorski and J.S zopa, J. Plant Physiol. 151 (1997) 689.
15. J. Szopa and K.M. Rose, J. Biol. Chem. 261 (1986) 9022.
16. H.E. Neuhaus and M. Stitt, Planta 182 (1990) 445.
17. A.L. Steiner, Methods in Enzymology 38 (1974) 96.
18. M.R. Stitt, R.M.C. Lilley, R. Gerhardt and H.W. Heldt, Methods in Enzymology, 174 (1989) 518.
19. J. Szopa and B. Mueller-Roeber, Plant Cell Rep. 14 (1994) 180.
20. G. Wilczyński, A. Kulma, I. Feiga, A. Wenczel and J. Szopa, Cell. Mol. Biol. Lett., 3 (1998) 75.
21. A. Takimoto., S. Kaihara, M. Shinozaki, J. Miura, Plant Cell Physiol. 32 (1991) 283.
22. A.M. Gilmore and O. Bjorkman, Planta, 192 (1994) 526.

Food Biotechnology
S. Bielecki, J. Tramper and J. Polak (Editors)
© 2000 Elsevier Science B.V. All rights reserved.

Transgenic plants as a potential source of an oral vaccine against *Helicobacter pylori* [*]

R. Brodzik[a], D. Gaganidze[a], J. Hennig[a], G. Muszyńska[a], H. Koprowski[b], A. Sirko[a#]

[a]Institute of Biochemistry and Biophysics, Polish Academy of Sciences, Pawińskiego 5A, 02-106 Warsaw, Poland

[b]Biotechnology Foundation Laboratories at Thomas Jefferson University, 1020 Locust St., Philadelphia, PA 19107, USA

Plants are one of several novel hosts that can be used for an inexpensive production of recombinant biopharmaceuticals such as vaccines. There is evidence that oral immunization with *Helicobacter pylori* urease, or its large subunit encoded by the *ureB* gene in the presence of oral adjuvants, is a feasible strategy for the development of a vaccine against this gastric pathogen. In this study we report on cloning and expression of the *H. pylori ureB* gene in a low alkaloid line of tobacco (*Nicotiana tabacum* cv. LA Burley 21) and in the cells of carrot callus (*Daucus carota* cv. Dolanka). The evaluation of the UreB level in transgenic tobacco plants and transformed carrot callus was accomplished and revealed expression levels sufficient for an immunological study in laboratory animals.

1. INTRODUCTION

Plants are attractive systems for the production of heterologous proteins. Many examples of plants producing transgenic proteins of industrial and pharmaceutical significance can be found in literature [1]. In plants, the upstream production costs are lower than in other systems. Most of the foreign proteins are not produced in plants on a very high level, however, there are many potential ways to increase the level of expression of transgenes *e.g.* by the use of appropriate promoters, enhancers and leader sequences, by the optimization of codon usage and removing of mRNA-destabilizing sequences and targeting proteins to different cellular compartments [2]. Downstream processing i.e. purification of proteins from plant biomass is assumed to be difficult and expensive, mostly because of the low ratio of recombinant protein to total biomass [3]. An attractive possibility to avoid costly a purification procedure is to use edible plant material as food and feed in order to deliver

* This research was financially supported by a grant from the State Committee for Scientific Research (KBN), project No 6P04B01112 and NATO Collaborative Research Grant No HTECH CRG 960885. We thank Dr. V. Yusibov from Biotechnology Foundation Laboratories, Thomas Jefferson University, Philadelphia, USA for collaborating with us on this project.
# for correspondence: tel. +4822 658 3848, fax: +48 39 121 623, e-mail: asirko@ibb.waw.pl

recombinant proteins to humans and animals. Such approach is especially useful for the development of plant-produced oral vaccines against bacterial and viral pathogens [4, 5]. In this case, in addition to such obvious advantages of the plant production system like safety (because of lack of contamination with animal pathogens), a relative ease of geneticmanipulation and cheap production, plant cells and plant viruses are expected to provide natural protection for the passage of antigens through the gastrointestinal tract which might enhance the mucosal immunogenicity of the tested antigen. Several important protein antigens, including binding subunit of *Escherichia coli* heat-labile enterotoxin [6], binding subunit of cholera toxin [7], hepatitis B surface antigen [8, 9], rabies virus glycoprotein [10] and Norwalk virus capsid protein [11] have been expressed in transgenic plants. This list is not yet impressively long, however, most of the data prove the usefulness of such an approach.

Helicobacter pylori is a causative agent for chronic active gastritis and peptic ulcer disease [12]. Since middle 1980s when direct evidence for the association between *H. pylori* infection and gastroduodenal disease was provided significant progress in our knowledge of mechanisms of pathogenesis [13] and immunopathology [14] of this organism has been made. The complete genome sequence of this pathogen has been established recently [15]. It is now possible to claim that vaccination against *H. pylori* is feasible [16]. There are several candidate antigens for prophylactic and even therapeutic vaccination against this gastric pathogen including urease and its large subunit - UreB [17,18,19]. Urease (EC 3.5.1.5., urea amidohydrolase) catalyses the hydrolysis of urea to ammonia and CO_2. This allows the bacterium to buffer its microenvironment when it is exposed to the low pH in the gastric lumen. In addition, hydrolysis of urea might provide bacteria with a source of nitrogen [20].

We decided to chose this *H. pylori* antigen as a first candidate for anti-*Helicobacter pylori* vaccine production in plants because it was the best known candidate for such a vaccine at the moment of starting our experiments. A recombinant, enzymatically inactive urease was successfully tested in different animal models [21] and recently underwent Phase II clinical trials in human volunteers [22]. It is worth to mention that urease is a protein frequently found in plants. It is present both in seeds [23] and in vegetative tissues [24]. Both types of urease (called embryo-specific and ubiquitous, respectively) were proposed to be involved in nitrogen recycling. Additional possibilities like its role in protection against plant pathogens has been also considered [25]. Moreover, UreB of *Helicobacter pylori* containing 3.6 mole percent methionine might be a good candidate for increasing methionine content and nutritive values of plant tissue when accumulated in the high amount.

We report production of *Helicobacter pylori* UreB protein in transgenic tobacco plants on a level sufficient for using such transgenic plants as a food supplement for laboratory animals. The content of *H. pylori* UreB in the analyzed transgenic plants is approximately 0.06%-0.1% of the total plant protein extract. It means that in 300 mg of lyophilized leaf mass there is approximately 50 µg of UreB. In attempts to obtain the UreB expression in edible crops we performed carrot transformation. The expression on a protein level was approximately 50 µg of UreB in 1 g of the wet transgenic carrot callus cells.

2. MATERIAL AND METHODS

2.1. Plant material

Low alkaloid line LA Burley 21 tobacco (*Nicotiana tabacum*) [26] (seeds obtained from G. Collins) was used for transformation experiments. *Nicotiana benthamiana* (seeds obtained from V. Yusibov) was used for the transient *in planta* assay of *ureB* expression. The plants were grown in MS medium or in soil in a growth chamber under controlled light and temperature conditions (16 hours light, 3000 Lx, 22°C and 8 hours darkness 18°C). For all experiments, 6 to 9 week-old plants were used.

2.2. Bacterial strains

Escherichia coli DH5αF' strain grown at 37°C in LB [27] was used for all cloning procedures. *Agrobacterium tumefaciens* strain LBA 4404 grown at 28°C in YEB medium (beef extract 5g/l, yeast extract 1g/l, peptone 5g/l, sucrose 5 g/l, MgSO$_4$ 2mM, pH7.2) was used for plant transformation and *in planta* transient assay experiments. The media contained appropriate antibiotics. Their concentrations were (μg/ml): for *E. coli*, ampicillin, 100; kanamycin, 25; for *A. tumefaciens*, kanamycin, 100; rifampicin, 10.

2.3. Oligonucleotides and PCR

The following oligonucleotides were used in this study: UREB1 (5'-GCG GTA CCC CAT GGA AAA GAT TAG CAG AAA AGA A-3'), UREB2 (5'-GCG TCT AGA CTA GAA AAT GCT AAA GAG TTG-3'). The typical PCR reaction for screening of the transformed plants was performed in a 1X Promega Reaction Buffer [10 mM Tris-HCl, pH 8.4, 50 mM KCl, 0.01% (w/v) gelatin] and contained about 50 ng of the total plant DNA , 200 nM primers, 1.5 mM MgCl$_2$, 25μM dNTPs and 0.2 units of Taq polymerase (Promega) in a 25 μl volume.

2.4. Plasmid constructions

All manipulations were done according to standard techniques [27]. Plasmid containing *ureA* and *ureB* genes of *Helicobacter pylori*, pHP902, [28] kindly provided by H. Mobley was used as a template in the polymerase chain reaction (PCR) with specific primers to *ureB* gene. Upstream UREB1 and downstream UREB2 primers possess on their 5'-ends 14-nt and 9-nt sequences, respectively. These sequences are not complementary to the template and contain restriction enzymes recogniction sites. Including these sequences enabled proper cloning of *ureB* into the plasmid vectors. The resulting PCR product (1.7 kb) containing the coding region of *ureB* was digested with *Kpn*I and *Xba*I, cloned into pUC19 and sequenced to confirm the accuracy of Taq polymerase. The resulting plasmid (pAR3) served as a source of DNA for further cloning experiments. In the first step the pAR8 was obtained by cloning of a (partially digested with *Nco*I) 1.7 kb *Nco*I-*Xba*I DNA fragment into pRTL2 [29]. Subsequently, a *Xho*I-*Pvu*II 2.2kb DNA fragment from pAR8, (partially digested with *Pvu*II) containing *ureB* proceeded by tobacco etch virus (TEV) leader and followed by transcriptional terminator, was cloned into pBISN1 [30] from which the coding region of *uidA* gene was removed by digestion with *Sal*I and *Sma*I. In the resulting plasmid, pAR14 (Figure 1), the *ureB* gene is expressed from the chimeric octopine-mannopine promoter. The coding *ureB* region is proceeded by TEV leader and followed by two signals of transcription

38

termination and polyadenylation, one originated from 35S RNA of cauliflower mosaic virus (CMV) and the other from *nos* gene of *Agrobacterium*.

2.5. Transient expression in *Nicotiana benthamiana* leaves

Fresh overnight culture of *Agrobacterium tumefaciens* containing the appropriate binary plasmid was incubated for 2 hours with 20 μM acetosyringone (Sigma). Bacteria were pelleted, washed in sterile water twice (to remove antibiotic) and suspended in 5-10 ml water at concentration 10^9cfu/ml. Plants of *Nicotiana benthamiana* 8 week-old grown in soil with fully expanded leaves (about 5 cm in diameter), were inoculated using a 5-ml syringe without needle by placing the syringe against the underside of the leaves and gentle pressing. Usually 2-3 leaves of one plant were inoculated with one bacterial culture. Leaves were harvested and analyzed for the transgene expression after 2 days.

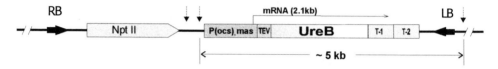

Figure1. Schematic map of a plant expression cassette used in this study. The *Eco*RV sites are indicated by vertical arrows and the 5-kb *Eco*RV-*Eco*RV fragment containing *ureB* is marked below the map. LB and RB, the left and right borders of the T-DNA; NptII, the expression cassette for kanamycin resistance; P(ocs)₃mas, the chimeric strong promoter; TEV, the non translated leader from tobacco each virus; UreB, the coding region of B subunit of *H. pylori* urease; T-1 signal for transcription termination and polyadenylation from 35S RNA of cauliflower mosaic virus (CMV); T-2, polyadenylation signal from *nos* gene. The size of mRNA containing *ureB* is indicated.

2.6. Transformation and regeneration procedure

Low alkaloid line LA Burley 21 [31] of tobacco (*Nicotiana tabacum*) seeds, kindly provided by G. Collins, were surface sterilized and grown in vitro on MS plates [32]. The 2-3 week-old seedling were transformed according to the published procedure [33].

The hypocotyls of two week-old dark grown seedlings of carrot (*Daucus carota* cv. Dolanka) were transformed according to the published procedure [34].

2.7. Molecular analysis of the transformed plants

The total plant RNA was extracted from the frozen leaf material using the cold phenol method as described previously [35]. Northern analysis of RNA (20 μg) separated on a 1.2% agarose was performed according to the standard procedure [27]. The digoxigenin-labeled PCR fragment obtained with UREB1 and UREB2 primers, and pAR8 as a template was used as a probe. RNA transferred to the nylon membranes was stained [36] and photographed prior to hybridization.

For protein analysis about 50 mg of tobacco leaves were homogenized in 200 μl of a standard Laemmli buffer [37] in an eppendorf tube using a mini mortar and a pinch of carborrundum. Samples were boiled for 5 minutes, chilled on ice for 2 minutes and microcentrifuged for 5 minutes. The same volume (20 μl) of each sample was loaded on a SDS-PAGE gel (12.5%) along with the appropriate controls. The separated proteins were transferred to the nitrocellulose membrane and the Western blot analysis was performed according to the standard procedures [27] using rabbit polyclonal antibodies against UreB.

Quantitative analysis of the results of Northern and Western blots was performed with LKB Ultrascan XL densitometer and the GelScan XL ver.2.1 (Pharmacia-LKB) and GelScan ver.1.12 (Kucharczyk TE) programs.

3. RESULTS AND DISCUSSION

3.1. Construction of the *ureB* expression cassette and its verification in a transient *in planta* assay

Before proceeding to a labor intensive stable plant transformation we decided to perform a transient assay of *ureB* expression from pAR14 (Figure 1) in order to check whether the whole construct is functional in plants and the signals are correctly recognized by the plant transcription/translation machinery. The results of the transient assay in *Nicotiana benthamiana* leaves confirmed that the plasmid was correctly designed and constructed (Figure 2). The protein of the expected size (about 60 kDa) recognized by rabbit polyclonal anti-UreB antibodies was detected in the analyzed material (Figure 2, lane 3). Such protein is absent in both controls i.e. leaves inoculated with water (Figure 2, lane 1) and leaves inoculated with *Agrobacterium* harboring vector pBISN1 (30) containing *uidA* gene instead of *ureB* (Figure 2, lane 2).

3.2. Analysis of ureB expression in transgenic tobacco plants

Tobacco was chosen as a transgenic system because of its ease of transformation, regeneration and transgene analysis. However, keeping in mind the possibility of future immunological (feeding) experiments we decided to transform a low alkaloid line of tobacco, (cv. LA Burley 21; [31]) containing 0.2-0.5% of alkaloids on a dry weight basis, to reduce harmful side effects of alkaloids present in the tested plant material. The 70 kanamycin-resistant plants were screened by PCR for the presence of the intact *ureB* gene using UREB1 and UREB2 primers. Only 28 plants gave the fragment of the expected 1.7 kb size (results not shown). Using the probe containing the coding region of *ureB* for the Northern blot analysis we were able to detect *ureB* mRNA in 19 (per 28 tested) kanamycin resistant plants containing intact *ureB* gene. The size of a hybridized transcript was in an agreement with the predicted size of *ureB* mRNA (2.1 kb). The results of the Northern analysis presented in Figure 3 are a compilation of four blots. Total plant RNA from transformant 1 served as an intensity control for each Northern blot. All transformants with detectable *ureB* mRNA were next probed for transgene expression on the protein level by using rabbit polyclonal anti-UreB antibodies. We were able to detect UreB in a total protein extract isolated from the leaf tissue of six analyzed plants (Figure 4). We did not notice any significant correlation between the level of mRNA and the amount of UreB protein detected. Plant 32, containing the highest level of UreB protein had a very moderate level of *ureB* mRNA. Most probably the lower amount of a transcript helps to avoid the post-transcriptional silencing [38] leading in consequence to the higher protein amount.

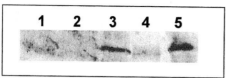

Figure 2. Results of a transient *in planta* assay of *ureB* expression in *Nicotiana benthamiana* leaves. UreB protein was immunodetected two days after inoculation using rabbit anti-UreB antibodies. Total soluble proteins from the *N. benthamiana* leaves inoculated with H_2O (lane1), and inoculated with *Agrobacterium* transformed with pBISN1 (lane 2) and pAR14 (lane 3) were loaded on a PAGE-SDS along with the protein extract from *E. coli* transformed with pHP902 (lane 4) and a plasmid producing UreB protein with additional His tag (lane 5). The expected size of UreB is 61.1 kDa while the expected size of His-UreB is 66 kDa, however, both proteins migrate slightly slower than expected.

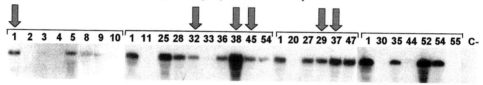

Figure 3. Northern blot analysis of transgenic plants using coding region of *ureB* as a probe. The plant material was harvested 3-5 weeks after transferring plants into soil. This is a compilation of four separate blots. The plant numbers are indicated and the brackets separate individual blots. 20 μg of total plant RNA was loaded on each well. RNA from the plant transformed with pBISN1 was used as a negative control (C-). The vertical arrows point to the plants producing detectable on immunoblots levels of UreB protein.

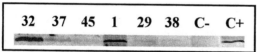

Figure 4. Western blot analysis of protein extract from selected (numbers indicated) transgenic plants using anti-UreB antibodies. The negative control (C-) is a kanamycin-resistant transgenic plant transformed with pBISN1. The positive control (C+) is a protein extract from *Escherichia coli* transformed with pHP902.

3.3. Carrot transformation and analysis of *ureB* expression in the carrot calli

In order to obtain an expression of *ureB* in a plant more appropriate then tobacco for a feeding experiments, the *Agrobacterium*-mediated carrot transformation was performed. Several obtained kanamycin-resistant callus lines were analysed for transgene expression on the protein level by using anti-UreB antibodies (Figure 5). According to our evaluation, in the case of the "most effective" callus lines, approximately 50 μg of UreB is present in 1g of the wet mass. The regeneration of transgenic carrot plants and their genetic and molecular analysis will be our aim in the nearest future.

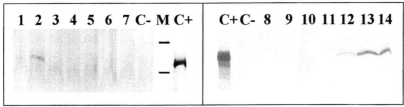

Figure 5. Western blot analysis of protein extract from transformed carrot callus lines using anti-UreB antibodies. C- is the protein extract from non transformed callus. C+ is a positive control as in Figure 4. M is the molecular weight marker with indicated positions of 97.4 and 66 kDa. Numbers refer to the callus tissues originated from the separate transformation events. The UreB protein was detected in 4 (nr 2, 12, 13, 14) from 14 analyzed callus lines.

4. CONCLUSSIONS

Transgenic plants producing a large subunit of urease (UreB) that is a strong *Helicobacter pylori* antigen were obtained. The transgene expression levels are sufficient for verification of the immunological potential of the obtained transgenic plant material in the animal model systems. To our knowledge it is the first report about the expression of *H. pylori* antigen in plants. In the future such transgenic plants or plant cells might be the source of cheap oral vaccines against this pathogen infecting majority of the developing countries.

REFERENCES

1. E. Franken, U. Teuschel, and R. Hain, Current Opinion in Biotechnol., 8 (1997) 411.
2. M.G. Koziel, N.B. Carozzi and N. Desai, Plant. Mol. Biol., 32 (1996) 393.
3. O.J.M. Goddijn and J. Pen, Trends Biotechnol., 13 (1995) 379.
4. H.S. Mason and C.J. Arntzen, Trends Biotechnol., 13 (1995) 388.
5. C.J. Arntzen, Nat. Biotechnol., 15 (1997) 221.
6. T.A. Haq, H.S. Mason, J.D. Clements and J.D. Arntzen, Science, 268(1995) 714.
7. T. Arakawa, D.K. Chong, J.L. Merritt and W.H. Langridge, Transgenic Res., 6 (1997) 403.
8. H.S. Mason, D.M.-K. Lam and C.J. Arntzen, Proc. Natl. Acad. Sci. USA, 89 (1992) 11745.
9. Y. Thanavala, Y.-F. Yang, P. Lyons, H.S. Mason and C. Arntzen, Proc. Natl. Acad. Sci. USA, 92 (1995) 3358.
10. McGaevey, P.B., Hammond, J., Dienelt, M.M., Hooper, D.C., Fu, Z.F., Dietzschold, B., H. Koprowski and F.H. Michaels, Biotechnology, 13 (1995) 1484.
11. H.S. Mason, J.M. Ball, J.-J. Shi, X. Jiang, M.K. Estes and C.J. Arntzen, Proc. Natl. Acad. Sci. USA, 93 (1996) 5335.
12. M.J. Blaser, Scientific American, 274 (1996) 92.
13. P. J. Falk, Inter. Med., 240 (1996) 319.
14. S.J. Czinn and J.G. Nedrud, Springer Semin. Immunopathol., 18 (1997) 495.

15. J.-F. Tomb, O. White, A.G. Kerlavage, R.A. Clayton, G.G. Sutton, R.D. Fleischmann, K.A. Ketchum, H.P. Klenk, S. Gill, B.A. Dougherty, K. Nelson, J. Quackenbush, L. Zhou, E.F. Kirkness, S. Peterson, B. Loftus, D. Richardson, R. Dodson, H.G. Khalak, A. Glodek, K. McKenney, L.M. Fitzegerald, N. Lee, M.D. Adams, E.K. Hickey, D.E. Berg, J.D. Gocayne, T.R. Utterback, J.D. Peterson, J.M. Kelley, M.D. Cotton, J.M. Weidman, C. Fuji, C. Bowman, L. Watthey, Wallin, W.S. Hayes, M. Borodovsky, P.D. Karp, H.O. Smith, C.M. Fraser and J.C. Venter, Nature, 388 (1997), 539.
16. J.L. Telford and P. Ghiara, Drugs, 52 (1996) 799.
17. R.L. Ferrero, J.-M. Thiberge, M. Huerre and A. Labigne, Infect. and Immun., 62 (1994) 4981.
18. M. Marchetti, B. Arico, D. Burroni, N. Figura, R. Rappuoli and P. Ghiara, Science, 267 (1995) 1655.
19. I. Corthesy-Theulaz, N. Porta, M. Glauser, E. Saraga, A.-C. Vaney, R. Haas, J.P. Kraehenbuhl, A.L. Blum and P. Michetti, Gastroenterology, 109 (1995) 115.
20. H.L.T. Mobley, M.D. Island and R.P. Hausinger, Microbiol. Rev., 59 (1995) 451.
21. S.J. Czinn, Gastroenterology, 113 (1997) 149.
22. P. Michetti, Ch. Kreiss, K.L. Kotloff, N. Porta, J.L. Blanco, D. Bachmann, M. Herranz, P. Saldinger, I. Corthesy-Theulaz, G. Losonsky, R. Nichols, J. Simon, M. Stolte, S. Ackerman, T.P. Monath and A.L. Blum, Gastroenterology, 116 (1999) 804.
23. L.E. Zonia, N.E. Stebbins and J.C. Polacco, Plant Physiol., 107 (1995) 1097.
24. M.E. Hogan, I.E. Swift and J. Done, Phytochemistry, 22 (1983) 663.
25. J.C. Polacco and M.A. Holland, Genetic control of plant ureases. In Setlow, J.K. ed. Genetic Engineering, Vol. 16, pp. 33-48. New York: Plenum Press, 1994.
26. P.D. Legg, J.F. Chaplin and G.B. Collins, J. Heredity, 60 (1969) 213.
27. J. Sambrook, E.F. Frisch and T. Maniatis, Molecular Cloning, A Laboratory Manual, 2nd ed., New York: Cold Spring Harbour Laboratory Press, 1989.
28. L.-T. Hu and L.T. Mobley, Infect. Immun., 61 (1993) 2563.
29. J.C. Carrington, D.D. Freed and A.J. Leinicke, Plant Cell, 3 (1995) 953.
30. M. Ni, D. Cui, J. Einstein, S. Narasimhulu, C.E. Vergara and S.B. Gelvin, Plant J., 7 (1995) 661.
31. P.D. Legg, J.F. Chaplin and G.B. Collins, J. Heredity, 60 (1969) 213.
32. T. Murashige and F. Skoog, Physiol. Plant., 15(1962) 473.
33. L. Rossi, J. Escudero, B. Hohn and B. Tinland, Plant Mol. Biol. Rep., 11 (1993) 220.
34. J.C. Thomas M.J. Guiltinan, S. Bustos, T. Thomas and C. Nessler, Plant Cell Reports, 8 (1989) 355.
35. H.J.M. Linthorst, F.T. Brederode, C. van der Does and J.F. Bol, Plant Mol. Biol., 21 (1993) 985.
36. D.L. Herrin and G.W. Schmidt, Biotechniques, 6 (1988) 196.
37. U.K. Laemmli, Nature, 227 (1970) 680.
38. T. Elmayan and H. Vaucheret, Plant J., 9 (1996) 787.

Food Biotechnology
S. Bielecki, J. Tramper and J. Polak (Editors)

Transgenic cucumber plants expressing the thaumatin gene

M. Szwacka[a], M. Krzymowska[b], M. E. Kowalczyk[a], A. Osuch[a]

[a]Department of Plant Genetics, Breeding and Biotechnology, Warsaw Agricultural University, Poland

[b]Institute of Biochemistry and Biophysics PAS, Warsaw, Poland

1. INTRODUCTION

Thaumatin occurs in the arils of fruit of *Thaumatococcus daniellii* Bentham a West African plant [1]. At least five different forms of thaumatin can be easily separated on the basis of their charge differences [2]. The two predominant forms (I and II) vary in quantities depending on the batch of fruit tested, which most probably reflects genetic variation among strains [3]. Both are made up of 207 amino acids in a single chain cross-linked with 8 disulphide bridges giving molecular weights of 22 209 Da and 22 293 Da [3]. Thaumatin is freely soluble in water. Its solutions taste sweet at concentrations as low as 10^{-8} M [4]. Thaumatin is at least 5000 times sweeter than sucrose at threshold concentrations and 2000 to 3000 times sweeter at the normal use of sucrose (6 to 10 %) [5]. Therefore, thaumatin is widely used as a flavour enhancer for various food, feed and pharmaceuticals.

Although thaumatin is incredibly sweet, its taste profile is quite different from that of sugar or the other high intensity sweeteners. There is some delay in the perception of sweetness and it leaves a liquorice-like aftertaste [4].

The structural gene of sweet-tasting protein from *T. daniellii* was cloned and expressed in *Escherichia coli* [3]. The primary translation product of this gene, preprothaumatin, contains extensions at both amino terminus and the carboxy terminus. The naturally occurring thaumatin represents the processed form [3].

Preprothaumatin II cDNA was used to prepare a plant shuttle vector, pWIT2, for *Agrobacterium rhizogenes*-mediated transformation of *Solanum tuberosum* [6]. Transformed hairy roots and regenerated whole plants had the characteristic taste and aftertaste of thaumatin. Dolgov et al. [7] have obtained successful transformation of apple, carrot, pear and strawberry with thaumatin II gene. The high level of thaumatin expression was observed in a callus tissue and lower in greenhouse plants of apple and strawberry.

In our earlier work the expression of preprothaumatin II cDNA under transcriptional control of constitutive promoter was shown in cucumber T_0 plants [8]. Here, we present data on the presence of thaumatin mRNA and protein in leaves and ripe fruits of two generations (T_1 and T_2) derived from cucumber T_0 plants bearing preprothaumatin II cDNA fragment under control of constitutive promoter 35S CaMV. On the basis of preliminary sensory evaluation of ripe fruits of T_2 cucumber plants and control variety fruits we concluded that the production of thaumatin protein improved cucumber fruit flavour.

This work was supported by grant No. 227 P06 96 11 from State Committee For Scientific Research.

2. MATERIALS AND METHODS

2.1. Plant material

The highly inbred cucumber line (line B) of monoecious variety Borszczagowski was used for the transformation experiments. For seed production cucumber plants were grown in a greenhouse in 16h/18h-day/night photoperiod (25°C-27°C in day/18°C-20°C in night). Light intensity in the greenhouse was about 1500 μmol m^{-2} s^{-1}.

The plants used for obtaining new generations were from seedlings after selection on modified MS medium [9], supplemented with 200mg/l kanamycin. Only the plants which exhibited normal growth and development in the presence of kanamycin were transferred to the greenhouse.

The fruits were harvested 6-7 weeks after self-pollination and stored (+4°C) for different periods of time before the sensory evaluation.

2.2. Vector plasmid

T. daniellii preprothaumatin II cDNA was provided by Dr. A.M. Ledeboer (Unilever, The Netherlands). *Agrobacterium tumefaciens* binary plasmid pROK2 [10] was used to construct the transformation vector pRUR528 [11].

2.3. RNA and protein gel blots

Total RNA was isolated from not fully expanded leaves and ripe fruit material using a method described by Linthorst et al. [12]. Ten micrograms of total RNA per line were separated on 1% agarose gel in 15 mM sodium phosphate, pH 6.5, blotted onto a Hybond N nylon membrane (Amersham), subsequently hybridized with the appropriate probes, washed at middle stringency (2xSSC / 0.1% SDS at RT and subsequently at 50°C) and exposed on autoradiographic film (Kodak) at –76°C for 67 to 140 h.

Protein fruit extracts were prepared in buffer containing 50 mM Tris pH 8.0 / 1 mM EDTA / 12 mM β-mercaptoethanol and 0.5% PMSF. Twenty five micrograms of protein extract were fractionated on 15% SDS-PAGE and subjected to an immunoblot analysis using specific anti-thaumatin rabbit polyclonal antibodies and anti-rabbit IgG goat antibodies conjugated with alkaline phosphatase (AP). Immunoblot was developed using the AP colorimetric detection kit (Boehringer Mannheim).

3. RESULTS

3.1. Transgenic cucumber plants that express thaumatin gene of *T. daniellii*

Northern blot analysis with thaumatin II cDNA probe was performed on total RNA isolated from leaves and some fruits of cucumber T_1 plants. Northern analysis revealed only two out of eleven T_1 plants, no. 219 and no. 235, showing very low levels of thaumatin II mRNA in leaf tissue, while containing the transgene cDNA of unchanged length as confirmed by Southern hybridization to genomic DNA (autoradiogram not shown). The plant no. 227 which showed no thaumatin II cDNA expression and no integration of transgene cDNA to the plant genome was treated as a negative control. We found variations in content of expressed thaumatin mRNA in leaf tissues (Fig. 1) but the levels of thaumatin expression in T_1 fruits were rather similar with the exception of fruit derived from plant no. 210 in which higher

104 212 235 227 217 210 215 224 233 219 225 C 207

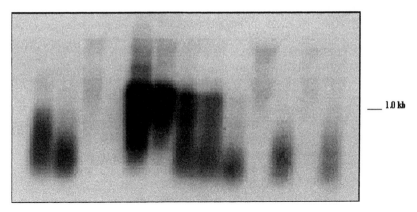

___ 1.0 kb

Figure 1. Expression of thaumatin gene in transgenic cucumber plants. Northern blot analysis with thaumatin II cDNA probe was performed on total RNA (10 μg) isolated from leaves of T_1 plants. As control (C) RNA from leaves of untransformed cucumber (line B) was assayed. Bar indicates the expected size of thaumatin mRNA (about 1.0 kb).

level of the thaumatin II mRNA was observed (autoradiogram not shown). We noticed a lack of marked differences between levels of thaumatin II mRNA in fruit and leaf tissues for five examined T_1 plants: no. 104, no. 210, no. 215, no. 224 and no. 225 (we only present data on expression of thaumatin gene in leaves - see Fig. 1). All T_1 plants were self-fertilized and produced fruits. The fruit development was normal, comparable with the development of control fruits. In another hybridization experiment we used samples of ripe cucumber T_2 fruits for Northern analysis. We noticed differences between levels of transgene-specific mRNA in fruit tissues for four examined T_2 lines (autoradiogram not shown).

3.2. Protein gel blots

Selected T_2 plants were screened for expression of the thaumatin II gene in fruits using SDS/PAGE immunoblot with anti-thaumatin anti-bodies. We estimated that the highest level of thaumatin protein was produced in the fruit of plant no. 212 01 which exhibited the highest (on 0-3 scale) sweetness intensity. Western blot analysis indicated presence of transgene-coded protein in a processed form (22 kDa) in eleven out of sixteen examined T_2 fruits (blot not shown). We observed a positive correlation between the level of thaumatin in fruit tissue and the sweetness intensity of fruit flesh (Table 1).

3.3. Sensory assessment of T_2 fruits

A total of 265 T_2 fruits from eleven T_2 plants were tasted to confirm a characteristic taste of thaumatin. The tastes of T_2 fruits were different from the taste of commercially available thaumatin (Sigma) but its taste profile was characteristic of that of thaumatin from *T. daniellii*. Some of the fruit samples exhibited the aromatic characteristics of cucumber fruits (3 out of 265) and some had a different aroma (3 out of 265). Control fruits were tasteless or exhibited different levels of acidity or bitterness. Sensory assessments were done after different periods of postharvest storage and during the longest storage (39 days) we could observe high sweetness (Table 2).

Table 1
Results of Western blot analysis and preliminary sensory assessment of ripe freshly harvested fruits derived from different T_2 plants

Plants	Intensity of			Flavour		Western blot analysis[*]
	sweetness (0 - 3)	acidity (0 - 3)	bitterness (0 – 3)	cucumber like	cucumber different	
210 03	2					+[**]
210 04	3					++
212 01	3					+++
14	2					+
	1					+/-
215 02				+		+
04			1			nd[***]
224 09	2				+	+
225 03	2					+
01		0				nd
07		3				nd
line B		1		+		-

0 (low) – 3 (high)
[*] content of recombinant thaumatin in 25 µg total protein extract estimated visually using the native thaumatin (Sigma) as a protein standard
[**] (content equivalent to 0.25 µg of protein standard)
[***] not determined

3.3. Sensory assessment of T_2 fruits

A total of 265 T_2 fruits from eleven T_2 plants were tasted to confirm a characteristic taste of thaumatin. The tastes of T_2 fruits were different from the taste of commercially available thaumatin (Sigma) but its taste profile was characteristic of that of thaumatin from *T. daniellii*. Some of the fruit samples exhibited the aromatic characteristics of cucumber fruits (3 out of 265) and some had a different aroma (3 out of 265). Control fruits were tasteless or exhibited different levels of acidity or bitterness. Sensory assessments were done after different periods of postharvest storage and during the longest storage (39 days) we could observe high sweetness (Table 2).

Table 2
Postharvest properties of various T_2 fruits after different periods of time before sensory assessment

Plants	Postharvest storage (days)	Intensity of		Western blot analysis[*]
		sweetness (0 - 3)	acidity (0 - 3)	
210 03	15	2		++
212 01	39	3		++
14	16	2		+[**]
224 09	27	2		+/-
225 01	32		0	nd[***]
07	11		3	nd
line B	21		1	-

0 (low) – 3 (high)
[*] content of recombinant thaumatin in 25 μg total protein extract estimated visually using the native thaumatin (Sigma) as a protein standard
[**] (content equivalent to 0.25 μg of protein standard)
[***] not determined

4. DISCUSSION

So far, the study on expression of recombinant thaumatin has been restricted to the primary transformants of different plant species. The results of experiments on transformation *S. tuberosum* with thaumatin II gene showed that this species has a new taste property: the characteristic taste of thaumatin [7]. Earlier we described the expression of the thaumatin II gene in cucumber T_0 progeny and preliminary sensory assessment of ripe cucumber T_1 fruits, indicating fruits with an improved taste [8]. In the present study we described the expression of thaumatin II gene in two subsequent generations (T_1 and T_2). The size of thaumatin protein accumulated in T_2 fruits was the same as that of naturally occurring thaumatin. This is consistent with the precursor protein (preprothaumatin) being processed correctly in the heterologous cucumber system. We examined the expression of thaumatin cDNA in leaf and fruit tissues of T_1 plants. The expression of transgene was observed in both kinds of transgenic cucumber tissues consistent with the regulatory properties of CaMV 35S promoter and not typical of the natural thaumatin gene [1]. We have observed the reproducibility of the thaumatin gene expression in T_1 progeny, which is indicative of no transgene rearrangements and silencing.

REFERENCES

1. G.E. Inglett and J.F. May, Econ. Bot. 22 (1968) 326.

2. H. van der Wel and K. Loeve, Eur. J. Biochem. 31 (1972) 221.
3. L. Edens, L. Heslinga, R. Klok, A.M. Ledeboer, J. Maat, M.Y. Toonen, Ch. Visser and C.T Verrips, Gene 18 (1982) 1.
4. M. Witty and J.D. Higginbotham (eds.), Developments in Sweeteners, Applied Science Publishers, 1979.
5. M. Witty, J.D. Higginbotham (eds.), Thaumatin, CRC Press, 1994, 47.
6. M. Witty, Biotechnol. Lett. 12 (1990) 131.
7. S.V. Dolgov, V.G. Lebedev, A.P. Firsov, T.V. Shushkova, G.B. Tukavin and S.A. Taran, Abstr. Conf. Papers IXth Intern. Cong. Plant Tiss. Cell Culture, Jerusalem, Israel, 14-19 June, 1998, 189.
8. M. Szwacka, M. Krzymowska, S. Malepszy, Proc. IXth Intern. Cong. Plant Biotechnology and In Vitro Biology in the 21st Century, Jerusalem, Israel, 14-19 June 1998, 619.
9. W. Burza and S. Malepszy, Plant Breed. 114 (1995) 341.
10. V.A. Hilder, A.M.R. Gatehouse, S.E. Sheerman, R.F. Barker and D. Boutler, Nature 300 (1987) 160.
11. M. Szwacka, M. Morawski and W. Burza, Genet. Pol. 37A (1996) 126.
12. H.J.M. Linthorst, F.T. Brederode, C. van der Does and J. F. Bol, Plant Mol. Biol. 21 (1993) 985.

Food Biotechnology
S. Bielecki, J. Tramper and J. Polak (Editors)
© 2000 Elsevier Science B.V. All rights reserved.

Diploidization of cucumber (*Cucumis sativus* L.) haploids by *in vitro* culture of leaf explant

N. M. Faris[a], M. Rakoczy-Trojanowska[b], S. Malepszy[b], K. Niemirowicz-Szczytt[b]

[a]Department of Botany, Faculty of Science, Al-Fateh University, P.O. Box 13228, Tripoli-Libya

[b]Department of Plant Genetics, Breeding and Biotechnology, Warsaw Agricultural University, Nowoursynowska 166, 02-787 Warszawa, Poland

Haploids provide valuable material for fundamental research and plant breeding. However, their utilization depends on the effectiveness of the chromosome doubling method. The use of two unconventional tissue culture methods from leaf explant contributed to the induction of doubled haploid.

The first method known as the long-term callus culture was based on the embryonic callus induction. The hypothesis is that it brings about endomitosis in callus cells. In this experiment 200 large leaf explants from four genetically different haploid clones were used. They were cultured on modified MS medium [1] with enhanced NH_4NO_3 without KNO_3, and 0.8 mg/l 2,4 D and 0.4 mg/l 2iP of growth regulators. The regeneration ability was estimated on the bases of the number of plantlets obtained per explant. The total of 36 explants exhibited regeneration and 10 produced viable plants. The rooted plants in the amount of 306 were transferred to greenhouse and 108 of them were tested for their ploidy level. As many as 74% of plants were diploid, 1% was triploid and 25% tetraploid

The second method employed the direct regeneration from the haploid leaf micro-explants (2-3 mm^2) from the first and second juvenile leaves. The explants were cultured on modified MS medium [2]. This method was applied to five genetically different haploid genotypes. In total, 21.1% of explants from the first leaf and 12.1% of the second leaf produced plantlets. As a result 388 plants from the first leaf and 210 from the second were obtained. Out of the first population 93 plants and 78 from the second were tested with respect to their ploidy level. All the plants obtained from the first leaf proved to be haploid, while in the population obtained from the second leaf there were 70.5% of haploids, 28.2% of diploids and 1.3% of mixoploids.

1. INTRODUCTION

Haploid plants can not be utilised for practical breeding because of their sterility. By doubling their chromosome number valuable homozygous lines can be obtained. The use of haploids depends on the effectiveness of diploidization method. In order to obtain doubled haploid plants conventional chromosome doubling techniques involving colchicine treatments are applied [3-5].

Direct somatic embryogenesis (direct regeneration) is another method by which doubled haploids were induced in a very short time. The ability to regenerate shoots and embryo is greatly dependent on the stage of tissue and the growth regulator content. Shoot regeneration can take place directly or by callus formation. The regeneration from organs, such as leaves, stems or roots, requires initiation of *de novo* meristems, which is controlled by the balance of cytokinines and auxines.

Shoot regeneration from cucumber leaf explants may take place either via callus [6] or directly [5]. There are also examples of direct regeneration from cucumber cotyledons [7].

The long-term culturing of haploid leaf explants and induction of callus allowed the development of *Nicotiana* doubled haploid plants [8-10]. In pot gerbera, 15.3% of diploid plants were obtained through callus culture of the ovule [11]. Endoduplication also took place during the two-step culture of wheat callus from anther, microspore and ovule [12].

The aim of this study was to obtain doubled haploids with the use of two unconventional tissue culture methods (the long term culture and direct regeneration from haploid leaf explants) and evaluate the plants produced for their ploidy level.

2. MATERIALS AND METHODS

Four haploid (16, 26 89, 99) derived from Polan F_1 variety were used for long term culture while the initial material for direct regeneration was derived from two cucumber forms: *C. sativus* var. Polan F1 (haploid No. 16) and *C. sativus* ssp. *rigidus* var. *sikkimensis* (haploids No. 748, 759a, 763 and 790). All these haploids were obtained by pollination with irradiated pollen grains and embryo rescue [13,14]. The haploid plants were maintained in jars under controlled conditions.

Long term culture
In this experiment 50 explants from the first and second leaves were cut into large explants ($0.5 \times 0.5 cm^2$) and cultured on modified MS medium. The medium consisted of 1/2 macro-elements with 1.7 x MS NH_4 NO_3, without KNO_3, full MS micro-elements, full MS vitamins, 250 mg/l edamine, 30 g/l sucrose and 7.5 g/l Diffco-bacto agar. The growth regulators amounted to 0.8 mg/l 2,4-D and 0.4 mg/l 2iP. The explants were incubated in dark condition at 25°C for four weeks and transferred to fresh medium every four weeks. The embryonic callus was transferred to hormone free medium for regeneration. The regeneration ability was estimated on the basis of the number of explants producing plantlets.

Direct regeneration
The first and second young leaves (0.50-$0.75 cm^2$) of the haploids were cut into small square micro-explants (2-$3 mm^2$) and cultured on modified MS medium. The medium consisted of 1/2 macro-elements with 1.7 x MS NH_4NO_3, without KNO_3, full MS micro-elements, full MS vitamins, 250 mg/l edamine, 30 g/l sucrose and 7.5 g/l Diffco-bacto agar. The growth regulators used included 2.4-D and 2iP at the concentration 0.4 mg/l each [2]. Explants were incubated for 11 to 15 days in dark and +25°C. After that time the explants were transferred onto hormone-free medium (1/2MS macro-elements, full MS micro-elements, full MS vitamins). Then the explants and shoots were transplanted into fresh

medium every 15 days. In the next stage plantlets were placed in jars containing 1/2 MS with 1 mg/l IAA for rooting. Some of the plants were planted in the greenhouse for further testing. The regeneration ability was estimated in the same way as in long term culture.

Ploidy level estimation

The ploidy level was determined at first phenotypically according to the habit of the plant, size, venation, texture and colour of the leaf and then by means of flow cytometry. Samples were prepared according to Galbraith *et al.* [15]. Fluorescence of DAPI was measured in a Partec CA II flow-cytometer (Partec GMbH, Germany).

3. RESULTS AND DISCUSSION

Long term culture

The embryonic callus induction was 100% for all genotypes, while the regeneration ability was rather low and genotype dependent. As shown in Table 1 only 36 explants were able to exhibit first regeneration symptoms. Ten explants produced viable plants. The genotype no. 99 proved to be the best for plant regeneration, while the genotype 26 did not produced any viable plantlets. Ploidy estimation is presented in Table 2. Out of 108 plants there were 80 diploids, 27 tetraploids and only one triploid.

Similar results were reported in anther derived haploid *Nicotiana*. Stem explant culture of *Nicotiana* haploids yielded in diploid plants similar to the diploid control, while the long lasting callus culture caused the formation of tetraploid [9-10]. The experiment described in this report also confirms the above statement. In fact more tetraploids were obtained after longer culture of calli. Grunewaldt and Malepszy [16] compared anther culture of five barley cultivars and found out that the number of polyploids and mixoploids increased with the duration of callus culture which was accompanied by the reduction of regeneration potentials.

Direct regeneration

Table 1
The regeneration ability from leaf explant of different haploid genotypes and the ploidy estimation of the plants obtained from long term culture

| Haploid symbol | No. of explants | | | | No. of rooted plants (average) |
	Placed	Explants with regeneration reaction	Explants produced plantlets	Explants which produced viable plants	
99	50	21	13	9	(30.2)
16	50	9	3	0	0
89	50	6	2	1	(43)
26	50	0	0	0	0
Total		36	18	10	(30.6)

During the induction period (11-15 days) in dark, on the edges of explants friable white shiny structures (pro-embryos) appeared. On the hormone-free medium, they developed into the bipolar somatic embryoids (of different stages) with shoot and root meristems.

Table 2
The different ploidy level of regenerants (R_0), analysed by flow -cytometry

Initial explant	Number of investigated plants	Ploidy level		
		2x	3x	4x
99/11	3	0	0	3
99/26	24	23	1	0
99/27	30	30	0	0
99/34	2	2	0	0
99/39	7	7	0	0
99/46	13	0	0	13
99/52	26	18	0	8
89/27	3	0	0	3
Total	108	80	1	27

Some of the plants had the adventitious shoots in the form a bunch of propagules (plantlets) which could be separated so as to give raise to regular plants. The induction ability was 100% for both the first and second leaves. The number of explants producing viable plantlets was 13 (28.9%) for genotype 790 and 7 (15.5%) for genotype 763 (Table 3).

Table 3
The direct regeneration ability of five different haploid cucumber genotypes

Haploid design	First leaf				Second leaf			
	No. of							
	Initial explant	Explants with (+) reaction	Explants producing plantlets	%	Initial explants	Explants with (+) reaction	Explants producing plantlets	%
16	15	9	9	60	15	0	0	0
748	15	3	0	0	15	0	0	0
759a	45	21	5	11.1	45	0	0	0
763	45	5	4	8.9	45	8	7	15.5
790	45	21	17	37.8	45	17	13	28.9
Total	165	59	35	21.2	165	25	20	12.1

The number of plants produced per explant from the first leaf was higher than that obtained from the second leaf. Out of five genotypes four regenerated from the first leaf and only two from the second. Similar tendencies were observed in plant development (Table 4).

Table 4
The regeneration and the ploidy estimation from the first and second leaves in different haploid genotypes

Leaf Position	Plant symbol	No. of explants with regeneration	No. of plants obtained	No. of plants evaluated	Ploidy level	
					n	2n
First leaf	16	9	203	31	31	0
	759	5	28	7	7	0
	763	4	18	7	7	0
	790	17	185	48	48	0
Total		35	434	93	93	0
Second leaf	763	10	51	21	19 (90.4%)	1+1* (4.8+4.8%)
	790	13	159	57	36 (62.2%)	21 (36.8%)
Total		23	210	78	55 (70.5%)	23 (29.5%)

* 1 plant mixoploid

The number of regular plants from the first leaf was greater than from the second leaf. All the plants from the first leaf were haploid and those from the second included haploids (70%), diploids (28.2%) and mixoploids (1.3%)

Similar observations were made on cucumber regeneration from the first diploid leaf (2). Plants regenerated from the first leaf did not change their ploidy level. Experiments with *Sorghum* culture indicated that haploid plants were regenerated from the very young panicles and haploid and diploid plants were obtained from more mature ones (17).

On the bases of the presented data and other reports it can be concluded that from very young leaves or other organs it is possible to obtain plants of the same ploidy level due to direct regeneration. Direct regeneration of older organs may give as a result a higher ploidy level. This phenomenon can be used for haploid doubling.

REFERENCES

1. T. Murashige and F. Skoog, Physiol. Plant., 15 (1962) 473.
2. W. Burza and S. Malepszy, Plant Breeding, 114 (1995) 341.
3. M. C. M. Iqbal, C. Mollers, G. Robbelen, J. Plant Physiol., 143 (1994) 222.

4. C. Mollers, M. C. M. Iqbal, G. Robbelen, Euphytica 75 (1994) 95.

5. V. Nikolova and K. Niemirowicz-Szczytt, Acta Soc. Bot. Pol., 65 (1996) 311.

6. S. Malepszy and A. Nadolska-Orczyk, Z. Pflanzenphysiol., 111 (1983) 273.

7. C. H. Colijin-Hooymans, J.C. Hakkery, J. Jansen, J.B.M. Custers, Plant Cell Tissue and Organ Culture, 39 (1994) 211.

8. J.P. Nitsch, Z. Pflanzenzuchtg., 67 (1972) 3.

9. M.J. Kasperbauer and G.B. Collins, Crop Sci., 12 (1972) 98.

10. K. Kochhar, P. Sabharwal, J. Engelberg, J. Hered., 62 (1971) 59.

11. K. Miyoshi and N. Asakura, Plant Cell Reports, 16 (1996) 1-5.

12. G. W. Schaefer, P. S. Baenziger, J. Worley, Crop Sci. 19 (1979) 697.

13. K. Niemirowicz-Szczytt and R. Dumas de Vaulx R., Cucurbit Genet. Coop., 12 (1989) 24.

14. J. Przyborowski and K. Niemirowicz-Szczytt, Plant Breeding, 112 (1994) 70.

15. D.W. Galbraith, K.R. Harkins, J.M. Maddox, N.M. Ayres, D.P. Sharma, E. Firoozabady , Science, 220 (1983) 1049.

16. J. Grunewaldt and S. Malepszy, Barley Genetics, III. Int. Barley Genet. Symp. Garching, PP. (1975) 367.

17. A. Elkonin, T.N. Godova, A.G. Ishin, U.S. Tyrnov, Plant Breed., 110 (1993) 201.

Food Biotechnology
S. Bielecki, J. Tramper and J. Polak (Editors)

Tomato (*Lycopersicon esculentum* mill.) transformants carrying *ipt* gene fused to heat-shock (*hsp70*) promoter

O. Fedorowicz, G. Bartoszewski, A. Smigocki[1], R. Malinowski, K. Niemirowicz-Szczytt

Departament of Plant Genetics, Breeding and Biotechnology, Warsaw Agricultural University
Nowoursynowska 166, 02-787 Warsaw, Poland

[1]U.S. Departament of Agriculture, 10300 Baltimore Av., Beltsville, MD 20705-2350, USA

Tomato *ls* mutant, characterised by suppressed lateral shoots, abnormal flowers and low level of endogenous cytokinins, was transformed with a *Agrobacterium tumefaciens* strain ACS101 carrying an *ipt* gene under a heat shock promoter. Of the 62 rooted shoots that were obtained, most exhibited unchanged ploidy levels. The PCR analysis confirmed that 76% of the plants were transgenic. The segregation of the selectable marker gene, *nptII*, in majority of the progeny was 3:1 on a kanamycin-containing medium. A two-hour heat shock treatment at 42°C increased *ipt* gene transcripts as analyzed by RT-PCR. Transcript levels decreased over time and after six hours could not be detected.

1. INTRODUCTION

The *lateral suppressor* (*ls*) mutant of tomato (*Lycopersicon esculentum* Mill.) fails to produce axillary buds in most of its leaf axils [1,2]. During vegetative growth, lateral meristems are not initiated. After the transition to floral development, buds are initiated in the axils of the two youngest leaves. The lateral bud in the axil of the leaf below the inflorescence becomes the main axis of the plant forming a sympodium, whereas the second axillary bud develops into a side-branch [2]. During flower development *ls* mutants fail to initiate petal primordia. The flowers are also characterised by reduced male and female fertility, which has a negative effect on fruit production [3]. The physiological basis for the pleiotropic effects of the *ls* mutation has not been described so far but it may be related to the imbalance of plant hormones. It has been shown that cytokinin levels in mutant shoots were reduced and auxin levels were increased as compared to a normal tomato plant [4,5,6]. According to Tucker [6,7] the reduced cytokinin levels are the primary cause for the absence of side shoot formation. Cytokinin applications to the leaf axils of *ls* plants initiate lateral buds in the axils above the point of treatment. It was also reported that axillary meristems could be induced by the application of cytokinin benzyladenine to *Stellaria media*, an herbaceous plant that frequently lacks axillary meristem formation [8].

An alternative approach to studying the role of cytokinins in meristem development, is the transformation of plants with the *ipt* gene in order to increase the endogenous level

This research was supported by the Polish State Committee for Scientific Research, Grand No. 5P06A02511.

of cytokinins. The *ipt* gene is one of the tumor-inducing genes of *Agrobacterium tumefaciens*, which codes for the enzyme isopentenyl transferase [9,10]. This enzyme catalyzes the condensation of isopentenyl pyrophosphate and 5'-adenosine monophosphate from isopentenyl-adenosine-5'-monophosphate, which is subsequently metabolised to isopentenyl-adenosine and isopentenyl-adenine, precursors to zeatin-riboside and zeatin, respectively.

The introduction of the *ipt* gene under a strong, constitutive promoter to petunia and tobacco brought about overproduction of cytokinins. As a result the plants had almost no root system, the shoots were darker green and they showed a lack of apical dominance [11,12].

In the transgenic *ls* mutant plants expressing the *ipt* gene under the native constitutive promoter the increase in endogenous zeatin riboside levels did not restore axillary meristem formation but in some cases bulbous structures were formed in leave axils [13].

The purpose of our experiments was to characterise the *ls* plants with elevating endogenous cytokinin and to study the effect of inducted *ipt* gene expression on the phenotype and the development of the plants more precisely, so the *ipt* gene under the heat-shock promoter was introduced to tomato mutant *ls*.

2. MATHERIALS AND METHODS

2.1. Plant material

Tomato *ls* mutant from our Department stock collection was used in all experiments.

2.2. *Agrobacterium* transformation

Agrobacterium transformation method described by McCormick [14] was used in this study. All the medium combinations used in the experiments were based on Murashige and Skoog [15]. Seeds of *ls* mutant were surface sterilised with a solution of sodium hypochloride and rinsed with sterile distilled water. Seeds were germinated, transferred to the jars with the medium and grown for 7-8 days.

The distal and proximal parts of the cotyledon were cut from 7-8 day old seedlings, the remaining fragment was cut in such a way that 2 fragments of similar length were obtained from each cotyledon. They were placed upside-down on the medium with a layer of *Nicotiana tabacum* feeder cells under a filter paper disc. The explants thus prepared were precultured for 2 days.

The binary vector, pHSCKn312, was used for the transformation. This plasmid contains the *Drosophila melanogaster hsp 70* gene promoter (456 bp) fused to the bacterial *ipt* gene (693 bp) (16). It also carries 35S-*nptII* selectable marker gene,. *Agrobacterium tumefaciens* strain EHA101 carrying the pHSCKn312 plasmid (ACS101) was grown to a density of 0.4 to 1.0 (OD_{600}), centrifugated and resuspended in liquid MS (containing 0.2% sucrose) to an OD_{600} of 0.4-0.5. Explants were submerged in the *Agrobacterium* suspension and returned to the feeder plates for 2 days of cocultivation. Explants were transferred to selective regeneration medium (MS + 2 mg/l zeatin, 75 mg/l kanamycin, 400 mg/l carbencilin). Regenerating explants were transferred to a fresh medium every 4 weeks. Shoots were transferred to the rooting medium (MS + 0.2 IAA, 50 mg/l kanamycin, 400 mg/l carbencilin). Rooted shoots were transferred to the greenhouse and grown in soil (16h light period, temperature 25°C and 20°C during the day and night respectively).

2.3. Ploidy level

The ploidy of regenerated shoots was determined by the flow cytometry method. A PARTEC II flow cytometer was used with a DAPI fluorescent stain. The softwere DAPC 2.0 (Partec 1992) was used to analyse the obtained data.

2.4. The PCR analysis

Primers for the *ipt* gene were designed in Oligo 4.0 program. The sequence of the primers was 5'AGT TAC CCG ACC AAG AGA CC 3' forward and 5'ACC TAA TAC ATT CCG AAC GG 3' reverse yielding a fragment of 355 bp. DNA was isolated from young leaves of putative transformants as described by Haymes [17]. PCR amplification was performed in a Perkin Elmer Thermocycler and the reaction products were resolved by electroporesis.

2.5. The RT-PCR analysis

Progeny of two transgenic plants with the integration of *ipt* gene and one non-transformed plant were chosen for the analysis. At 8-day, seedlings were heat shocked at 45°C for 2 h. Tissues were collected before, just after, 2 h and 6 h after high temperature treatment. RNA was isolated with QIAGEN Rneasy Plant Mini Kit, treated with DNAseI and quantified. The RT-PCR analysis was conducted using 1 μg of RNA with Access RT-PCR System (Promega).

3. RESULTS

Three independent transformation experiments were conducted. Of the 436 explants inoculated with *Agrobacterium tumefaciens* about 95.5% produced callus after 4 weeks of selection and 42.3% regenerated shoots. Shoot regeneration was first noticed after 6 weeks of culture and after 8 to 12 weeks of culture. Shoots were excised and transferred to the rooting medium containing 50 mg/l kanamycin. A total of 62 shoots formed roots on kanamycin (Table 1).

Table 1
Effectiveness of transformation

Explants	Used for transformation	Producing callus	Regenerating shoots	Developing rooting shoots
Number	436	416	184	62
%	100	95.5	42.3	14.2

The ploidy level of the putative transgenic plants was estimated by the flow cytometry method. Most of the rooted shoots (84%) showed unchanged ploidy level. Only diploid plantlets were transferred to the greenhouse where 59% survived, flowered and produced fruits and seeds. Most of the plants showed the *ls* mutant type morphology. The plants did not produce axillary shoots and showed certain abnormalities of the flowers. Four plants with normal phenotype were also regenerated.

The PCR analysis confirmed that 76% of the plants transferred to the greenhouse were transgenic with an integrated chimeric *ipt* gene. Finally, seeds were collected from 26 diploid plants.

The segregation of a *nptII* gene (kanamycin resistance gene linked with *ipt*) in the R_1 progenies was tested on a medium with kanamycin (Table 2). Most of the progeny (76%) showed Mendelian segregation 3:1. Progeny of one plant segregated 15:1 ratio. One was susceptible and did not segregate. The remaining 19.2% segregate in non Mendelian way.

The segregation of *ipt* gene was tested by the PCR analysis of kanamycin resistant progenies. The analysis proved that both genes, *ipt* and *nptII*, segregated with each other.

Table 3
Segregation ratio in R_1 progeny

Type of segregation	Mendelian segregation		Non Mendelian	Not segregating (susceptible)	Total
	3:1	15:1			
Number	19	1	5	1	26
%	73.0	3.8	19.2	3.8	100

The RT-PCR analysis of *ipt* transcript was carried out on RNA isolated from plants treated with high temperature. The analysis indicated that the *ipt* transcripts were not detected or detected at background levels as demonstrated in lanes 6-11 in plants and appeared after the heat shock. The level of *ipt* transcripts decreased over time and almost disappeared after six hours (Figure 1).

The R_1 generation of transgenic plants was grown in a greenhouse. It was noticed that high temperatures (over 40°C) applied during the last few weeks of growing period did not affect plants' morphology.

Seedlings of R_2 generation were tested on medium containing kanamycin. Homozygous transgenic plants were selected. Finally 9 transgenic lines were chosen that showed a single locus of transgene integration.

The progeny (R_2) of the lines, which showed the reversion of the *ls* mutation to normal morphology did not express *ls* gene effects.

3. DISCUSSION

The experiments confirmed that the decreased level of cytokinin in *ls* mutant shoots did not influence the regeneration ability of the *ls* mutant of tomato defective in the initiation of axillary shoot meristem. The *ls* line showed a high regeneration ability from the cotyledon fragments cultured under the standard transformation conditions. Nearly 100% of the explants produced callus after 4 weeks of culture and 42.3% regenerated shoots. Majority of the regenerated plants were stably transformed. They showed the integration of both the *ipt* and *nptII* genes and had unchanged ploidy levels. The analysis of the segregation of both genes in the progeny indicated that as a rule they integrated in a single locus, as expected.

Most of the plants transferred to the greenhouse had a phenotype typical for *ls* mutant. Only four plants were phenotypicaly different. They showed a reversion of all the *ls* mutant features. They produced axillary shoots, developed normal flowers and had larger fruits.

The transgenic plants of tobacco, petunia and *Arabidopsis* with the *ipt* gene under the heat shock promoter even before the high temperature induction showed a lack of apical dominance and an intensive side shoot development [18]. The PCR analysis of these plants did not reveal the presence of the *ipt* gene in their genome. However, the segregation of the progeny on the kanamycin medium suggested the presence of the *nptII* gene. The negative result of the PCR analysis could be caused by rearrangements in the *ipt* gene that could have occurred during or after the integration of the T-DNA fragment into the plant genome.

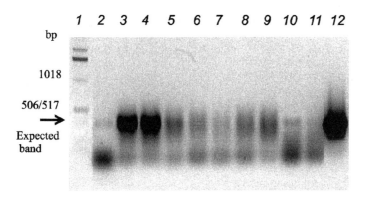

Figure 1.
RT-PCR analysis of some transgenic plants carrying the *ipt* gene
lane 1. 1kb ladder (Gibco)
lane 2. Lack of *ipt* expression in *ls* transformants before heat-shock
lane 3. Expression of *ipt* gene in *ls* transformants after heat-shock
lane 4. Ekspression of *ipt* gene in *ls* transformants 2 hours after heat-shock
lane 5. Lack of *ipt* expression in *ls* transformants 6 hours after heat-shock
lanes 6-9. Lack of *ipt* gene expression in non-transformed *ls* plants after heat shock treatment (respectively as above)
lanes 10-11. Control without reverse transcriptase: RNA from transgenic and non-transgenic plants before heat shock treatment
lane 12. Control of Access RT-PCR System (Promega)

The basis of the phenotype reversion of the *ls* mutation in these plants has not been determined and will require further studies.

The analysis of the expression of the *ipt* gene indicated that the *hsp70* promoter from *Drosophila melanogaster* is recognised by transcriptional factors of tomato. In addition, heat treatment induced transcription of the *ipt* gene. The expression of *ipt* gene in transgenic tobacco plants after the heat shock induction was studied by Smigocki [16]. As a result of high temperature induction the *ipt* transcripts appeared in leaves and shoots. The largest amount of transcripts was observed after 2 hours after the heat shock. After 46 hours *ipt* transcripts were detectable only in the leaves. The high temperature treatment conducted either in *in vitro* culture or in a greenhouse had no influence on transgenic *ls* plants' morphology.

The results obtained do not provide a sufficient basis for concluding whether the increase of the level of endogenous cytokinin can cause the reversion of *ls* mutation. Further research will be done on these regenerated transgenic lines.

REFERENCES

1. A.G. Brown, Rep. Tomato Genetics Cooperative, 5 (1955) 6-7.
2. J.C. Malayer and A.T. Guard, Am. J. Bot., 51 (1964) 140-143.
3. S.P.C. Groot, W. de Ruiter and J.J.M. Dons, Sci. Horti., 59 (1994) 157-162.
4. R. Maldiney, F. Pelrse, G. Pilate, L. Sossountzov and E. Miginiac, Physiol. Plant, 68 (1986) 426-430.
5. L. Sossountzov, R. Maldiney, B. Sotta, I. Sabbagh, Y. Habricot, M. Bonnet and E Miginiac, Planta, 175 (1988) 291-304.
6. D.J. Tucker, New Phytol., 77 (1976) 561-568.
7. D.J. Tucker, Ann. Bot., 48 (1981) 837-843.
8. H. Tepper, J Plant Physiol., 140 (1992) 241-243.
9. D.E. Akiyoshi, H. Klee, R.M. Amassino, E.W. Nester and M.P. Gordon, Proc. Natl. Acad. Sci. USA, 81 (1984) 5994-5998.
10. G.F. Barry, S.G. Rogers, R.T. Fraley and L. Brand, Proc. Natl. Acad. Sci. USA, 81 (1984) 4776-4780.
11. H. Klee, R. Horsch and S. Roger, Ann. Rev. Plant. Physiol., 38 (1987) 467-486.
12. A.C. Smigocki and L.D. Owens, Plant Physiol., 91 (1989) 808-811.
13. S.P.C. Groot, R. Bouwer, M. Busscher, P. Lindhout and H.J. Dons, Plant Growth Regulation, 16 (1995) 27-36.
14. S. McCormick, Plant Tissue Culture Manual, B6 (1991) 1-9.
15. T. Murashige and F. Skoog, Physiologia Plantarum, 15 (1962) 473-497.
16. A.C. Smigocki, Plant Mol. Biol., 16 (1991) 105-115.
17. K.M. Haymes, Plant Mol. Biol. Reporter, 14 (1996) 280-284.
18. J. Medford, R. Hogan, Z. El-Sawi and H. Klee, Plant Cell, 1 (1989) 403-413.

Food Biotechnology
S. Bielecki, J. Tramper and J. Polak (Editors)

Regulation of carbon catabolism in *Lactococcus lactis*.

T. Aleksandrzak,[a] M. Kowalczyk[a], J. Kok[b] and J. Bardowski[a*]

[a] Department of Microbial Biochemistry, IBB PAN, Pawinskiego 5a, 02-106 Warsaw, Poland

[b] Department of Genetics, University of Groningen, Kerklaan 30, 9751 NN Haren, The Netherlands

The *Lactococcus lactis* IL1403 is a lactose negative, plasmid free strain. Nevertheless, it is able to hydrolyze lactose in the presence of cellobiose.

In this work we describe identification of a gene involved in this process. The gene was found to be homologous to the sugar catabolism regulator, *ccpA*. The complete DNA sequence and analysis of the region encoding the *ccpA* gene is also presented.

1. INTRODUCTION

Lactic Acid Bacteria (LAB) are able to utilize a great number of sugars in fermentation processes. The most important carbon source for LAB grown in milk is lactose, because it is a main sugar in this medium. To be catabolized lactose has first to enter the bacterial cell. Lactose can be transported into the cell in two ways. One way, which is energetically preferable, is the PEP:PTS system (phosphoenolpyruvate-dependent phosphotransferase system). Using this system lactose is transported and simultaneously undergoes phosphorylation. Therefore, it enters the cell as phosphorylated molecule. Then, inside the cell, it is hydrolyzed by P-β-galactosidase into glucose and galactose-6-P [1]. The second way of lactose internalization occurs with the use of the permease and without phosphorylation of lactose. Inside a cell a non-phosphorylated lactose is cleaved into glucose and galactose by β-galactosidase.

The whole lactose catabolic pathway is one of the best known metabolic systems in LAB. Lactose operon consists of *lacABCDEFGX* genes [2], among which there are those coding for lactose transport and phosphorylation (*lacEF*), hydrolysis of lactose-6-P (*lacG*) and for tagatose pathway enzymes (*lacABCD*). Expression of the *lac* operon is regulated by the product of *lacR* gene, which is the transcriptional repressor. The *lacR* gene constitutes a monocistronic operon, which is divergently orientated to *lac* operon. A function of LacR protein depends on inducers such as tagatose-6-P and an intermediate of tagatose-6-P pathway. In the absence of lactose the LacR protein represses transcription of *lac* operon. In the presence of lactose inducers inactivate the LacR protein and therefore alleviate repression of *lac* genes.

---

* corresponding author
This work was partly supported by KBN Grant 6 P04B 025 10

Some of the LAB, mainly from the genus *Lactococcus*, are also able to assimilate β-glucosides, which are sugars found in plants. This catabolic potential probably results from the fact that the natural niche of lactococci is plant environment.

The regulatory system involved in catabolism of β-glucosides relies on *bglR* gene function. It was shown that BglR protein, which belongs to the BglG / SacY family of regulators is required for utilization of β-glucosides by *Lactococcus lactis* [3].

This protein acts as an antiterminator, which prevents RNA polymerase from stopping the transcription of β-glucosides operon. A *bglR* mutant is not able to grow neither on salicin, arbutin, nor on esculin but is still able to grow on cellobiose [4].

In this work we show that on reach M17-agar medium supplemented with cellobiose and X-gal the wild type *Lactococcus lactis* IL1403 strain forms blue colonies. Since this strain normally does not grow on lactose as the only source of carbon, we suggest that the ability to hydrolyze X-gal is a cellobiose-inducible phenotype. Moreover, it seems that a system involved in cellobiose catabolism is different from that involved in arbutin, salicin and esculin assimilation. This observation led us to concentrate our work on identification and characterization of the gene responsible for cellobiose inducible lactose catabolism.

2. IDENTIFICATION OF THE GENE INVOLVED IN CELLOBIOSE-INDUCIBLE LACTOSE CATABOLISM

To identify the gene responsible for cellobiose inducible lactose catabolism, the plasmid integration mutagenesis system was used [5]. This system is efficient for lactococci and other Gram-positive bacteria. The plasmid pGhost that is used in this system is characterized by a thermosensitive replication, contains the IS*S1* sequence and the erythromycin resistance marker which is used for selection (Figure 1). Integration of the plasmid occurs at non-permissive temperature (37°C) when plasmid replication is blocked. This random integration results in duplication of the IS*S1* sequence. As a result of this duplication plasmid pGhost is flanked by two IS*S1* sequences. Presence of unique restriction sites HindIII and EcoRI in the pGhost molecule is very useful for DNA cloning. The HindIII or EcoRI digestion of a DNA with integrated pGhost9:IS*S1* plasmid leads to excision of the pGhost9:IS*S1* with its right or left flanking fragment of chromosomal DNA respectively. The nucleotide sequences of pGhost9:IS*S1* downstream or upstream regions can next be determined with primers corresponding to IS*S1* or pGhost sequences.

This plasmid integration mutagenesis system was applied to isolate mutants of *Lactococcus lactis* IL1403 strain that do not exhibit cellobiose-inducible β-galactosidase positive phenotype. The selection was carried on reach M17-agar medium supplemented with cellobiose, X-gal and erythromycin. We expected to obtain white colonies of transformants, which had lost their ability to cellobiose-iducible lactose catabolism and blue colonies for all other transformants. The antibiotic selection excluded non-transformed *Lactococcus lactis* cells from growing.

After mutagenesis of *Lactococcus lactis* IL1403 strain, performed by random integration of pGhost9:IS*S1* into its chromosomal DNA, 2000 colonies were obtained on the selective M17-agar medium with cellobiose and X-gal. Out of 2000 erythromycin resistant integrants 6 grew as white colonies. Integration sites in all white mutants were analyzed by Southern hybridization.

Southern hybridization patterns showed that 6 clones could be divided into 5 classes. After plasmid rescues from 5 integrants, pGhost9:ISS1 with flanking fragments of chromosomal DNA were introduced into *Escherichia coli* EC1000 strain. The chromosomal DNA fragments from all 6 white integrants were subsequently sequenced.

Analysis of the partial nucleotide sequences obtained revealed that in two mutants the same gene was mutated. The two, above-mentioned sequences show the homology to the CcpA protein from *Bacillus subtilis*.

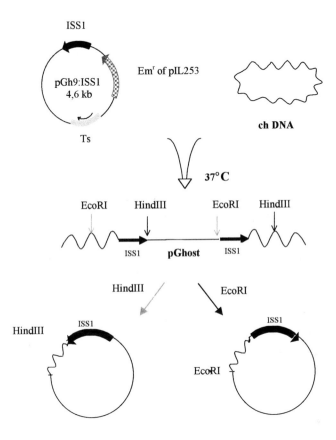

Figure 1. Scheme of DNA mutagenesis and DNA cloning with pGhost9:IS*S1* integration plasmid (modified from [6]).

COMPLETE DNA SEQUENCE OF *LACTOCOCCUS LACTIS* CcpA PROTEIN

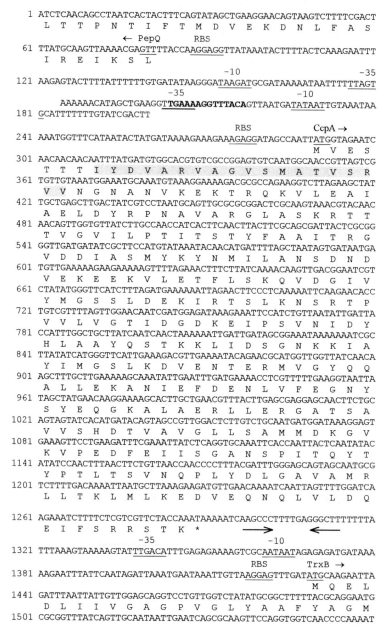

Figure 2. The nucleotide and amino acid sequences of the *ccpA* gene. Potential ribosome binding site (RBS), -10 and −35 promoter elements, and terminator (inverted repeats) are underlined. A 14-bp sequence with homology to *cre* consensus sequence is bolded and HTH (helix-turn-helix) element is indicated in gray.

The nucleotide sequence of *ccpA* gene (Gene Bank accession number AF106673) contains 999 bp that corresponds to 333 amino acids (Figure 2). The gene is preceded by a putative ribosome binding side RBS (AGAGG) located 10 bp upstream of the methionine start codon. The putative promoter consists of -10 box (TATAAT) at 226 bp and a -35 box (TTGAAA) at 203 bp. We also identified a putative *cre* sequence, which differs by 1 bp from the *cre* consensus sequence. (Table 1). This sequence is upstream of putative *ccpA* and is overlapping the promoter region. The CcpA protein contains also HTH (helix-turn-helix) element, which is the DNA binding domain. The *ccpA* gene is followed by a transcriptional terminator (inverted repeats) at 1297 bp.

The location of *ccpA* gene in chromosomal DNA was further characterized. It was found that there is the *pepQ* gene coding for peptidase immediately upstream of the *ccpA* gene. The *pepQ* gene is divergently oriented in respect to *ccpA*.

The putative *pepQ* promoter consists of -10 box (TAGAAT) at 153 bp and a -35 box (GTGATT) at 175 bp. The *trxB* gene coding for thioredoxin reductase, transcribed in the same direction, follows *ccpA*. The putative *trxB* promoter was identified. It consists of -10 box (AATAAT) at 1361 bp and -35 box (TTGACA) at 1338 bp.

Table 1
Comparison of *Lactococcus lactis ccpA cre* sequence with the *cre* consensus

	cre sequence
	T A
Consensus sequence	TGA* A* CG* T*TCA
Lactococcus lactis ccpA	TGAAAAGGTTTACA

3. CONCLUSIONS

We identified a gene implicated in cellobiose inducible lactose catabolism. The gene that was found is homologous to the *ccpA* gene from *Bacillus subtilis*. CcpA protein from *Lactococcus lactis* reveals 96% identity and 97% similarity with the CcpA protein from *Lactococcus lactis* subsp. *cremoris* [7]. The high homology (more then 48% amino acids identity) is also observed with CcpA proteins from: *Lactobacillus casei, Streptococcus mutans, Listeria monocytogenes, Bacillus subtilis, Bacillus megaterium, Staphylococcus xylosus* and *Thermoactinomyces sp.*

Some other proteins, which are homologous to CcpA from *Lactococcus lactis*, are members of the LacI Family of regulators, like: DepA - transcriptional regulator from *Bacillus subtilis*, RbsR - transcriptional regulator from *Thermatoga maritima*, RbsR - ribose operon represor from *Bacillus subtilis*.

These results strongly suggest that the CcpA protein play a role in cellobiose-inducible lactose catabolism in *Lactococcus lactis*. Moreover, it is possible to speculate that the CcpA from *Lactococcus lactis* is involved in carbon catabolite repression.

Three elements had been reported to be necessary for carbon catabolite repression of *amyE* gene in *Bacillus subtilis*. One was the CcpA protein, another was P-Ser-HPr, which interacts with CcpA. The third one - an operator-like, cis acting element in front of the regulated gene is catabolite responsive element, *cre* [8]. It is already known that CcpA protein binds to the *cre*, which is present in most operons sensitive to carbon catabolite repression in Gram-positive bacteria. This interaction occurs in the presence of P-Ser-HPr. Therefore binding of CcpA to the *cre* elements is probably mediated by an allosteric interaction between P-Ser-HPr and CcpA [9]. HPr can be phosphorylated at Ser-46 by the ATP-dependent, metabolite activated protein kinase (HPr kinase). The activity of HPr kinase is positively regulated by fructose-1,6-bisphosphate (FBP).

The P-Ser-HPr is formed when the cell takes rapidly metabolisable carbon source. The role of CcpA protein in carbon catabolite regulation in *Lactococcus lactis* is still not well elucidated. However, we demonstrated that CcpA is involved in growth of *Lactococcus lactis* on lactose in the presence of cellobiose. The *ccpA* sequence analysis shows that the CcpA protein consists of 333 amino acids. Similarly to its counterparts from other Gram-positive bacteria the lactococcal CcpA protein contains the HTH motif, which is essential for binding to the DNA of regulated genes and interacts with the *cre* sequences [10]. The presence of the *cre* sequence upstream of the *ccpA* and overlapping the putative promoter region suggests the autoregulation of the *ccpA* gene. Moreover, location of the *cre* between *ccpA* and *pepQ* genes hints that the expression of the latter gene can be regulated by CcpA protein.

The results presented in this work can be promising for future applications in respect to the possibility of improvement of dairy starters through genetic modifications. The *ccpA* gene takes part in the carbon catabolite regulation and increasing or decreasing of carbon catabolism could become an industrial, biotechnological advantage.

REFERENCES

1. W.M de Vos, Lactic Acid Bacteria:Genetics, Metabolism and Application – Proceedings of the Vth Symposium held in Veldhoven, (1996) 223.
2. R .J. van Rooijen, Ph.D. Thesis, Wageningen, The Netherlands, 1993.
3. J. Bardowski, S.D. Ehrlich, A. Chopin, J. Bacteriol., 176 (1994) 5681.
4. J. Bardowski, S.D. Ehrlich, A. Chopin, Dev. Biol. Stand. Basel, Karger, 85 (1995) 555.
5. E. Maguin, J. Bacteriol., 178 (1996) 931.
6. E. Maguin, H. Prevost, A. Gruss, Lait, 76 (1996) 139.
7. E.J. Luesink, Mol. Microbiol, 30 (4) (1998) 789.
8. J. Deutscher, C. Fischer, V. Charrier, A. Galinier, C. Lindner, E. Darbon and V. Dossonnet, Folia Microbiol., 42 (1997) 171.
9. Y. Fujita, Y. Miwa, A. Galinier and J. Deutscher, Mol. Microbiol., 17 (1995) 953.
10. M.J. Weickert and A.Sankar, J. Biol Chem., 267, (1992) 15869.

Food Biotechnology
S. Bielecki, J. Tramper and J. Polak (Editors)

Production and genetic regulation of an amylase in *Lactococcus la....*

M.Domań [b], E.Czerniec [a], Z Targoński [b] and J.Bardowski [a]*

[a] Department of Microbial Biochemistry, Institute of Biochemistry and Biophysics Polish Academy of Sciences, Pawińskiego 5a, 02-106 Warsaw, Poland

[b] Department of food Technology, Agricultural University, Skromna 8, 20-950 Lublin, Poland

Over 100 lactococcal strains isolated from their natural niche cow-milk were tested for production of amylolytic enzymes. Two of the tested strains were found to produce an amylase. One of them, *Lactococcus lactis* IBB500 was used for biochemical and genetic studies. We observed that *L. lactis* IBB500 strain started to produce the extracellular amylase in the BHI broth at the end of the logarithmic phase of growth. The maximal amount of the enzyme was detected at the early stationary phase. It demonstrated wide temperature optimum with at least 50% of its activity found in the range between 25°C and 60°C. The highest activity was observed at 50 °C. Moreover, production of the amylase was induced to different levels by various sugars e.g.: starch, maltose, maltotriose, pullulan etc. When glucose was added to the medium repression of the enzyme production was observed. Genetic studies showed that amylase gene was located on a 30 kb plasmid. This gene was cloned and expressed in *Lactococcus lactis*. Analysis of a partial DNA sequence of the region containing the gene indicated that it could encode a polypeptide homologous to various amylases like izoamylase from *Pseudomonas amyloderamosa*, pullulanase from *Thermus sp.* IM6501, pullulanase precursor from *K. pneumonie*, termostable pullulanase from *Bacillus stearothermophilus* and pullulanase from *Bacteroides thetaiotaomicron*.

1. INTRODUCTION

Starch is composed of mixture of linear and branched polysaccharides consisting of α-1,4-glucosyl residues and α-1,6-glucosyl branch points. The enzymatic saccharification of starch is catalyzed by several enzymes, classified according to the linkages hydrolyzed and their mode of action. α-Amylases (EC 3.2.1.1) attack exclusively α-1,4-glucosidic linkages in an endo-fashion yielding dextrins, oligosaccharides and maltose. Pullulanases (EC 3.2.1.4)

*Corresponding author. Tel.:+48 22 6594419; fax +48 22 6584636
This work was partly supported by the KBN grant 5 PO6G 032 14

hydrolyze the α-1,6-linkages of pullulan, a fungal polysaccharide composed of maltotriosyl units linked by α-1,6-glucosidic bonds. Most pullulanases also cleave α-1,6-linkages in amylopectin, and are therefore said to have debranching activity. Glucoamylases (EC 3.2.1.3) are exoglucanases that consecutively split α-1,4-glucosidic bonds of polymeric α-glucans to yield glucose α-1,6-linkages, which are also attacked, but with a rate of hydrolysis lower than that of α-1,4-linkages.

Amylases have several applications in the starch industry, mainly in the production of maltose syrups and high purity glucose and fructose. In these processes, pullulanases are used in tandem with other glycanolytic enzymes that hydrolyze amylopectin [1].

Amylolytic enzymes are produced by different Gram-positive bacteria e.g.: *Bacillus subtilis* [2, 3], *Bacillus amyloliquefaciens* [4], *Bacillus acidopulluliticus* [1], *Streptococcus bovis* [5] etc.. Moreover, enzymes degrading starch are produced by various thermophilic bacteria that belong to the genera: *Clostridium* [6] and *Thermoanaerobium* [7].

Amylolytic activity has been detected in some strains of the lactic acid bacteria, for example in: *Lactobacillus plantarum* [8] or *Lactobacillus fermentum* [9].

So far, there has not been any data on occurrence of amylolytic strains from the genus of *Lactococcus*. *Lactococcus lactis* belongs to the group of lactic acid bacteria, which play an important role in making fermented food (dairy products) and fodder. Plant environment is the natural niche for lactococci. Therefore, we supposed that some of these bacteria could produce enzymes degrading one of the main storage carbohydrates in plants i.e. starch.

This paper presents results of our research on production of an amylase by *L. lactis* IBB500 strain isolated from cow-milk environment.

2. RESULTS AND DISCUSSION

2.1. Identification of amylolytic strains

Over one hundred lactococcal strains isolated from cow-milk were tested for the amylolytic phenotype using a simple plate-test. The strains were dropped onto a rich BHI-agar medium supplemented with starch and after incubation at 30°C covered with Lugol solution (J_2 in KJ). In the case of strains producing an amylase, a halo of unstained, degraded starch was clearly visible, while undegraded starch became blue. After 3 days of incubation at 30°C two of the tested strains were found to produce an amylase. One of them, *Lactococcus lactis* IBB500 was used for further studies.

2.2. The effect of temperature on amylase activity

Amylase activity was assayed by measurement of the amount of the iodine-starch complex formed [10]. The clear, cell-free supernatant that contained an amylase enzyme was obtained after centrifugation. The enzymatic activity was determined at different temperatures within the range of 25°C – 60°C and at various starch concentrations in the BHI medium (Figure 1).

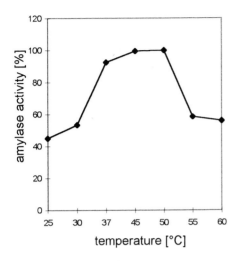

Figure 1. The effect of temperature on amylase activity

The enzyme demonstrates wide temperature optimum with at least 50% of amylase activity found in the range between 25°C and 60°C. Maximum activity was observed at 50°C.

2.3. The effect of glucose and starch on amylase activity

Production of the enzyme can be modulated by some sugars when present in the culture medium (Figure 2).

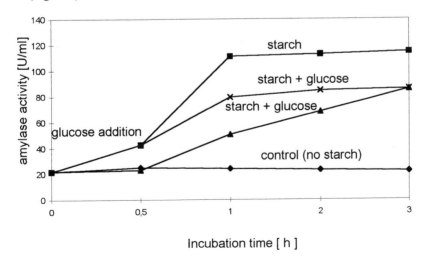

Figure 2. The effect of glucose and starch on amylase activity

We observed that glucose (0.5%) almost completely inhibited the enzyme biosynthesis. However, when starch was added together with glucose to the culture medium, activity of the enzyme decreased by about 20%. This suggests that production of the amylase by IBB500 strain undergoes glucose repression. The addition of starch to the BHI medium activates the enzyme biosynthesis. After 1 h cultivation of the producer strain amylase activity detected in the culture medium increased from 20 U/ml to 100 U/ml.

2.4. Effect of various starch concentrations on amylase production

Production of the amylase increases with the increase of starch concentration in the BHI medium (Figure 3).

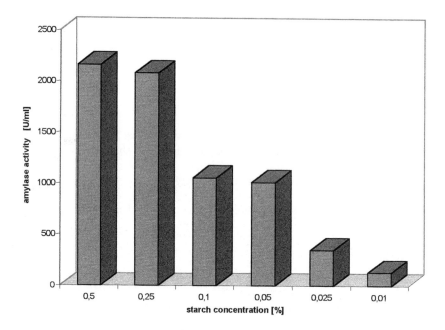

Figure 3. The effect of starch concentrations on amylase production

The highest amount of the amylase was detected at 0.5% of starch in the culture medium. Amylase activity was 4 times higher in the BHI medium + 0.25% starch than in the BHI medium + 0.025% starch. These results indicate that starch is a very good inducer for amylase production.

2.5. The effect of various sugars on amylase production

Influence of various sugars (isomaltose, glycogen, cyclodextrin, maltoheptaose, maltopentaose, maltotriose, pullulan, cellobiose, maltooligosaccharide, amylose, amylopectin, starch) on amylase production by *L. lactis* IBB500 strain was tested (Table 1). The strain was cultivated for 18 h in BHI medium containing various sugars in the final concentration of 0.5%. Then amylase activity was assayed as previously described [10].

Table 1
The effect of various sugars on amylase production

Sugar	Amylase activity [U/ml]
No sugar	5.25
Isomaltose	15.95
Glycogen	61.8
α-Cyclodextrin	83.5
β-Cyclodextrin	62.05
γ-Cyclodextrin	47.35
Maltoheptaose	53.0
Maltopentaose	3,15
Maltotriose	55.85
Maltohexaose	39.25
Maltotetraose	63.5
Maltose	54.85
Pullulan	64.45
Cellobiose	22.6
Maltooligosaccharide	26.55
Amylose	1120
Amylopectin	1223.6
Starch soluble acc. to Zulkowsky	1217.7
Starch from potato	1186.9
Starch from rye	1135.6
Starch from triticale	1136.9

We observed that various types of starch, as well as amylose and amylopectin, were good inducers for amylase production. Besides that maltose, maltopentaose and other oligosaccharides were also shown to be inducers. However, they exerted lower levels of induction comparing to starch.

2.6. Genetic studies on the amylase from *L. lactis* IBB500

The IBB500 producer strain was found to contain several plasmids. We used plasmid-curing strategy to find out if the amylase gene is plasmid located. It was observed that cultivation of the strain at elevated temperature (37°C) or its long-time incubation in the stationary phase of growth resulted in appearance of amylase-negative clones with the frequency of 3 %. The clones unable to produce the amylase were also devoid of a 30 kb plasmid. These experiments demonstrated that amylase gene was located on a plasmid in *Lactococcus lactis* IBB500 strain.

In cloning strategy, total plasmid DNA was separately cut with various restriction endonucleases. It was then ligated to pIL253 plasmid vector linearized with the same restriction enzyme and finally introduced to *L. lactis* MG1363 host through electroporation.

One amylase positive transformant, containing an insert of about 25 kb was found. Subcloning of the insert into smaller fragment through restriction-and-deletion procedure finally resulted in its diminishing to 8 kb fragment. We demonstrated that this plasmid, called pIBB502, fully expressed the amylase in the lactococcal host.

DNA sequence analysis of the region containing the gene indicated that it could encode an Orf with high similarity to the amino acid sequences of izoamylase from *Pseudomonas amyloderamosa*, pullulanase from *Thermus sp.* IM6501, pullulanase precursor from *K. pneumonie*, termostable pullulanase from *Bacillus stearothermophilus* and pullulanase from *Bacteroides thetaiotaomicron*. DNA sequence of the *L. lactis* IBB500 amylase revealed four highly conserved boxes characteristic for amylase family enzymes.

In conclusion, we demonstrated that *L. lactis* IBB500 strain produces an amylase. This enzyme is secreted into the culture medium. Production of the amylase can be easily modulated to different extend by various sugars. We found starch strongly induces while glucose represses biosynthesis of the amylase.

REFERENCES

1. J.P. Jensen, B.E.Norman, Process. Biochem., 19 (1984) 129.
2. N.Y. Ensari, B. Otludil, M.C. Aytekin, Starch/Starke, 47 (1995) 315.
3. E. Ferrari, A. Jornagin, B. Schmidt, Commercial production of extracellular enzymes, ASM, Washington, D.C. (1993) 917.
4. C. Groom, A. Daugulis, B. White, Appl. Microbiol. Biotechnol., 28 (1988) 8.
5. E. Satoh, T. Uchimura, T. Kudo, K. Komagata, Appl. Environ. Microbiol., 63 (1997) 4941.
6. H.H. Hyun, J.G.Zeikus, J. Bacteriol., 164 (1985) 1146.
7. A.R. Plant, S. Parrat, R.M. Daniel, H.W. Morgan, Biochem. J., 255 (1988) 865.
8. A. Fitzsimons, M. O'Conell, FEMS Microbiol. Lett., 116 (1994) 137.
9. V. Agati, J.P.Guyot, J. Morlon-Guyot, P. Talamond, D.J. Hounhouigan, J. Appl. Microbiol., 85 (1998) 512.
10. W.L. Nicholson, Y.K Park, T.M. Henkin, M. Won, M.J. Weicker, J.A. Gaskell, G.H. Chambliss, J. Mol. Biol., 198 (1987) 609.

Food Biotechnology
S. Bielecki, J. Tramper and J. Polak (Editors)
© 2000 Elsevier Science B.V. All rights reserved.

Introducing the killer factor into industrial strains of *S. cerevisiae* as a marker

Gniewosz M., Bugajewska A., Raczyńska-Cabaj A., Duszkiewicz-Reinhard W., Primik M.

Department of Food Biotechnology, Agricultural University, ul. Rakowiecka 26/30, 02-528 Warsaw, Poland

The presence of the killer factor in yeast may be a good marker for the identification of a strain because of the ease of its detection under laboratory conditions. The aim of the present research was the introduction of the killer factor into the cells of distillery and wine yeast using protoplast electrofusion. As a result of the performed protoplast electrofusion of the laboratory strain *S. cerevisiae* ATCC 42300 K_1 and *S. cerevisiae* of the S.o./1 strain; hybrids with the killer factor, whose stability did not change during two years of storage, were obtained.

1. INTRODUCTION

The killer toxin produced by the *S. cerevisiae* yeast may kill cells of sensitive yeast strains. This property may be utilized in fermentation in order to inhibit undesirable wild type yeast. On the other hand this factor is easy to detect under laboratory conditions [1,2].

Electrofusion is one of the methods of introducing the killer factor into industrial strains of the yeast *S. cerevisiae* [3,4]. This process allows for joining of cells with different degrees of ploidy. This is significant in the case of genetic modifications of industrial yeast strains, which are often polyploid [5]. It thus seemed justified to undertake research on the construction of industrial wine or distillery yeast which have a killer phenotype.

2. MATERIALS AND METHODS

2.1 Yeast strains

Wine yeast: Bordeaux, Bingen, Steinberg, Laureiro, Sauternes, Tokay, Riesling, mead strains 33 and 34, Malaga, Portwein, Madeira, Sherry, Burgund, Bratyslawa 1, S.o./l species *Saccharomyces cerevisiae* from the Pure Culture Collection of the Department of Food Biotechnology of the Agricultural University in Warsaw and distillery yeast B-4, B-4b, B-4i, Bc-16a, Bc-16ab, D-2, As-4, R-15, Ja, O-11 *Saccharomyces cerevisiae* from the Central Laboratory of Agriculture in Bydgoszcz were used. Yeast for checking the killer phenotype: *S. cerevisiae* ATCC 42300 α ade 2 thr 1 ski 2-1 [kill-K_1], *S. cerevisiae* ATCC 44069 a ura 1 MTKJ$^+$ [kill-K_2] [NEX-O], *S. cerevisiae* strain sensitive to the killer factor from the Department of Genetics, Institute of Biochemistry and Biophysics, Polish Academy of Sciences in Warsaw. Yeast for determining the mating type of monosporic populations:

S. cerevisiae PMY 274 (MAT a his 3 - Δ 200 leu 2-3, 112 ura 3-52 lys 2-801 trp 1-1),
S. cerevisiae PMY 275 (MAT α his 3 - Δ 200 leu 2-3, 112 ura 3-52 lys 2-801 trp 1-1).

2.2. Determination of killer phenotype of the strains

The killer phenotype was analyzed according to Rose et al. 1990 [6].

2.3 Determination of the neutral and sensitive phenotype of the strains

The sensitive and neutral phenotype of the strains was evaluated according to Cansado et al. 1991 [7].

2.4 Tetrad analysis

The distillery strains were grown on the YPD medium (2% glucose, 2% peptone, 1% yeast extract) and subsequently plated in duplicate on the McClary medium (1% potassium acetate, 0.25% yeast extract, 1% glucose, 1.5% agar) and the KAc medium (1% potassium acetate) in order to induce sporulation. The presence of sporulation asci was analyzed under the microscope several times during the 14 days of culture. The efficiency of sporulation was expressed as the percentage of all asci in the population of 300 counted cells [8]. The cells taken from sporulation media were suspended in a solution of the enzyme β-glucuronidase (Sigma, G-7770, 90000 units/ml, from *Helix pomatia*) in order to digest the cell walls of the asci. Single spores were isolated by means of a micromanipulator (Singer MSM System Series 200, Singer Instruments) on solid YPD medium. The plates were then incubated at 30°C for 7 days. The survival of the spores was determined as the percentage of isolated spores which formed macroscopically visible colonies (monosporic populations).

These populations were again plated on McClary medium in order to confirm the lack of sporulation. The mating type of non-sporulating monosporic populations was determined using *S. cerevisiae* PMY 274 and PMY 275 tester strains according to Rose et al. 1990 [6].

2.5 Protoplastization of the strains

The protoplastization of yeast strains was performed in 1 ml of the SorTrisCa solution (1 M sorbitol; 10 mM CaCl₂; 10 mM MgCl₂; 100 mM Tris; pH 7.5) supplemented with 250 μl β-glucuronidase (Sigma) and 0.015 g the enzymatic mixture of *T. harzianum* (Sigma) or the lytic enzymes of (*Rhizoctonia solani*, Sigma) dissolved in sodium acetate buffer pH 5.0 (30-60 min, 37°C) according to Rose et al. 1990 [6].

2.6 Protoplast electrofusion

In order to perform the fusion protoplasts of the two strains were mixed in a ratio of 1:1. After mixing 50 μl of the suspension (about 10^4-10^5 cells/ml) was taken and was placed between two electrodes of a frame cell. The process of protoplast electrofusion was observed under the microscope and fixed on an SVHS tape. After finishing the fusion process the post-fusion suspension was transferred to selective MMGS medium [5,9,3] supplemented with 0.5% chloramphenicol and 0.5% kanamycin. Incubation was performed for 7-14 days at 28°C.

2.7 Checking the killer phenotype of the isolated hybrids

The killer phenotype of the isolated hybrids was checked according to Rose et al. 1990 [6].

3. RESULTS AND DISCUSSION

The analysis was initiated by checking the killer phenotype of the wine and distillery strains. Among 16 wine and 10 distillery strains none was found to have produced the killer factor. The literature provides different information on the subject of the frequency of occurrence of the killer factor in yeast isolated from the fermentation media. According to Hidalgo and Flores [10] this is a frequently occurring factor. Among 270 strains derived from 11 wineries as many as 42.6% had the killer phenotype. However, among 38 winery yeast strains from the collection of the Forschunganstalt Geisenheim [11] only 4 had the killer phenotype.

Table 1
Phenotype of *S. cerevisiae* strains

Strain	Killer factor			
	K_1		K_2	
	Phenotype			
	Sensitive S	Neutral N	Sensitive S	Neutral N
Wine yeast				
Bordeaux	+[1]	-[2]	+	-
Bingen	+	-	+	-
Steinberg	+	-	+	-
Laureiro	+	-	+	-
Sauternes	+	-	+	-
Tokay	+	-	+	-
Riesling	+	-	+	-
No 33	+	-	+	-
No 34	+	-	+	-
Portwein	+	-	+	-
Malaga	+	-	+	-
Madeira	+	-	+	-
Burgund	+	-	+	-
Sherry	+	-	+	-
Bratyslawa 1	+	-	+	-
(S.o./1)	+	-	+	-
Distillery yeast				
B-4	+	-	-	+
B-4b	+	-	+	-
B-4i	+	-	+	-
Bc-16a	+	-	+	-
Bc-16ab	+	-	+	-
D2	+	-	+ weak	-
As-4	+	-	+	-
R-15	+	-	-	+
Ja	+	-	+ weak	-
O-11	+	-	-	+

[1] Zone of growth inhibition was observed
[2] Zone of growth inhibition was not observed

Thus in the present work all strains were checked for their sensitivity and neutrality in respect to the killer toxins of the K_1 and K_2 type produced by the *S. cerevisiae* ATCC 42300 (K_1) and ATCC 44069 (K_2) strains (Table 1).

All wine and distillery strains showed sensitivity to K_1 toxin, but differences in sensitivity to K_2 killer factor were observed. Five distillery strains B-4b, B-4i, Bc-16ab and As-4 had high sensitivity to this factor, which was indicated by a broad zone of growth inhibition. Two other distillery yeast strains D-2 and Ja showed a weak sensitivity to K_2 toxin. In this case the zone of inhibition did not exceed 2-3 mm. Among the remaining three distillery strains B-4, R-15 and O-11 no zone of growth inhibition was observed, which indicates their neutral character in respect to the killer toxin of the K_2 type.

We started the work on obtaining the haploid forms by checking the ability of selected distillery (Table 2) and wine (Table 3) strains to sporulate. Four wine strains: Tokay, Bordeaux, Burgund and S.o./l did not produce spores on any of the two used sporulation media. The Sherry strain produced a small amount of sporulation asci only on KAc medium, and the distillery strain B-15 only on McClary medium. The remaining strains formed sporulation asci with an efficiency permitting their easy detection and separation using a micromanipulator. The distillery yeast D-2 and the wine strains Riesling, Bingen and Malaga had a distinctly higher yield of sporulation on KAc medium.

Spore survival was evaluated after their isolation from 6 of the best sporulating wine and distillery strains. According to Pretorius and van der Westhuizen [12] most of the wine yeasts are homothalic and have higher spore survival than distillery yeasts. The present work appears to confirm this rule - distillery strains had a low spore survival rate from 0.6 to 25%. Observations by other authors analyzing Polish cultures of industrial yeasts also indicate the problem of the low survival of the spores. Of the two distillery strains capable of sporulation analyzed by Szopa et al. [13] one did not produce any spores capable of growth and in the case of the second one the spore viability was 26%. Sałek and Arnold [14] screened about 30 industrial yeast strains and observed poor sporulation and the colonies obtained from single spores grew slowly and were small.

Table 2
Frequency of sporulation distillery yeast (%)

Strain	Frequency of sporulation	
	McClary medium	Kac medium
Ja	74	76
D-2	34	62
O-11	58	60
B-4	60	59
B-4i	48	56
B-4b	42	53
Bc-16a	9	9
R-15	0.5	0

Table 3
Frequency of sporulation wine yeast (%)

Strain	Frequency of sporulation	
	McClary medium	Kac medium
Riesling	25	80
Bingen	5	60
Madeira	35	30
Steinberg	25	25
Malaga	2	10
Portwein	3	5
Sherry	0	5
Tokay	0	0
Bordeaux	0	0
Burgund	0	0
S.o./1	0	0

After analyzing the selected monosporic populations most of the wine strains were found not to possess a mating type. The ability to mate was only found in the case of a few colonies obtained from spores of the Bingen and Riesling strains.

In the case of monosporic populations obtained from distillery strains the ability to mate was observed frequently. These results appear to be encouraging as homothality observed in many monosporic populations obtained from industrial strains [14] does not appear to be a common phenomenon among the cultures obtained in the present work. Many of the obtained monosporic populations may thus become an attractive material in programs of genetic enhancement of technological properties of the strains, also in the method of protoplast electrofusion.

For further investigations two yeast strains were chosen: a distillery strain Ja with mating type a and the polyploid wine strain S.o./l, which because of its extraordinary properties (osmotolerance, alcohol tolerance) acquired by means of prolonged environmental selection [15] is being prepared for testing on industrial scale. The introduced killer factor will be a marker permitting the control of stability of these strains under industrial conditions.

The next stage of the work was the adaptation of conditions of protoplastization of *S. cerevisiae* strains Ja, S.o./l and ATCC 42300 with the killer phenotype K_1. In the investigations the suitability of three lytic enzymes was checked: β-glucuronidase, a lytic enzymes from *R. solani* and from *T. harzianum*. On the basis of the degree of protoplastization and regeneration of individual yeast strains the best conditions of the protoplastization process using β-glucuronidase (approx. 40 min., 37°C) and an enzymatic mixture composed of two preparations β-glucuronidase and enzymes from *T. harzianum* (approx. 40 min., 37°C) were selected (Figures 1 and 2).

The degree of protoplastization using lytic enzymes from *R. solani* and from *T. harzianum* turned out to be insufficient for this process.

Dielectrophoresis of the killer strain K_1 with the wine strain S.o./ l and also the killer strain with the haploid form of the distillery yeast Ja strain was performed using the changing field voltage of 15-20 V and the frequency of 2000 kHz. The fusion of protoplasts arranged in chains was obtained using 1 - 3 pulses lasting for 40 - 60 μs and the intervals between the pulses from 0.5 to 1 s. The value of the voltage of the rectangular pulse causing the joining of protoplasts was at the level of 100 to 140 V. The obtained electrofusion parameters are close to those given by Sałek and Arnold [14] and Aarnio and Suikho [16]. These authors obtained

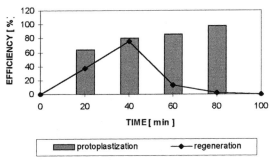

Figure 1. Efficiency of protoplastization and regeneration of ATTC 42300 K_1 strain with β-glucuronidase and enzymes from *T. harzianum*.

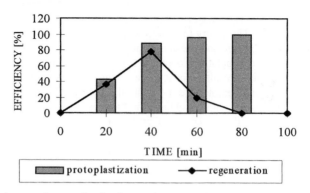

Figure 2. Efficiency of protoplastization and regeneration of *S. cerevisiae* S.o./1 wine strain with β-glucuronidase and enzymes from *T. harzianum*.

an alignment of yeast protoplasts using changing field voltages from 12 to 20 V and frequencies of 2000 kHz. They also used a similar number of pulses 2 to 4 and a similar time of their duration 40 - 60 μs but higher values of amplitudes of the rectangular pulse as they were 140 - 160 V.

As a result of the performed electrofusion of protoplasts of the laboratory strain *S. cerevisiae* ATCC 42300 K_1 and *S. cerevisiae* of the S.o./1 strain hybrids were obtained which had the killer factor and their stability did not change during two years of storage at 4°C. However, the obtained hybrids of the K_1 killer strain with the distillery strain Ja turned out to be non-permanent and lost the killer factor after 6 months of storage at 4°C.

REFERENCES

1. S. Goto, K. Kitano, T. Shinohara, J. Ferment. Bioeng., 73 (1991) 70.
2. S. Hara, Y. Uimura, H. Oyama, T. Kozeki, K. Kitano, K. Otsuka, Am. J. Enol. Vitic., 32 (1980) 28.
3. T. Yamazaki, H. Nonomura, J. Ferment. Bioengin., 4/ 72 (1991) 300
4. U. Zimmermann, J. Vienken, J. Halfman, C.C. Emeis, Adv. Biotechnol. Proces., 4 (1985) 79.
5. H.J. Halfmann, C.C. Emeis, U. Zimmermann, FEMS Microbiol. Lett., 20 (1983) 13.
6. M.D. Rose, F. Winston, P. Hieter, Methods in yeast genetics. A laboratory course manual. Cold Spring Harbor Laboratory Press, 1990.
7. J. Cansado, E. Longo, P. Calo, C. Sieiro., I.B. Velazquez, T.G. Villa, Appl. Mikrobiol. Biotechnol., 34 (1991) 643.
8. N. Nakazawa, T. Ashikari, N. Goto, T. Amachi, R. Nakajima, S. Harashima, Y. Oshima, J. Ferment. Bioeng., 73 (1992) 265.
9. N. Urano, H. Sahara, S. Koshino, Enzyme Microbiol. Technol., 15 (1993) 959.
10. P. Hidalgo, M. Flores, Food Microbiol., 11 (1994) 161.
11. K. Shimizu, T. Adachi, K. Kitano, T. Shimazaki, A. Totsuka, S. Hara, H.H. Dittrich, J. Ferment. Technol., 63/ 5 (1985) 421.
12. S. Pretorius, T. J. van der Westhuizen, S. Af. J. Enol. Vitic., 12 (1991) 3.

13. J.S. Szopa, B. Gańczyk, K. Kowal, G. Zwoliński, Acta Alimentaria Polonica, 16/50 (1990) 97.
14. A.T. Sałek, W. Arnold, Chem. Mikrobiol. Technol. Lebensm., 16 (1994) 165.
15. A. Bugajewska, W. Wzorek, Przemysł Fermentacyjny i Owocowo – Warzywny, 6 (1995) 20. (In Polish)
16. T. H. Aarnio, M. L Suihko, Appl. Biochem. Biotech., 1 / 27 (1991) 65.

Food Biotechnology
S. Bielecki, J. Tramper and J. Polak (Editors)

The development of a non-foaming mutant of *Saccharomyces cerevisiae*

E. Kordialik-Bogacka[a] and I. Campbell[b]

[a] Technical University of Lodz, Poland
Institute of Fermentation Technology & Microbiol., 90-530 Łódź, ul. Wólczańska 173

[b] Heriot-Watt University, Edinburgh, United Kingdom
Department of Biological Sciences, Riccarton, EH14 4AS

The non-foaming mutant of the brewery production top-fermenting ale strain *Saccharomyces cerevisiae* was induced by serial repitching of bottom-cropped yeast. The concentration of total protein and hydrophobic polypeptide in wort fermented with the non-foaming mutant and its parent strain was determined in order to elucidate their effect on the formation of a yeast head during beer fermentation. A slight difference in the quantity of total protein in beer fermented with the non-foaming mutant and its parent strain was found. In turn, there were not any systematic differences in hydrophobic polypeptides. Simultaneously, organoleptic properties of green beer fermented with the non-foaming mutant and the original strain were checked and significant variations were observed. Large variations were also found in attenuation of these beers and in the viability of the bottom-cropped non-foaming mutant and the top-cropped parent strain collected after successive fermentations.

1. INTRODUCTION

The production of foam in fermentation vessels is a disadvantage to the brewer. Overfoaming can result in (a) fouling of the carbon dioxide off-take pipe with foam and compromise the hygiene or the operation of the fittings at the top of the vessel, (b) the loss both of beer and of hop bitterness and (c) possible damage to the foaming characteristics of the final beer due to loss of some important foam components. A large number of yeast cells is included in the froth and it appears that these cells adhere to the surface of carbon dioxide bubbles and contribute to stabilising the froth-head. This means that cylindroconical vessels may only be filled to 80 to 90% of their capacity in the case of lager worts. For faster fermenting beers such as ales or stouts, the fill level could be as low as 60%. The capital cost implication of these reduced fill levels is the expenditure on a greater number of fermenters than would otherwise be required. There already exist several methods for controlling foam levels in a fermenter, among which the most common is the use of antifoam, mainly silicone and fatty acid esters such as sorbitan monolaurate.

---

* The project has been carried out with the support of the European Commission within the framework of the Tempus Phare Scheme.

However, all these methods either compromise the foam quality of the final beer or pose a significant hygiene hazard for the process. Therefore, alternatives for limiting foam in fermentation vessels are being sought. One of them is the use of non-head forming yeast strains.

The aim of this work was to develop the non-foaming mutant of *S. cerevisiae* and check whether foaminess of its parent strain was mainly influenced by total protein and/or hydrophobic polypeptide present in fermented wort by comparing the concentration of total protein and hydrophobic polypeptide in wort fermented with the non-foaming mutant and its parent strain. Simultaneously, organoleptic properties of green beer fermented with the non-foaming mutant were checked in comparison with organoleptic properties of green beer fermented with its parent strain.

2. MATERIALS AND METHODS

2.1. Yeast
The brewery production top-fermenting ale strain *S. cerevisiae* was employed. The non-foaming mutants were induced by serial repitching of bottom-cropped yeast (until generation 27). Laboratory fermentations were performed in 2 litre beakers with 12°P autoclaved hopped wort from one batch with 30% addition of glucose at 17°C. When the froth reached the maximum height i.e. at 50 hours of fermentation time both the non-foaming mutant (cropped from the bottom of the fermentation vessel) and its parent strain (skimmed off) was selected and used as pitching yeast for the next fermentation.

2.2. Beer analysis
Samples of green beer collected after successive fermentations were analysed for total protein, hydrophobic polypeptide, flavour compounds and specific gravity. The protein content and the hydrophobic polypeptide concentration of beer were measured using the method developed by BRF [1]. Flavour compounds were analysed by gas chromatography with a flame ionisation detector (FID). The specific gravity was measured on a PAAR DMA 46 calculating digital density meter.
Each sample was assayed in triplicate.

2. 3. Viability assay
Viability was determined by staining with methylene blue according to Pierce method [2].

3. RESULTS AND DISCUSSION

Dixon and Kirsop [3] showed that top-fermenting strains of *S. cerevisiae* stabilised bubbles of fermented wort foam by binding surface-active materials whereas those of bottom-fermenting *Saccharomyces uvarum* did not. The foaminess of yeast strains is thus irretrievably tied up with the content of surfactants in fermented wort. However, yeast can also release foam-promoting substances itself. Tybussek *et al.* [4] have reported that foam formation of *S. cerevisiae* is mainly controlled by extracellular proteins which can act as collectors and thus are able to influence the yeast cell enrichment in the foam. Bahr *et al.* [5]

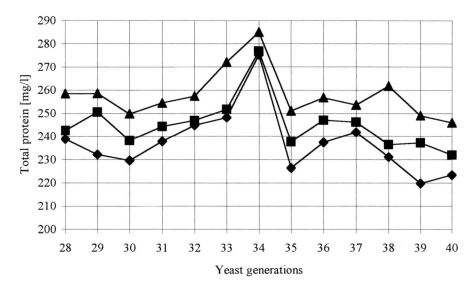

Figure 1. Comparison of total protein content in beer fermented with the non-foaming mutant (■) and with its parent strain (♦) in successive fermentations; total protein content in wort before fermentation (▲).

have found similar relationship with regards to *Hansenula polymorpha*. Bahr *et al.* have shown that the flotation of the cells *H. polymorpha* is also strongly dependent on extracellular protein composition. This was not confirmed for *S. cerevisiae* by Tybussek *et al.*, who noted only a slight effect of extracellular proteins on the flotation of *S. cerevisiae*.

In this study the concentration of total protein in wort was on average 5.2% higher than in beer fermented with the non-foaming mutant and 8.8% higher than in beer fermented with the original strain (Fig. 1). Thus, there is, although slight (3.6%), a difference in the quantity of total protein in beer fermented with the non-foaming mutant and its parent strain. However, on account of these results, it is not possible to assess the impact of total protein on foaminess of *S. cerevisiae*. It may be assumed that there is accumulation of protein on the yeast cell wall and that more protein adhere to the yeast cell surface in the case of the foaming parent strain. It is likely that these interactions influence the foaminess of the fermented wort. It can be presumed that the cells of the foaming parent strain acquire an additional surface-active coating of proteins, enabling them to the larger extent to adhere to bubbles of carbon dioxide, which carry them to the surface. However, it must be stressed that these results cannot lead to a definite conclusion because the non-foaming mutant and its parent strain can also exert different quantities of exoproteins contributed foaminess.

No systematic differences in hydrophobic polypeptides in beer fermented with the non-foaming mutant and its parent strain were found (Fig. 2).

Significant variations in the organoleptic properties of green beer fermented with the non-foaming mutant in relation to its parent strain were observed. Green beer fermented with the non-foaming mutant contained on average by 165% lower concentrations of acetaldehyde

84

than green beer fermented with its parent strain (Fig. 3). In turn, the non-foaming mutant produced higher amounts of fusel alcohols (n-propanol – 25%; isobutanol – 25%; 2-methyl-butanol – 27%; 3-methyl-butanol – 27%).

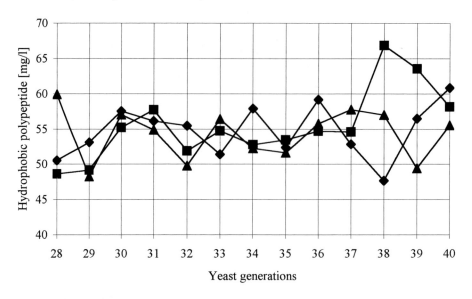

Figure 2. Comparison of hydrophobic polypeptide content in beer fermented with the non-foaming mutant (■) and with its parent strain (♦) in successive fermentations; hydrophobic polypeptide content in wort before fermentation (▲).

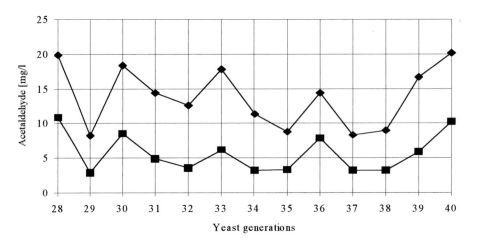

Figure 3. Comparison of acetaldehyde content in beer fermented with the non-foaming mutant (■) and with its parent strain (♦) in successive fermentations.

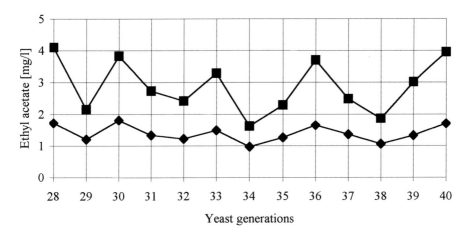

Figure 4. Comparison of ethyl acetate content in beer fermented with the non-foaming mutant (■) and with its parent strain (♦) in successive fermentations.

The non-foaming mutant also produced higher amounts of esters (ethyl acetate – 104% (Fig. 4); iso-amyl acetate – 135%; ethyl hexanoate – 123%). From the presented results it can be seen that the change of foaming properties of yeast entailed the change of organoleptic properties of fermented beer.

A difference was also found in attenuation of beer fermented with the non-foaming mutant and its parent strain, with higher final concentration of ethanol by the former (Fig. 5).

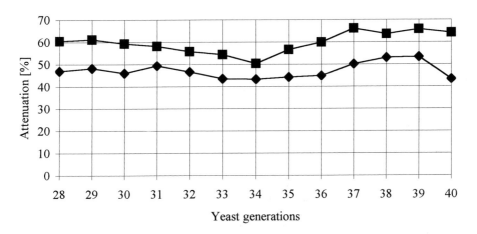

Figure 5. Comparison of attenuation of beer fermented with the non-foaming mutant (■) and with its parent strain (♦) in successive fermentations.

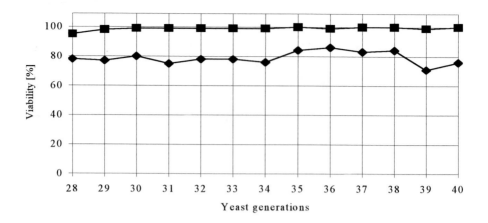

Figure 6. Comparison of viability of the original strain (♦) and the non-foaming mutant (■) after successive fermentations.

Moreover, it was observed that viability of the bottom-cropped non-foaming mutant was consistently high, above 99%, whereas viability of the top-cropped parent strain varied from 71% to 86% (Fig. 6). The reasons for this are not known.

4. CONCLUSIONS

Non-foaming brewing yeast can be obtained easily from pure culture strain by serial repitching of bottom-cropped yeast. Little difference in the concentration of total protein and no systematic differences in the concentration of hydrophobic polypeptide in beer fermented with the non-foaming mutant and its parent strain were found. The change of foaming properties of yeast resulted in the change of organoleptic properties of produced beer as well as the increase of both beer attenuation and viability of yeast.

REFERENCES

1. C.W. Bamforth, The Brewer, 81 (1995) 396.
2. J.S. Pierce, J. Inst. Brew., 76 (1970) 442.
3. I.J. Dixon and B.H. Kirsop, J. Inst. Brew., 75 (1969) 200.
4. R. Tybussek, F. Linz, K. Schűgerl, N. Moses, A.J. Léonard, P.G. Rouxhet, Appl. Microbiol. Biotechnol., 41 (1994) 13.
5. K.H. Bahr, H. Weisser, K. Schűgerl, Enzyme Microb. Technol., 13 (1991) 747.

Food Biotechnology
S. Bielecki, J. Tramper and J. Polak (Editors)
© 2000 Elsevier Science B.V. All rights reserved.

Genetic transformation of mutant *Aureobasidium pullulans* A.p.-3 strain

J. Kuthan –Styczeń[a], M. Gniewosz[b], K. Strzeżek[a], E. Sobczak[b]

[a]Institute of Biotechnology and Antibiotics, ul. Starościńska 3 ,Warsaw, Poland
[b]Department of Food Biotechnology, Agricultural University, ul. Rakowiecka 26/30, 02-528 Warsaw, Poland

Protoplasts of the low yield Mx-3 mutant were transformed with native DNA isolated from the high yield mutant Dy-17. Three transformants were obtained which utilized galactose as a carbon source and produced pullulan at the same level as the high yield mutant Dy-17. The kinetics of pullulan production, cell biomass content, saccharose utilization and changes in pH were compared with batch culture of transformants and parental strains performed in a shaker at 28°C for 96 hours. The analyzed parameters of kinetics for the transformants proved to be similar to the kinetics of the Dy-17 mutant.

1. INTRODUCTION

Pullulan – polysaccharide produced by *A. pullulans* strains [1] can be used as low calorie food additive or substrate for the production of biodegradable packagings. Thin layers of pullulan are impenetrable to oxygen and lipids. Due to these characteristics it may serve to coat food products protecting them from oxidative processes [2]. Chemically modified pullulan can also be used as material for manufacturing bags and cartons or boxes [3].

As a result of mutagenesis of *A. pullulans* A.p.-3 strain, the mutants with high and low yields of pullulan production were obtained [4]. Finding a selective marker, which would enable quick isolation of clones with pullulan gene expression, was the aim of the studies.

2. MATERIALS AND METHODS

2.1. Organism and growth conditions

The mutants Dy-17 and Mx-3 of *A. pullulans* A.p.-3 from the collection of the Department of Food Biotechnology, Agricultural University, Warsaw, Poland, were used. Cultures were transferred from slants to liquid media containing: 6% saccharose, 0.75% K_2HPO_4, 0.15% NaCl, 0.07% $(NH_4)_2SO_4$, 0.04% $MgSO_4 \cdot 7\ H_2O$, 0.04% yeast extract (bioMerieux). The medium pH was 6.0. The cultures were shaken (150 rpm) at 28°C for 24 h. 1 ml cultures were then transferred into the fresh medium and cultivated in the same conditions as above for 18 h.

This work was financially supported by the KBN grant No 5 PO6G 007 10.

2.2 DNA isolation

DNA of the Dy-17 *A. pullulans* mutant was isolated according to a previously described procedure [5]. Electrophoresis was performed in 1% agarose, applying 10 μl of the sample [6]. DNA obtained from a *Saccharomyces cerevisiae* strain was used as the standard.

2.3 Preparation of protoplasts

Mycelium of the Mx-3 *A. pullulans* mutant was centrifuged at 8000 rpm for 15 min at 40°C and washed with water. Subsequently the mycelium was transferred to a sterile flask and suspended in a solution (0.6 M KCl, 10 mM phosphate buffer pH 6.0). The lytic enzymes from *Trichoderma harzianum* (Sigma) was added to a concentration of 2.5 mg/ml. Digestion was performed in a rotary shaker at 150 rpm at 37°C for 30 minutes. The protoplasts were filtrated through a sterile Schott funnel and washed twice with 0.6 M KCl. Subsequently the protoplasts were suspended in a sorbitol solution (1.2 M sorbitol, 50 mM $CaCl_2$, 10 mM Tris pH 7.5) and centrifuged at 4000 rpm for 10 minutes. The pellet was suspended in the same solution to a density of 10^7 cfu/ml.

2.4 Protoplast transformation

10 μl of the native DNA of the Dy-17 strain was added to 100 μl of protoplasts of the Mx-3 strain, mixed and incubated on ice for 5 minutes. Subsequently 15 μl of PEG solution (25% PEG 6000, 50 mM $CaCl_2$, 10 mM Tris pH 7.5) was added. The mixture was incubated on ice for approximately 30 minutes and 1 ml of PEG solution was added. After 5 minutes at room temperature 3 ml of a solution containing: 1.2 M sorbitol, 50 mM $CaCl_2$, 10 mM Tris pH 7.5 was added and mixed carefully. Subsequently, the sample was transferred to a dry test tube and suspended in 4 ml of double concentrated selective medium containing: NH_4NO_3 6.0 g/l; galactose 10 g/l, KCl 0.52 mg/l, Mg_2SO_4 0.52 mg/l, $KHPO_4$ 1.52 mg/l in 2.4 M sorbitol (dissolved and kept in a water bath at the temperature of 56°C). After mixing the contents were spread onto 15 ml of a selective medium. The results of the transformation were observed after 4-5 days.

2.5. Analytical methods

The measurements of kinetic parameters of the process were done every 24 h. The fungus biomass was centrifuged (14 000 rpm, 15 min at 4°C) from a sample of batch culture, washed three times with distilled water and dried to constant weight at 105°C. Pullulan was precipitated from the samples (after cell removal) with 96% ethanol and the precipitate was centrifuged (14 000 rpm for 20 min at 4°C), re-washed with ethanol and dried to constant weight at 105°C. The dry sample was then hydrolized with 1 M HCl for 2 h at 100°C and monosaccharides were determined by the phenol-sulphuric acid method [7]. Saccharose was determined in the supernatants using the phenol-sulphuric acid method [7] after separation of cells and pullulan.

3. RESULTS AND DISCUSSION

The search for high yield *A. pullulans* mutants for pullulan production is related to time-consuming investigation. The research is based on performing several fermentation trials for a large number of isolated and randomly selected clones (after mutagenization) and assaying the content of the pullulan released by them into the environment [8,9,10]. Additionally, the mechanism of action of many mutagenic factors on filamentous fungi has not been yet sufficiently understood, which often makes success a matter of chance. Therefore, an attempt was made to find a selective marker making it possible to shorten this tedious selection procedure considerably, and at the same time increasing the probability of obtaining high-yield *A. pullulans* mutants for pullulan production.

The kinetics of pullulan production, the increase in cell biomass, the utilization of saccharose and the change in pH for high (Dy-17) and low yield (Mx-3) mutants were analyzed. These mutants were obtained as a result of coupled mutagenization (ethyleneimine together with UV) of the *A. pullulans* strain A.p.-3 [4]. Significant differences in pullulan production were found between these strains already after the second day of cultivation (Fig. 1). With the same increase in cell biomass of the analyzed strains a two times more intensive pullulan production by the high yield mutant was obtained. After 96 hours of cultivation the Dy-17 mutant produced 3.57 g pullulan/100 ml of the substrate whereas the low yield mutant only 1.01 g/100 ml. A more intense utilization of saccharose was obtained during the second and third day of cultivation for the strain Dy-17, whereas the changes in pH were similar.

A characterization of the biochemical properties of wild type strain *A. pullulans* A.p.-3 and twenty of each high and low yield mutants was performed using API-tests (bioMerieux). Differences were found between these mutants based on the fact that all low yield mutants were unable to utilize galactose as the sole source of carbon. All high mutants and wild strain *A. pullulans* A.p.-3 were able to utilize galactose as the sole source of carbon. An attempt was made to demonstrate that pullulan production by a strain is linked to its ability to utilize galactose.

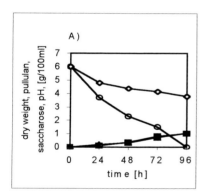 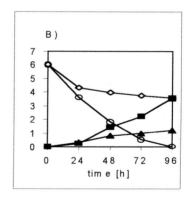

Figure 1. Time course of pullulan biosynthesis, saccharose utilisation and pH changes in cultures of parental strains of *A. pullulans*: A) Mx-3, B) Dy-17.
Dry weight ▲, pullulan ■, saccharose O, pH ◇ .

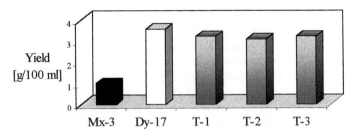

Figure 2. Biosynthesis of pullulan of low-yield mutant Mx-3, high-yield mutant Dy-17 and T-1, T-2, T-3 transformants.

Protoplasts of the low yield mutant Mx-3 were transformed with native DNA isolated from the high yield mutant Dy-17. After transformation the protoplasts were plated on a selective medium with galactose. Three transformants were obtained utilizing galactose as the sole source of carbon. Next the ability to synthesize pullulan of these transformants was checked and they were found to produce pullulan at the same level as the Dy-17 mutant i.e. almost three times more than the Mx-3 mutant (Fig. 2).

For the obtained transformants the time course of pullulan production, the increase in cell biomass, saccharose utilization and pH changes during 96 hours of cultivation was checked (Fig. 3). The changes of these parameters was very similar for all transformants. After 96 hours of cultivation 0.98 - 1.02 g of dry weight/100 ml medium was found. Changes were observed in pH from an initial value of 6.0 to about 4.0 after 24 hours and 3.43-3.52 after 96 hours of cultivation. The production of pullulan and the utilization of saccharose were at the same level, as for the parental high yield mutant Dy-17.

The comparison between media containing saccharose and galactose led us to conclusion that on the medium with galactose mutants with high pullulan productivity were obtained twice as frequently [data not shown].

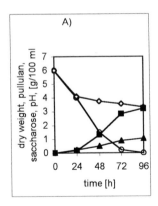 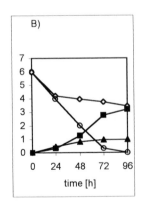 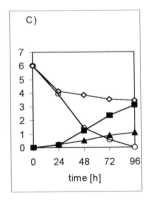

Figure 3. Time course of pullulan biosynthesis, saccharose utilisation and pH changes in cultures of *A. pullulans* transformants: A) T-1, B) T-2, C) T-3.
Dry weight ▲ , pullulan ■, saccharose O , pH ◇ .

4. CONCLUSION

The results of the experiments led us to the conclusion that galactose, as the selective marker, can be used for fast isolation of the high-yield *A. pullulans* mutants.

REFERENCES

1. H. Bender, J. Lehman, K. Wallenfels, K. Biochim. Biophys. Acta, 36 (1959) 309.
2. S. Yuen, Process Biochem., Nov. (1974) 7.
3. J. Rice, Food Procesing, 10 (1991) 34.
4. M. Gniewosz, E. Sobczak, D. Wojciechowska, J. Kuthan-Styczeń, Pol. J. Food Nutr. Sci., 8/49, No 2 (1999) 1.
5. M. M.Yelton, J. E Hamer. and W.Timberlake, Genetics, 81 (1984) 1470.
6. J.Sambrook, F.F. Fritsch, T. Maniatis(2nd ed.), Cold Spring Harbor Labotatory Press, Cold Spring Harbor, New York,1989.
7. K. A. Dubois, J.K. Gilles , P.A. Hamilton, F. Rebers and Smith , Anal. Chem., 28 (1956) 350.
8. T.P.West and B. Reed-Hamer, FEMS Microbiol. Lett., 124 (1994) 167.
9. L. Tarabasz-Szymanska and E. Galas, Enzyme Microb. Technol., 15 (1993) 317.
10. P.J. Kelly and B.J. Catley, J.General Microbiol., 102 (1977) 249.

FOOD PROCESSING AND FOOD PRODUCTS

Food Biotechnology
S. Bielecki, J. Tramper and J. Polak (Editors)
© 2000 Elsevier Science B.V. All rights reserved.

Enzymes in food and feed: past, present and future

Gert S.P. Groot, Marga A. Herweijer, Arthur L.M. Simonetti, Gerard C.M. Selten and
Onno Misset

DSM Food Specialties R&D,
P.O. Box 1, 2600 MA Delft, The Netherlands

Don't waste clean thinking on dirty enzymes [ref.1]

1. INTRODUCTION

Enzymes act as catalysts to carry out chemical reactions. In the living world each chemical reaction is catalyzed by its own enzyme and therefore enzymes exhibit a high specificity and are able to discriminate between slightly different substrate molecules.

Compared with chemical catalysts, advantages of enzymes are, besides this high specificity, their ability to efficiently operate at moderate conditions with respect to temperature and pressure. Moreover, they function efficiently in aqueous solutions. It is therefore no surprise that enzymes are more and more used in the food and feed preparation.

2. PAST

Enzymes have been, unwittingly, used in the preparation of food already for centuries. The best known example is the use of yeast, which by amongst others producing enzymes leavens bread, a practice known a long time before the start of our era. Equally old is the use of yeast for the manufacturing of beer and wine. In fact the word enzyme is derived from the Greek words εν ζυμη, meaning: (present) in yeast. Also the use of stomachs (and thus, we know now, chymosin) for the preservation of milk in the form of cheese was known in the ancient days. Finally, numerous forms of fermented foods are known for centuries from the Orient.

3. PRESENT

In addition to the direct consumption of vegetables, fruits and animals, man is consuming more and more processed food products, in the production of which enzymes play an important, though rather empirical role. For the production of these processed foods we can distinguish two main routes. The first one is direct production of food products from plant raw material by using enzymes or intact microorganisms producing enzymes. Examples of this are: bread from wheat flour, sugar or sugar syrups from starch, wine from grapes, beer from barley or other cereals, fruit juices from apples, pears or other fruits, etc.

Moreover enzymes and microorganisms are used to produce food ingredients such as taste ingredients, like fermented and/or hydrolyzed proteins and texture ingredients, like modified pectin and modified starch.

The alternative route uses domestic animals as an intermediate. Here the plant raw materials are processed in animal feed and man consumes the animal derived products, like milk, eggs, meat and processed derivatives like cheese, sausages, etc. Here, as in the production of feed the actions of enzymes or intact microorganisms are required. Table 1 summarizes the various enzyme activities and their fields of application.

Table 1
Enzyme activities used in the food and feed application areas

Enzyme	1. Baking	2. Starch processing	3. Brewing	4. Wine/Fruit juices	5. Dairy	6. Meat	7. Taste	8. Texture	9. Agro	10. Confectionery
Carbohydrate modifiers										
amylase		2	3						9	
glucoamylase	1	2	3							
glucose isomerase		2								
invertase		2								10
cellulase complex	1	2	3						9	
xylanase	1	2	3						9	
lactase					5					
pectinase complex				4				8		
glucanase			3						9	
protease										
endoprotease	1	2	3		5	6	7	8		
exopeptidase	1		3		5		7			
lipase/esterase										
lipase	1				5					
phospolipase	1									
esterase					5					
miscellaneous										
phytase		2							9	
glucose oxidase	1			4						
FDase/deaminase							7			
lysozyme					5					
transglutaminase						6				

The use of enzyme (preparations) in the different food and/or feed applications used to be rather empirical and based more or less on intuition or experience. However, the increased knowledge of substrate and enzyme structures by sophisticated technology like X-ray crystallography, 2D-3D NMR, mass spectrometry and the development of rec-DNA helped to put a more rational basis under the enzyme applications.

4. FUTURE

One of the areas in which this approach is very successful is that of the research and development activities in cell wall degrading enzymes. The reason is that, especially in the food and feed industry, plant cell walls play an important role since they are part of the raw materials used. Enzymatic breakdown or modification of them is required in order to improve processes or products.

4.1. Pectin structure and enzymatic modification

Pectin is a major constituent of the primary cell wall of most flowering plants. It is a polysaccharide of complex structure and composition (see Figure 1). Because of extensive research activities more details about its structure and composition, together with the enzymatic activities acting on it, are still being elucidated every day. Technology coupled with appropriate screening and application tests helped to put a more rational basis under the enzyme applications.

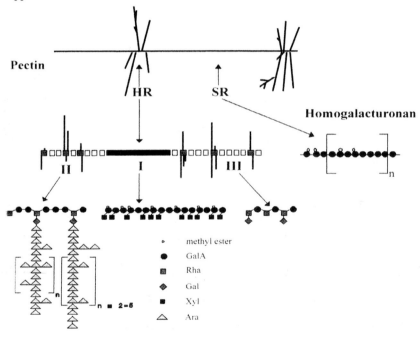

Figure 1. Hypothetical structure of apple pectin. **SR** = smooth region, **HR** = hairy region. GalA = galacturonic acid, Rha = rhamnose, Gal = galactose, Xyl = xylose, Ara = arabinose. From ref. 2.

Smooth region

Up till recently, pectin was considered to be a "simple" molecule, consisting of what is now referred to as the smooth region. This smooth region is a homogalacturan i.e. a linear polymer of galacturonic acid that is methylated in varying degrees, depending on the source.

Hydrolysis in this region can be achieved in two ways:

- Polygalacturonase hydrolyses only demethylated pectin and therefore additionally requires the action of pectin methyl esterase
- Pectin lyase cleaves the main chain between methylated galacturonic acid residues. Pectate lyase catalyzes the same reaction on low-methylated pectin.

The calcium-dependent gelling properties of pectin can be influenced by pectin methyl esterase alone because, through the action of this enzyme alone, more negatively charged galacturonic acid groups are formed.

Hairy region or rhamnogalacturonan

The hairy region differs from the smooth region in that it possesses polysaccharide side chains attached to the main chain. Figure 2 gives a schematic overview of the structure of these regions and the enzymes acting on them.

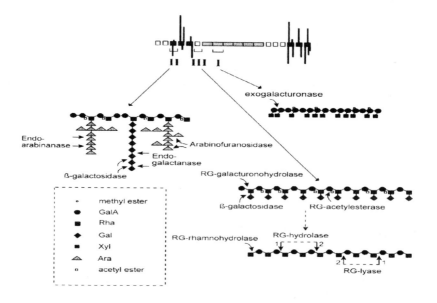

Figure 2. Schematic overview of enzymes acting on the hairy regions of pectin. **I**, **II** and **III** are different subunits; GalA, Rha, Gal, Xyl and Ara are as indicated in Figure 1. From ref 2 and 3.

Together with the elucidation of the structure of these hairy regions, many more enzymes were discovered that act on these regions. These enzymes and their site of action are indicated in Figure 2.

Increased knowledge of the substrate structure, especially of those substrates that are related to the technical effect one wants to establish during the application, will facilitate the choice of the enzyme activity for the application.

4.2. Improvements of PluGbug ^{TM}

In order to test the potential applications of these enzymes in the food and feed area sufficient amounts of them should be available. This used to be a serious bottleneck, but due to improvements of rec-DNA technologies, solutions are now at hand.

In order to be able to produce substantial amounts of a particular enzyme, we have started with classically improved *Aspergillus* production strains for glucoamylase. These strains have undergone a series of rounds of mutation/selection and are optimized for high production levels of this enzyme. Although the classical strain improvement is in essence a "black box" approach, in which the nature of the improvement cannot be predicted, one of the reasons for the higher production level of glucoamylase in this strain is the multiplication of a region of the DNA, containing amongst others the promoter and coding sequence of the gene: the *gla*A-loci.

Secondly, we have "emptied" these loci using the *amd*S forward and backward selection system. Forward selection, i.e. selection of strains containing this gene, is based on the fact that the transformant can use acetamide as a nitrogen source and can be selected for. Backward selection, i.e. strains cured of the gene, is based on the fact that they are resistant to fluoro-acetamide. Using a proper replacement vector and a double cross-over event (see ref. 4) strains with "empty" *gla*A-loci, but still containing the 5'- and 3'- flanking sequences can be constructed.

Thirdly, the loci can be "filled" again by using an expression cassette containing the *amd*S gene, with proper flanking regions, and the "gene of interest" under the control of the *gla*A promoter. Selection on acetamide, followed by counterselection on fluoro-acetamide leads than to the desired strain. Compared to the original strains they are completely identical except for the fact that the "gene of interest" replaces the glucoamylase genes. Substantial amounts of the "protein of interest"can than isolated in a relatively pure form.

5. CONCLUSION

Due to these recent developments we no longer have *"to waste clean thinking on dirty enzymes"*, but can use this thinking for defining clever application experiments using defined substrates and enzyme (mixtures).

REFERENCES

1. E. Racker, Mechanisms in Bioenergetics, Academic Press, New York, 1965.
2. M. Mutter, Ph.D.-thesis, Agricultural University Wageningen, The Netherlands, 1997.
3. H.A. Schols and A.G.J. Voragen in J. Visser and A.G.J. Voragen (eds.), Progress in Biotechnology 14: Pectins and pectinases, Elsevier, Amsterdam, 1996.
4. EP application No. 94201896.1 (1995), US Patent No. 5,876,988 (1999)

Food Biotechnology
S. Bielecki, J. Tramper and J. Polak (Editors)

Functional foods with lactic acid bacteria: probiotics - prebiotics - nutraceuticals

W. Kneifel

Department of Dairy Research & Bacteriology, University of Agricultural Sciences,
Gregor Mendel-Str.33, A-1180 Vienna, Austria

Functional foods are claimed to have several health-specific advantages. In addition to their basic nutritive value and naturalness, they contain a proper balance of ingredients which will help us to function better and more effectively, including aiding in the prevention and treatment of illnesses and diseases. Within this category, products containing lactic acid bacteria have become of high importance. Although the recognition of beneficial effects of dairy products containing these bacteria has its roots in the beginning of this century, this trend has re-emerged particularly in the early nineties when probiotics, prebiotics and synbiotics have appeared on the market. This essay describes and discusses some of the features and drawbacks of these products and is aimed at giving an overview of the current situation with functional foods with special regard to dairy products.

1. INTRODUCTION

The area of functional foods has been designed to provide beneficial ingredients and effects, besides their normal composition and properties. Within this novel category of food products, the application potential of lactic acid bacteria (LAB) has been extended markedly due to recent developments in food technology, microbiology, biotechnology and hygiene. LAB are known to comprise a heterogeneous group of bacteria of rods and cocci which produce lactic acid. Geni most relevant for probiotics are *Lactobacillus* and *Bifidobacterium* (although the latter is phylogenetically assigned to the group of *Actinomycetes*). Traditionally, LAB have been used in several kinds of food fermentation leading to products with sensory attractiveness and improved keeping quality. A very modern issue of LAB is their use as beneficial live components in so-called „healthy eating". The central aim behind this idea is to modulate the human gut microflora by administering health–relevant bacteria or by stimulating their growth by providing specific substrates of oligo- and polysaccharidic nature which cause an increase of their population. Since there is some evidence that humans are daily confronted with several factors negatively influencing the well-being, the optimization of the composition of the gastrointestinal microflora has become of major importance (for survey of main criteria see Table 1).

Under normal conditions, the resident bacterial flora of the gut comprises up to more than 10^{12} cells per g dry weight faeces. It has been recognized that lactic acid bacteria and bifidobacteria play a major role in this complex environment since they may contribute considerably to the microbiological balance.

Table 1
Possible factors influencing the well-being via the gastrointestinal tract

NEGATIVE	POSITIVE
- Illness / Disease	- Good general status / Fitness
- Ageing	- Balanced nutrition
- Stress	- Regular consumption of
- Allergies	fermented milk products
- Malnutrition	- Probiotics, prebiotics, synbiotics
- Antibiotic therapy	- Nutraceuticals
- Environmental factors	
- Jet-lag	
- Others	

In 1907, Elie Metchnikoff reported for the first time on the positive effects of fermented dairy products via their influence on well-being. Even almost one hundred years later, this area is about to develop towards one of the cardinals of microbiological research. Buzzwords such as *probiotics, synbiotics, biotherapeutics, HEFFI's (health enhancing functional food ingredients), FOSHU's (foods for specific health use), PARNUT's (foods for particular nutrition, mainly used in clinical nutrition)* etc. were created by scientists as well as advertising strategists or even by legal experts. These terms may on the one hand have stimulated common interest in this topic, but on the other hand also caused confusion among consumers and researchers. All these products exhibit a great diversity which makes it difficult to classify them according to the still differing legal situation of the countries. Beyond the long-term experience with all kinds of traditional foods, functional foods are now being studied with regard to their special beneficial effects, and numerous reports dealing with prophylactic and therapeutic claims of pro-, pre- and synbiotics have appeared and also provoked research activities aiming at the clarification of the function of these products.

2. DEFINITIONS AND PROPERTIES

2.1 Probiotics

Several definitions have been made for „probiotics" since 1965, ranging from quite restricted explanations to a more general meaning [1]. Briefly, *probiotics* are live microbes which influence the well-being of their host through their effect on the intestinal microflora [2]. A broad array of beneficial effects has been attributed to probiotics, however not all of them have been confirmed on a reliable scientific basis [3,4]. Moreover, there is some evidence that only some of the properties can be fulfilled by some of the bacterial strains. At present, approximately ten to fifteen bacterial strains have passed extensive investigations for some of the probiotic criteria (Table 2). However, many other strains possessing the same taxonomic properties have benefited from those results despite lacking specific in-depth studies. Although probiotic bacteria usually act in the gastrointestinal tract, positive effects can also be initiated in other regions of the human body [5].

Originating from the intestinal content of healthy humans, most probiotic bacterial strains have been selected for their specific properties in extensive screening procedures. Owing to their human origin, probiotic bacteria are expected to display a proven adaptation to the human gastrointestinal environment which is of importance with regard to their safety criteria. Based on the large number of reports in the literature an attempt was made to evaluate the most relevant beneficial criteria attributed to probiotic strains (Table 3).

In general, probiotic effects are only observed when the bacteria are applied during long-term administration periods. However, probiotic strains gradually vanish from the intestinal microbial ecology after their administration has been stopped. From these experiences we may conclude that only those foods should be considered as being sensible vectors for such strains that are eaten frequently and regularly (e.g., fermented dairy products, dairy-based drinks, cereals). Nevertheless, also other products containing probiotics (sausage, sweets, chocolate bars, ice cream etc.) have now emerged on the market. According to current market statistics, within the category of dairy products, probiotics comprise approximately 10 to 20 percent. Increasing portions are expected for the future. From the evaluation of effects listed in Table 3 it is also evident that some of the claimed beneficial properties still lack scientific evidence. Particularly, features such as the role in cancer prevention or as hypocholesteremic effects are not yet clarified.

Table 2
Some probiotic strains with scientific documentation on beneficial effects [6,7]

Lactobacillus johnsonii LA1
Lactobacillus rhamnosus GG
Lactobacillus acidophilus NCFB 1748
Lactobacillus acidophilus 74-2
Lactobacillus gasseri ADH
Lactobacillus reuteri ATCC 55370
Lactobacillus paracasei subs. paracasei Shirota
L. paracasei subsp. paracasei Actimel
Lactobacillus plantarum DSM 9843
Streptococcus thermophilus and
L. delbr. subsp. bulgaricus
Bifidobacterium lactis BB-12
Bifidobacterium longum BB-536

Table 3
Reported effects of probiotic bacteria: evidence proven (P), partly proven (PP), questionable (Q)

Probiotic effect	Evidence
Reduced duration of diarrhea	P
Antagonistic effects against pathogenic micro-organisms	P
Regulation of intestinal motility	P
Immunomodulation (in vitro)	P
Support of the immune system (in vivo)	PP
Reduction acitivities of cancer-relevant enzymes	P
Cancer prevention	Q
Cholesterol-lowering effect	Q
Improved lactose digestion	P
Improved Calcium resorption	P
Provision of water-soluble vitamins	P

Some conclusions may therefore lead to misinterpretation problems, even when considering advertising arguments or particular properties declared on the label. In many cases, studies referred to such functions are either based on in vitro models or on animal testing and this makes it difficult to predict possible effects in the human body. Moreover, the legal situation

104

remains still unclear, since no official regulation exists defining the demands and prerequisites which make a „normal" strain become a probiotic one. Hitherto there is no official viable count limit which should be fulfilled by a probiotic product. Although a microbial population of at least one million viable bacteria per gram of sample has usually been recommended as some kind of standard value (this may guarantee a dose of approximately 100 million cells per meal portion), only dose-response studies can provide some indication of the specific requirements which may be expected to be strain-dependent.

2.2 Prebiotics

More recently, so-called prebiotics have been discovered to constitute an interesting alternative to the probiotics (for examples of products see Table 4), providing specific nutrients to beneficial bacteria hosted in the colon. *Prebiotics* belong to the functional group of dietary fibers which are of oligo- and polysaccharidic nature. Most products presently marketed in Europe are mainly of plant or whey origin [8], however, many other compounds of different origin, technological basis and composition have been introduced and established in Japan and may now stimulate the situation worldwide.

Prebiotics are increasingly incorporated into foods aiming at the same beneficial effects as probiotics by supporting the microbial balance of the intestine. These substances are known to be bifidogenic which means that they are capable of promoting the growth of bifidobacteria in the colon [9]. Also the population of some lactobacilli can be increased in the colon by prebiotics.

Table 4
Some examples for commercially available prebiotic sugar compounds [8,10]

Category	Brand name	Supplier
Galacto-oligosaccharide	Elix´or	Borculo Domo Ingredients
	Oligomate 55	Yakult
Fructo-oligosaccharide	Raftilose	Orafti
	Actilight	Beghin-Meiji
	Frutasun	Sensus-Frutall
	Meioligo	Meiji
	Nutraflora	Golden Technologies
Inulin	Raftiline	Orafti
	Frutafit	Sensus-Frutall
Lactulose	Laevolac	Laevosan/Fresenius
	Bifiteral	Solvay
	MLS/P/C	Morinaga

From this shift in the microbial composition of the intestine secondary positive effects are induced, in analogy to the reported probiotic features. The most important prerequisite for finally acting in the large intestine is the ability of prebiotic sugars to reach this intestinal region without being hydrolysed through their passage of the upper gastrointestinal tract.

In addition to their health relevant properties, prebiotics may also improve sensory and textural attributes when added to foods. Moreover, they are of low caloric value. Because of the diversity in the composition of prebiotics, different recommendations are given regarding the doses necessary to exert positive effects. For instance, in the case of fructo-oligosaccharides, microbiologically and medically relevant effects were observed ranging from 1 g up to 18 g per day [11-13]. On the other hand, undesirable side effects such as abdominal discomfort, flatulence, borborygmus and even diarrhea may occur, depending on the individual sensitivity [14].

While probiotic products are mainly dairy based, prebiotic sugars can be easily applied to many other categories of food products. Factors such as the hydrolysis properties under acidic conditions as provided by some foods and the specific behaviour under thermal processing have to be considered, but in general most of the preparations are known to exhibit good stability.

2.3 Synbiotics

It was quite obvious to combine pro- with prebiotics in so-called synbiotics which may be expected to cause a twofold beneficial effect. *Synbiotics* are defined as mixtures of pro- with prebiotics which influence positively the well-being of the host by supporting the survival of probiotics and the colonisation of the colon with beneficial bacteria [9].

Probiotic +	**Prebiotic** =	**Synbiotic**
Stability problems of probiotic bacteria during storage of the product	Good stability of oligo-saccharides during food processing and storage	Stability problems of probiotic bacteria during storage of the product, partly corrected by the prebiotic
Stability problems of bacteria during passage of the upper gastrointestinal tract	Good stability of oligo-saccharides during passage of the upper gastrointestinal tract (exceptions possible)	Stability problems of bacteria during passage of upper gastrointestinal tract
Surviving bacteria exert beneficial effects	Oligosaccharides reach the colon and stimulate benefical resident micro-organisms	Surviving bacteria exert beneficial effects, in addition stimulation of resident beneficial microflora
NO SIDE EFFECTS		POSSIBLE SIDE EFFECTS

Figure 1. Comparison of the most relevant characteristics of pro-, pre- and synbiotic products [8].

It has been shown that not each probiotic strain is in perfect agreement with any prebiotic sugar to form a sensible synbiotic combination [15]. Moreover, it can be expected that the probiotic component acts alone and, in addition, the prebiotic substrate may stimulate the resident colonic microflora. In Figure 1, the main criteria of pro-, pre- and synbiotics are compared, outlining the specific adavantages and disadvantages of these components.

At present some interesting synbiotic dairy products containing certain probiotic strains and prebiotic sugars (mainly fructo-oligosaccharide) are available on the market. Mainly, the doses of prebiotic sugars in these products are around 3 to 4 g per portion (beaker).

2.4 Nutraceuticals

A currently appearing novel trend coming primarily from the American market is the creation of pharmaceutical-like food products (*nutraceuticals*). This is a very heterogeneous group of products of varying efficacy which still causes severe problems of classification, since some can be considered as foods others as pharmaceuticals, partly not meeting the high requirements of pharmaceutical regulations. Nutraceuticals contain biologically active ingredients (e.g., lipophilic materials, saponins, flavonoids, prebiotic oligosaccharides, vitamins, antioxidants, minerals, trace elements and also LAB). Because of the unclear and interdisciplinary position there is a considerable need to deal with this novel group of products. It is quite obvious that they offer beneficial therapeutic effects, much more than any other health-relevant products. Although some authors state that they are some kind of foods and no drugs [16], their appearance markedly resembles that of pharmaceuticals. There is also some evidence that they should be mainly used supplementarily to regular nutrition, either as powders, tablets, capsules or as syrups. These drug-like properties may be misleading to the consumer, since only in rare cases they have something common with normal food products.

Quite differing from the situation in Europe, in the United States some health claims have already been allowed for such products, according to the Nutrition Education and Labelling Act (NLEA) or to the Dietary Supplement and Health Education Act (DHSEA) [17]. Hence, the term nutraceutical has often been used as a synonym for foods that can prevent and treat diseases. In Europe it is not allowed to declare medically relevant properties on the product labels, and the consumer must not be misled by any information about therapeutic properties on the food packaging.

3. FUTURE PERSPECTIVES

It is quite obvious that current developments in the area of functional foods have considerably stimulated the advertising activities in the public media. This trend has even resulted in a more or less legally permitted presentation of health claims, therapeutic effects and other arguments of medical relevance. Although this situation has on one hand caused some difficulties in terms of product declaration and also in the understanding by the consumer due to lacking sufficient information, we have, on the other hand, also to consider the positive effect arising from the intensified research in the area of functional foods. Dealing with actual questions such as „how healthy is our food ?" or „how does our food influence the gastrointestinal microflora and the common well-being ?" has given new insights into the understanding of food composition and the digestion process.

Moreover, the interdisciplinary involvement of several research fields such as medicine, microbiology, food technology and biochemistry, has demonstrated that the complexity behind the relationship between health and food can only be clarified if these disciplines cooperate in a synergistic way. It also seems to be positive that the food manufacturers are now increasingly recognizing biologically relevant and natural ingredients to be incorporated into food products.

REFERENCES

1. W. Kneifel and C. Bonaparte. Ernaehrung/Nutrition, 22 (1998) 357.
2. G. Schaafsma. Newsletter of the International Dairy Federation no. 145 (1996) 23.
3. B.R. Goldin. Brit. J. Nutr., 80 (Suppl. 2) (1998) S203.
4. A.C. Ouwehand and S. J. Salminen. Int. Dairy J., 8 (1998) 749
5. R. Sieber and U.T. Dietz. Ernaehrung/Nutrition, 23 (1999) 401.
6. W. Kneifel, T. Mattila-Sandholm and A. von Wright. Probiotics In (R. K. Robinson, C.A. Blatt and P.D. Patel, eds.) Encyclopedia of Food Microbiology, (1999) Academic Press, London, in press.
7. T. Mattila Sandholm, J. Mättö, M. Saarela. Int. Dairy J., 9 (1999) 25.
8. C. Bonaparte and W. Kneifel. Ernaherung/Nutrition, 23 (1999) 152.
9. G. R. Gibson & M.B. Roberfroid. J. Nutr., 125 (1995) 1401.
10. M.J. Playne & R. Crittenden. Bull. International Dairy Federation, 313 (1996) 10.
11. H. Hidaka et al. Bifidobacteria Microflora, 5 (1986) 37.
12. H. Tomomatsu. Food Technol., (1994) October, 61.
13. Y. Bouhnik et al. Eur. J. Clin. Nutr., 50 (1996) 269.
14. R. Hartemink and F. Rombouts. p.57 In Proc. Symp. Non digestible oligosaccharides: healthy food for the colon ? (1997) Wageningen, The Netherlands.
15. W. Kneifel, A. Rajal and K.D. Kulbe. Microb.Ecol.Health Dis., in press.
16. I. Goldberg. p.3 In Functional Foods, Designer Foods, Nutraceuticals (I. Goldberg, ed.). (1994) Chapman and Hall, New York.
17. K.W. Rychlik and C.G. Grennwald. p.13 In New Technologies for Healthy Foods and Nutraceuticals (M. Yalpani, ed.), (1997) ATL Press Inc., Shrewsbury MA, USA.

Food Biotechnology
S. Bielecki, J. Tramper and J. Polak (Editors)

Microbial production of clavan, an L-fucose rich exopolysaccharide

P.T. Vanhooren and E.J. Vandamme

Dept. Biochemical and Microbial Technology
Faculty of Agricultural and Applied Biological Sciences
University of Gent, Coupure links 653, B-9000 Gent, Belgium

INTRODUCTION

Certain microbial extracellular polysaccharides (EPS), best known for their thickening, gelling or emulsifying properties, are an easy source of the unusual 6-deoxyhexose, L-fucose [1, 2].

L-Fucose can be used as a substrate in the chemical synthesis process of furanone-flavouring compounds. Furthermore, L-fucose and L-fucose containing oligosaccharides have potential application in the medical field in preventing tumour cell colonisation of the lung (anticancer effect), in controlling the formation of white blood cells (anti-inflammatory effect) and in cosmeceuticals as skin moisturizing agent.

L-Fucose production via chemical synthesis is laborious and suffers from low yield, while direct extraction from brown algae is costly and subject to seasonal variations in supply volume and quality. Chemical or enzymatic hydrolysis of L-fucose rich microbial EPS opens up a new route towards efficient L-fucose production.

1. *CLAVIBACTER MICHIGANENSIS* EPS AS A SOURCE OF L-FUCOSE

Growth and EPS production of the *Clavibacter michiganensis* strains LMG 3675, 3676, 7323, 3690, 3694, 5604 and RPZBC 3108, reported to contain L-fucose in their EPS [3, 4], were investigated in detail. The first three strains belong to the subspecies *insidiosus*, while the remaining four belong to the subspecies *michiganensis*.

In preliminary experiments, *Clavibacter michiganensis* subsp. *michiganensis* LMG 5604 and *C. michiganensis* subsp. *insidiosus* LMG 3675 yielded the highest viscosity. The EPS of these strains was isolated and hydrolysed and the nature of the constituting monosaccharides was determined using gaschromatography (Figures 1 and 2) with xylitol as the internal standard. De monosaccharides detected are L-fucose, D-glucose and D-galactose. Trace amounts of mannose were also detected, originating from mannan, a component present in the culture medium yeast extract which coprecipitated during EPS isolation.

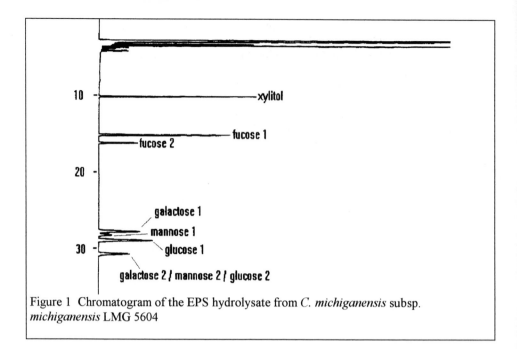

Figure 1 Chromatogram of the EPS hydrolysate from *C. michiganensis* subsp. *michiganensis* LMG 5604

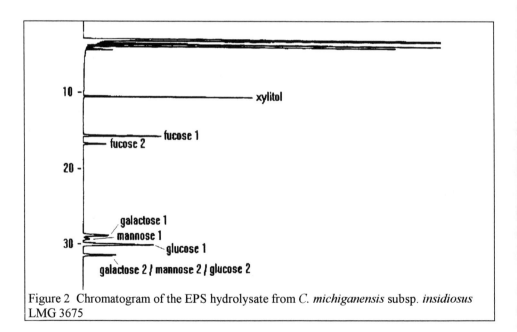

Figure 2 Chromatogram of the EPS hydrolysate from *C. michiganensis* subsp. *insidiosus* LMG 3675

Cultivation of these two strains on fermentor scale (2 litre) resulted in the selection of strain LMG 5604 for further optimisation of the EPS production since strain LMG 3675 did not produce any EPS on larger scale culture due to its shear stress sensitivity. The EPS, produced by the selected *Clavibacter* LMG 5604 strain was named clavan [5].

Subsequently, the production of clavan was studied via submerged fermentation of *Clavibacter* LMG 5604 (Figure 3). When a virulence analysis of this normally phytopathogenic strain was performed on tomato plants, no disease symptoms developed. The required safety and containment precautions during clavan producing fermentation processes could therefore be substantially reduced.

The nature and concentration of carbon and nitrogen source, the carbon/nitrogen ratio, the addition of tomato juice to the growth medium, the initial pH of the growth medium, the K_2HPO_4, KCl, NaCl, Tween 80, antifoam agent and organic acids concentration of the growth medium were optimised; as well as the cultivation temperature, dissolved oxygen concentration and inoculum age. Fermentation processes were performed on erlenmeyer, 2 litre and 60 litre fermentor scale. Optimisation of the L-fucose recovery resulted in an L-fucose yield of **7.4 g/kg**, corresponding to a clavan yield of **19.7 g/kg** culture broth.

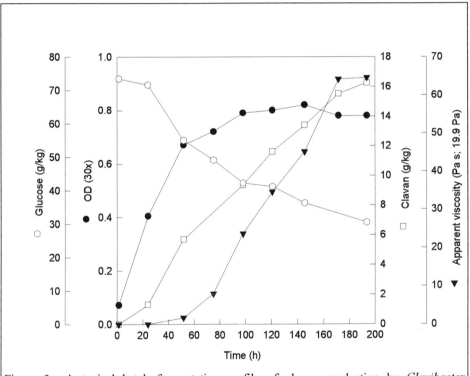

Figure 3   A typical batch fermentation profile of clavan production by *Clavibacter michiganensis* LMG 5604

2. RHEOLOGICAL PROPERTIES

The rheological properties of fermentation broths were studied in detail for clavan producing *Clavibacter michiganensis* LMG 5604, using a controlled stress Bohlin CS10 rheometer. An adequate model, describing the observed non-Newtonian flow behaviour, the Herschel-Bulkley model, was selected (Figure 4) and the influence of shear history, polysaccharide concentration, temperature and pH on this behaviour was investigated.

The obtained results demonstrated that the fermentation broth does not exhibit any *time dependency*. When the rheological characteristics of broths with different *clavan concentrations* were studied, it was found that broths with a higher EPS concentration are more viscous and pseudoplastic due to increased entanglement of the polymeric chains.

The broth becomes less viscous and displays a more Newtonian behaviour as the *temperature* increases. This trend is due to increased Brownian motion which lowers the viscosity and the pseudoplasticity through decreased polymer interactions and decreased shear induced molecular orientation, respectively. A rather sharp drop in the apparent viscosity appeared when the *pH* value exceeds 7.0, probably indicating a structural alteration of the clavan.

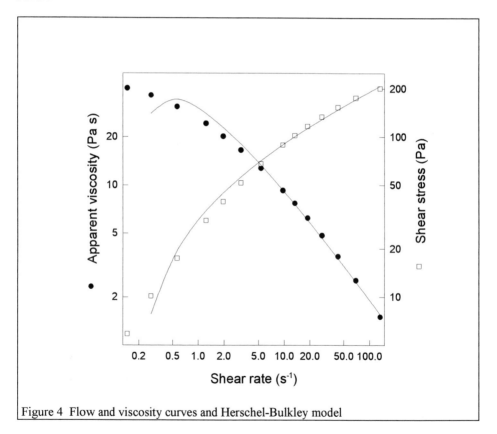

Figure 4 Flow and viscosity curves and Herschel-Bulkley model

3. *CLAVIBACTER MICHIGANENSIS* MUTANTS WITH ALTERED EPS OR L-FUCOSE LEVELS

A first goal here was the selection of bacitracin resistant mutants, since such mutants can contain elevated levels of isoprenoid lipids resulting in EPS hyperproduction [6]. The selection and use of such a strain, resistant to 50 µg/l bacitracin, in fermentation processes indeed resulted in an increase of the clavan yield by **38 %** as compared to the wild type.

Secondly, a UV-mutation programme was set up in order to select strains, still producing free L-fucose or L-fucose containing oligosaccharides. Six non-mucoid strains could be selected. Further experiments performed with these strains revealed that most of them did not produce clavan anymore, but no free L-fucose either. Only one mutant was still producing clavan, albeit in very low yields. When fermentation processes were performed with this strain, no viscosity build-up of the medium occurred, while **266 mg/l** L-fucose could be detected in the broth after acid hydrolysis. Whether this L-fucose originates from clavan oligo- or polysaccharides could not be determined, since oligosaccharides as well as low polysaccharide concentrations could be responsible for the observed low broth viscosity.

4. CONCLUSIONS AND PERSPECTIVES

Fermentation optimisation, both nutritional and physico-chemical, can substantially increase the yield of L-fucose rich EPS during submerged fermentation. In the case of *Clavibacter michiganensis* subsp. *michiganensis* LMG 5604 a final clavan yield of **19.7 g/kg** culture broth could thus be obtained. This corresponds to a L-fucose yield of the culture broth of **7.4 g/kg**.

A very important, if not the most important, drawback during these submerged fermentations is undoubtedly the high viscosity build-up of the fermentation broths. The non-Newtonian, pseudoplastic flow behaviour does not only affect the downstream processing, but also strongly hampers the mass transfer and mixing phenomena during the submerged EPS fermentations.

A promising alternative to achieve a higher L-fucose yield, thereby avoiding high viscosity, is the selection of mutant strains, displaying less polymerisation of the EPS constituents. As such, free monosaccharides or oligosaccharides should be formed in the medium. However, further investigation is needed to optimise this approach.

ACKNOWLEDGEMENTS

The authors are grateful to the "Fonds Wetenschappelijk Onderzoek-Vlaanderen" (FWO-Vl) for financial support. Part of the results mentioned here are from the doctoral thesis of P.T. Vanhooren, FWO-Vl-bursist.

114

REFERENCES

1. P.T. Vanhooren, S. de Baets, G. Bruggeman, E.J. Vandamme, Biologi Italiani, 28(1), (1998), 7–11.
2. P.T. Vanhooren, E.J. Vandamme (1998). Cosm'ing 98. October 1998.
3. R.W. van den Bulk, L.P.T.M. Zevenhuizen, J.H.G. Cordewener, J.J.M. Dons, Phytopathology, 81, (1991), 619-623.
4. L.D. Varbanets, Mikrobiologicheskii Zhurnal, 54, (1992), 16-26.
5. P.T. Vanhooren (1997), Ph.D. thesis. Universiteit Gent.
6. I.W. Sutherland (1977), Academic Press, New York, San Fransisco, pp 472.

Food Biotechnology
S. Bielecki, J. Tramper and J. Polak (Editors)
© 2000 Elsevier Science B.V. All rights reserved.

Glucansucrases: efficient tools for the synthesis of oligosaccharides of nutritional interest

P. Monsan, G. Potocki de Montalk, P. Sarçabal, M. Remaud-Siméon and R.M. Willemot

Centre de Bioingénierie Gilbert Durand, UMR CNRS 5504, UR INRA 792, DGBA, INSA, 135 avenue de Rangueil, 31077 Toulouse Cedex 4, France

Besides their traditional role as energy source and sweeteners, oligosaccharides are more and more regarded as informative biological molecules. Particularly, their ability to specifically stimulate the beneficial intestinal flora and the digestive immune system is of increasing nutritional interest (prebiotics). Besides the obtention of such oligosaccharides by extraction from plant sources or by hydrolysis of plant polysaccharides, their direct enzymatic synthesis from a widely available agricultural raw material, sucrose, can be catalysed by glucansucrases of microbial origin.

1. INTRODUCTION

Glucansucrases are transglucosidase enzymes (EC: 2.4.1.-) which catalyse the synthesis of α-D-glucopyranosyl homopolymers and oligomers using such a simple molecule as sucrose as substrate (α-D-glucopyranosyl donor) and as unique energy source. No activated glucosyl-intermediate or cofactor is needed. The reaction proceeds through the formation of a glucosyl-enzyme covalent intermediate, from which the D-glucopyranosyl unit is transferred onto an acceptor compound.

In the presence of sucrose alone, high molecular weight polymers are obtained [1], which contain a majority of:

- α-1,6 linkages with dextransucrase (dextran);
- α-1,3 linkages with mutansucrase (mutan);
- alternating α-1,6/α-1,3 linkages with alternansucrase (alternan);
- α-1,4 linkages with amylosucrase (amylose).

In the presence of both sucrose and an efficient α-D-gucopyranosyl acceptor, like maltose or isomaltose for example, glucansucrases catalyse the synthesis of low molecular weight oligosaccharides [2]. It is possible to control the structure and the size of such oligosaccharides by controlling the type of glucansucrase, the type of acceptor and the reaction conditions (donor/acceptor ratio).

In addition, the isolation of the genes coding for such glucansucrases allows to approach the understanding of their mechanism of action at the molecular level. The identification of the amino acid residues involved both in the specific recognition of the D- glucosyl donor and acceptor, and in the control of the regioselectivity of the D-glucosyl transfer reactions, opens the way to the possibility to design a new range of glucansucrases with controlled specifity and regio selectivity.

2. DEXTRANSUCRASES

The soil microorganism *Leuconostoc mesenteroides* was identified by van Thiegem [3] in 1874 as the cause of the gelification of canne sugar syrups. This gelification is due to the production of an extracellular polysaccharide, named dextran because of its positive rotary power.

The name dextransucrase was given by Hestrin et al. [4] to the extracellular enzyme produced by *Leuconostoc mesenteroides*, which catalyses the synthesis of dextran polysaccharides from sucrose with D-fructose as co-product. Dextransucrase (sucrose: 1,6 α-D-glucan 6-α-D-glucosyltransferase, EC 2.4.1.5) is an extracellular enzyme, which is specifically induced by sucrose [5].

Dextran is a D-glucopyranosyl polymer which contains more than 50% α-1,6 osidic bonds. Its molecular mass ranges from 0.5 to 6.10^6 Da [6]. According to the origin of dextransucrase, the type and degree of branching of dextran can vary very significantly [7-9]. Dextrans have thus been classified into three groups according to the main type of branching present in their structure: group A for α-1,2 branching, group B for α-1,3 branching and group C for α-1,4 branching [10].

Dextran produced by *L. mesenteroides* NRRL B-512F is a very highly linear polysaccharide, containing 95% α-1,6 osidic bonds. It was one of the first biopolymers to be produced on an industrial scale in 1948 and to find applications in the medical field, as blood plasma substitute and as raw material for the production of iron carriers, and in the field of separation media (Sephadex®) [11]. In addition, dextran is a food grade texturing agent.

Besides *Leuconostoc sp.*, dextransucrase is also produced by two other types of lactic bacteria, *Streptococcus sp.* and *Lactobacillus sp.* [6]. Streptococcal glucosyltransferases synthesize from sucrose glucan polymers which are playing a key role in cariogenesis phenomena: they are involved in the adhesion mechanism of bacteria on the teeth surface to form the dental plaque [12,13].

Very interestingly, the structure of the genes coding for the dextransucrases from *L. mesenteroides* is very similar to the structure of the genes coding for the streptococcal glucosyltransferases [14-16]. In fact, they successively contain:

- a N-terminal signal peptide allowing protein excretion [17],
- a variable region [18],
- a highly conserved N-terminal catalytic domain, which contains the catalytic amino acid residues [18],
- a C-terminal glucan binding domain, which is involved in both dextran and oligosaccharide synthesis mechanism [19].

The combination of biochemical studies, sequence alignment studies, hydrophobic cluster analysis and structure prediction has shown that these glucansucrases share many mechanistic and structural features [20]. In addition, they also present very strong mechanistic and structural analogies with glucoside hydrolases, and particularly with those belonging to the family 13 [21] according to Henrissat's classification [22], which possess a classical catalytic $(\beta/\alpha)_8$ barrel domain: α-amylase, CGTase, pullulanase, etc...

The understanding at the molecular level of the structure/function relationships for glucansucrases, combined with random mutagenesis and molecular evolution will very probably result in the design of new enzymes with controlled specificity towards both substrates and acceptors, and controlled regioselectivity in D-glucopyranosyl unit transfer [21]. It has been shown, for example, that the substitution of the threonine residue T589 of the glucosyltransferase GTF-S from *Streptococcus mutans* by an aspartic acid residue results in an increase by 30% of the amount of α-1,3 osidic linkages in the synthesised glucan [23]. Similarly, the substitution of the threonine residue T667 in the dextransucrase DSR-S from *L. mesenteroides* NRRL B-512F results in a 13% increase of the content in α-1,3 osidic bonds in the synthesised dextran [21].

Our interest has been focused on the dextransucrase enzyme from *L. mesenteroides* NRRL B-1299. This strain is known to produce a highly α-1,2 branched dextran that contains 27% to 35% α-1,2 branch linkages, as well as a limited amount of α-1,3 branch linkages [24-27]. The dextransucrase activity of *L. mesenteroides* NRRL B-1299 is mainly insoluble [28,29], but both insoluble and soluble dextransucrases produce dextran polysaccharides and oligosaccharides that contain α-1,2 osidic linkages [30,31]. Such oligosaccharides are efficiently obtained when dextransucrase is reacted in the presence of sucrose and maltose (D-glucopyranosyl acceptor)[31]. A mixture of three families of oligosaccharides is then obtained [32], which contains a maltose residue at the reducing end and either (i) only α-1,6 linkages (serie OD), or (ii) α-1,6 linkages and one α-1,2 linkage at the non-reducing end (serie R), or (iii) α-1,6 linkages and one α-1,2 linkage on the penultimate D-glucosyl residue (serie R'): Figure 1.

The presence of such α-1,2 linkages results in a very high resistance of these oligosaccharides to the attack of the digestive enzymes of humans and animals [33]. This is the reason why they were originally designed and developed as low-calory food bulking agents [30], in view to be formulated in combination with intense sweeteners. It was demonstrated that α-1,2 containing glucooligosaccharides (GOS) are not metabolised by germ-free rats, i.e. without any bacterial intestinal flora [33]. It was also shown that such glucooligosaccharides are specifically metabolised by bifidobacteria, lactobacilli and bacteroides [34,35], while they are poorly used as substrates by potentially detrimental strains, namely clostridia, eubacteria, enterobacteria and coliforms. This introduces the concept of prebiotic oligosaccharides, which has been originally initiated in Japan [36], and is now developing in Europe with the use of fructooligosaccharides obtained either by enzymatic synthesis or by controlled hydrolysis of chicory inulin.

GOS are fermented by the intestinal flora to short-chain fatty acids. In heteroxenic rats inoculated with a complex human intestinal flora, the short-chain fatty acids profile in the caecum is changed, with a decrease in butyric, isobutyric and isovaleric acid proportions ($p < 0.01$) and an increase in the proportion of caproic acid ($p < 0.05$) [33]. When compared to fructooligosaccharides and galactooligosaccharides, GOS are not bifidogenic, but they promote the growth of the cellulolytic intestinal flora. More important is the fact that they induce a broader range of glycolytic enzymes, without any significantly increased side-production of gases, and thus any detrimental effect [37].

GOS are presently marketed as prebiotics used in human food complements, in formulations combining probiotics (living microorganisms) and B family vitamins. Their main application is the control of the intestinal transit.

118

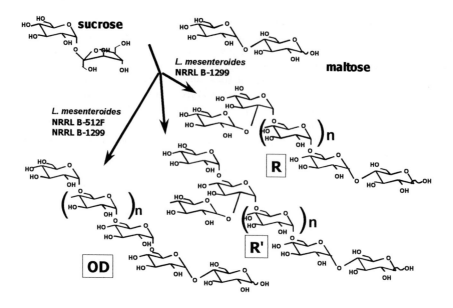

Figure 1. *Leuconostoc mesenteroides* NRRL B-1299 dextransucrase reaction in the presence of sucrose (D-glucosyl donor) and maltose (D-glucosyl acceptor): structure of the three types of glucooligosaccharides obtained. Maltose residue is located at the reducing end. In addition, serie OD only contains α-1,6 linkages, serie R contains a α-1,2 linkage at the non-reducing end, and serie R' contains a α-1,2 branching on the penultimate D-glucosyl residue.

It is also possible to take advantage of the prebiotic effect of GOS in animal feeding, for increasing the mean daily weight gain. Such effect, which can be attributed to the broad induction of glycolytic enzymes in the colon of mongastric animals, has been demonstrated for piglets, broilers and calves [35]. Besides such direct effects on the intestinal flora, GOS induce a biospecific immunostimulation effect: the simple addition of 1.5 g of GOS per kg of young calves feed results in a 20% decrease of the veterinary cost [35].

The same concept of prebiotic effect can be applied to the skin microbial flora. In fact, promoting the growth of beneficial lactic bacteria results in a very significantly decreased growth of detrimental microbial strains involved in acne, dandruff or bad smell problem [38]. GOS is thus widely used for dermocosmetic applications: BioEcolia ®.

The application of *L. mesenteroides* NRRL B-1299 dextransucrase to glucooligosaccharide synthesis from sucrose and maltose mixtures has been optimised [39] and scaled-up by BioEurope (Solabia Group, France). Dextransucrase is obtained by fed-batch cultivation of *L. mesenteroides* NRRL B-1299, using sucrose as carbon source and specific enzyme inducer [40,41]. As sucrose is also the substrate of the extracellularly obtained dextransucrase, dextran is synthesised during microbial growth and D-fructose is produced. D-fructose presence results in a repression of dextransucrase production. This repression effect can be suppressed in the presence of D-glucose. The simultaneous fed-batch addition of sucrose and D-glucose thus allows to increase by 100% the dextransucrase activity obtained in the culture medium and to reach a final activity of 9.7 U/ml [42]. Very fortunately, more than 90% of this dextransucrase activity is associated to the cells and the surrounding dextran slime. It can thus be very easily recovered from the culture medium by centrifugation.

For developing a continuous process for glucooligosaccharide synthesis, dextransucrase immobilisation is needed. This can be easily obtained, thanks to the strong association of dextransucrase with dextran slime and cells, by simply entrapping the catalytic activity within calcium alginate beads. The immobilisation yield is 93%, and a specific activity of 4.1 U/ml of gel can be obtained [41]. The present industrial production is of 40 tons per year, and is obtained by operating the immobilised dextransucrase in a 1-cubic meter continuous packed-bed reactor. The key parameter for efficient operation is the sucrose to maltose ratio, which controls both the glucooligosaccharide yield and size distribution [41]. Sucrose conversion at the outlet of the reactor is 100%.

3. AMYLOSUCRASE

Amylosucrase catalyses the transfer of the D-glucopyranosyl unit of sucrose to yield a polysaccharide containing α-1,4 osidic linkages, similar to amylose. This enzyme was first described by Hehre and Halmiton from *Neisseria perflava* [44]. It was later described from non-pathogenic *Neisseria polysaccharea* strains isolated from throats of healthy children [44].

The gene coding for the amylosucrase from *N. polysaccharea* has been isolated, sequenced, and cloned in plants by the group of Kossmann [45]. But some sequencing errors were identified by our group [46]. This allowed to put in evidence the very high level of homology between amylosucrase and α-amylases: sequence analysis showed a $(\beta/\alpha)_8$-barrel fold, interrupted by a separate folding module homologous to a calcium-binding domain, and a C-terminal β-barrel. This means that amylosucrase can be regarded as a member of the family 13 of the glucoside hydrolases [46]. Sequence alignement has also allowed to identify key amino acid residues involved in the catalytic mechanism: the amino acids D-294, D-401 and E-336 correspond to the catalytic triad involved in glucosidic bond cleavage in α-amylases. Their mutagenesis into N-294, N-401 and Q-336 respectively results in the obtention of totally inactive mutants [47]. From an evolutionary point of view, it appears that amylosucrase is very far from the other sucrose utilising glucosyltransferases [46].

Amylosucrase has been purified to homogeneity as a fusion protein with Glutathion-S-Transferase, using Glutathion-Sepharose-4-B affinity chromatography [46]. Its molecular mass is 70 kDa.

In the presence of sucrose alone, amylosucrase catalyses simultaneously sucrose hydrolysis, maltose and maltotriose synthesis (using D-glucose as acceptor), and high molecular weight glucan synthesis (containing only α-1,4 linkages). Very surprisingly, amylosucrase is activated by high sucrose concentrations: while Vmax and Km values are respectively equal to 510 U/g and 2 mM when sucrose concentration are lower than 20 mM, they are respectively equal to 906 U/g and 26 mM when sucrose concentrations are higher than 20 mM [47].

In the presence of sucrose and glycogen, amylosucrase is highly activated: at 105 mM sucrose, the addition of 30 g/l glycogen results in a 100-fold increase in amylosucrase activity [47]. This activator effect decreases when increasing sucrose concentration, suggesting a competition between sucrose and glycogen. This reaction results in the obtention of modified glycogen, which contains extended linear chains with an average DP of 75 D-glucosyl units according to iodine-complex characterisation [46].

Amylosucrase can thus generate a wide variety of oligosaccharides and polysaccharides of potential interest as food ingredients, due to their structure similar to that of amylose.

4. CONCLUSION

Glucansucrases are able to efficiently synthesise a wide variety of carbohydrate structures using sucrose as substrate, ranging from polysaccharides to oligosaccharides and glucoconjugates. The applications of the corresponding products are also very broad: from food texturing agents to prebiotic compounds and biospecific signals.

These spectra will be surely enlarged in the coming years, as a result of both the screening of new microbial strains from non-conventional biotopes and the possibility to modify and control at the molecular level the structure of the genes encoding such enzymes through the mutagenesis and molecular evolution techniques.

This will result in the possibility to produce as well new polysaccharides with controlled physico-chemical properties, as new oligosaccharides with highly specific biological properties.

The industrial development of such innovative products will take advantage both of the fact that the reaction substrate, sucrose, is a widely available agricultural raw material, and that D-glucosyl conjugates are totally biodegradable and are hardly suspect of any xenobiotic effect.

AKNOWLEDGMENT
This work was supported by the European Union within the frame of the project "Alpha-Glucan Active Designer Enzymes" (PL970022).

REFERENCES

1. E.J. Hehre, Methods Enzymol., 1 (1955) 178.
2. H.J. Koepsell, H.M. Tsuchiya, N.N. Hellman, A. Kasenko, C.A. Hoffman, E.S. Sharpe, R.W. Jackson, J. Biol. Chem., 200 (1953) 793.
3. P. van Tieghem, Ann. Sci. Nat. Bot. Biol. Veg., 7 (1878) 180.
4. S. Hestrin, S. Avireni-Shapiro, M. Aschner, Biochem. J., 37 (1943) 450.
5. W.B. Neely, J. Nott, Biochem., 1 (1962) 1136.
6. K.H. Ebert, G. Schenk, Adv. Enzymol., 30 (1968) 179.
7. R.L. Sidebotham, Adv. Carbohydr. Chem. Biochem., 30 (1974) 371.
8. M.D. Hare, S. Svensson, G.J. Walker, Carbohydr. Res., 66 (1978) 254.
9. A. Jeanes, W.C. Haynes, C.A. Williams, J.C. Rankin, E.H. Melvin, M.J. Austin, J.E. Cluskey, B.E. Fischer, H.M. Tsuchiya, C.E. Rist, J. Am. Chem. Soc., 76 (1954) 5041.

10. F.R. Seymour, R.D. Knapp, Carbohydr. Res., 81 (1980) 105.
11. E.I. Garvie, Methods Microbiol., 16 (1984) 147.
12. H. Muzaka, H.D. Slade, Infect. Immun., 8 (1973) 555.
13. S. Hamada, H.D. Slade, Arch. Oral Biol., 24 (1979) 399.
14. R.R.B. Russell, Arch. Oral Biol., 35 (1990) 53S.
15. C.L. Simpson, P.M. Giffard, N.A. Jacques, Infect. Immun., 63 (1995) 609.
16. M.M. Vickerman, M.C. Sulavik, J.D. Nowak, N.M. Gardner, C.W. Jones, D.B. Clewell, DNA Seq., 7 (1997) 83.
17. M. Wilke-Douglas, J.T. Perchorowicz, C.M. Houck, B.R. Thomas, PCT Patent No. 8912386 (1989).
18. V. Monchois, R.M. Willemot, M. Remaud-Siméon, C. Croux, P. Monsan, Gene, 182 (1996) 23.
19. V. Monchois, A. Reverte, M. Remaud-Siméon, P. Monsan, R.M. Willemot, Appl. Environ. Microbiol., 64 (1998) 1644.
20. V. Monchois, R.M. Willemot, P. Monsan, FEMS Microbiol. Rev., 23 (1999) 131.
21. M. Remaud-Siméon, R.M. Willemot, P. Sarçabal, G. Potocki de Montalk, P. Monsan, J. Mol. Catalysis B in press.
22. B. Henrissat, A. Bairoch, Biochem. J., 316 (1996) 695.
23. A. Shimamura, Y.J. Nakano, H. Musaka, H.K. Kuramitsu, J. Bacteriol., 176 (1994) 4845.
24. M. Kobayashi, Y. Mitsuhishi, K. Matsuda, Biochem. Biophys. Res. Com., 80 (1976) 306.
25. M. Kobayashi, K. Matsuda, Agr. Biol. Chem., 41 (1977) 1931.
26. F.R. Seymour, R.D. Knapp, E.C.M. Chen, A. Jeanes, S.H. Bishop, Carbohydr. Res., 71 (1979) 231.
27. Y. Mitsuhishi, M. Kobayashi, K. Matsuda, Carbohydr. Res., 127 (1984) 331.
28. E.E. Smith, FEBS Lett., 12 (1970) 33.
29. M. Kobayashi, K. Matsuda, Biochim. Biophys. Acta, 370 (1974) 441.
30. F. Paul, A. Lopez Munguia, M. Remaud, V. Pelenc, P. Monsan, US Patent No 5 141 858 (1992).
31. M. Remaud-Siméon, A. Lopez Munguia, V. Pelenc, F. Paul, P. Monsan, Appl. Biochem. Biotechnol., 44 (1994) 101.
32. M. Dols, M. Remaud-Siméon, R.M. Willemot, M. Vignon, P. Monsan, Carbohydr. Res., 305 (1998) 549.
33. P. Valette, V. Pelenc, Z. Djouzi, C. Andrieux, F. Paul, P. Monsan, J. Sci. Food Agric., 62 (1993) 121.
34. T. Kohmoto, F. Fukui, A. Takaku, T. Mitsuoka, Agr. Biol. Chem., 55 (1991) 2157.
35. P. Monsan, F. Paul, in Biotechnology in animal feeds and feeding (R.J. Wallace, A. Chesson, eds.), VCH, Weinheim, 1995, pp 233.
36. H. Hidaka, M. Hirayama, Biochem. Soc.Trans., 19 (1991) 561.
37. Z. Djouzi, C. Andrieux, British J. Nutrition, 78 (1997) 313.
38. J.P. Lamothe, Y. Marchenay, P. Monsan, F. Paul, V. Pelenc, PCT Patent No.WO 9300067 (1993).
39. M. Dols, M. Remaud-Siméon,P. Monsan, Proceedings Second European Symposium on Biochemical Engineering Science (S. Feyo de Azevedo, E.C. Ferreira, K. Ch.A.M. Luyben, P. Osseweijer, eds.), Fac. Eng. Univ. Porto, 1998, pp 86.
40. M. Dols, M. Remaud-Siméon, P. Monsan, Enzyme Microb. Technol., 20 (1997) 523.
41. M. Dols, M. Remaud-Siméon, P. Monsan, Appl. Biochem. Biotechnol., 62 (1997) 47.

42. M. Dols, W. Chraibi, M. Remaud-Siméon, N.D. Lindley, P. Monsan, Appl. Environ. Microbiol., 63 (1997) 2159.
43. E.J. Hehre, D.M. Hamilton, A.S. Carlson, J. Biol. Chem., 166 (1946) 77.
44. J.Y. Riou, M. Guibourdenche, M.Y. Popoff, Ann. Microbiol. (Inst. Pasteur), 134B (1983) 257.
45. V.Büttcher, T.Welsh, L. Willmitzer, J. Kossmann, J. Bacteriol., 179 (1997) 3324.
46. G. Potocki de Montalk, M. Remaud-Siméon, R.M. Willemot, V. Planchot, P. Monsan, J. Bacteriol., 181 (1999) 375.
47. G. Potocki de Montalk, P. Sarçabal, M. Remaud-Siméon, R.M. Willemot, V. Planchot, P. Monsan, in R.R.B. Russel (ed.) Newcastle, England. Recent advances in Carbohydrate Bioengineering, in press.

Food Biotechnology
S. Bielecki, J. Tramper and J. Polak (Editors)
© 2000 Elsevier Science B.V. All rights reserved.

Oligosaccharide synthesis with dextransucrase
Kinetics and reaction engineering

K. Demuth, B. Demuth, H.J. Jördening, K. Buchholz

Chair for Carbohydrate Technology
Technical University, Langer Kamp 5, D 38106 Braunschweig, Germany

1. INTRODUCTION

Dextransucrase is a glucosyltransferase, and more precisely a transglucosidase, named sucrose: 1,6 α-D-glucan 6-α-D-glucosyltransferase. Its Enzyme Commission nomenclature number is: EC 2.4.1.5. The only natural substrate of dextransucrase (DS) is sucrose, which serves as a high energy glucosyl donor for polysaccharide and oligosaccharide synthesis; the main product is dextran.

In a secondary reaction, the so-called acceptor reaction, the glucosyl moiety is transferred from sucrose to the acceptor molecule instead of the growing dextran chain to produce the acceptor molecule elongated by one single glucosyl unit as the primary product. Usually the new bond formed is an α-1,6-glucosidic bond. In most cases, the product can itself serve as an acceptor, so that a homologeous series of glucosylated oligosaccharides is formed [1,2].

Acceptor properties have been reported for a wide variety of mono-, di- and oligosaccharides (e.g. glucose, fructose, maltose, raffinose [3-7], but also for some functionalized saccharide derivatives - as for example the sugar alcohols D-sorbitol and D-glycerol or the 5-O-α-D-glucopyranosyl-arabonic acid, or even glucal, an unsaturated sugar [3,8,9]. This secondary reaction therefore is of special technical interest: It offers perspectives for the synthesis of new oligosaccharides, which are not accessible otherwise by feasible technical routes.

Maltose and isomaltose are the most efficient acceptors. In most acceptors, the glucose from sucrose is transferred to the 6-hydroxyl group of the nonreducing-end glucose residue to give a series of isomaltooligosaccharides of degree of polymerization of 2–7 [6]. A detailed kinetic analysis revealed a nearly equally high acceptor strength for the di- to tetrasaccharides in the reaction mixture [10].

1.1 Application

The most widely used dextran is produced by the dextransucrase of the strain *L. mesenteroides* NRRL B-512F, which synthesizes a very highly linear polysaccharide containing 95% α-1,6 bonds.

The dextransucrase from *L. mesenteroides* NRRL B-1299 is known to catalyse the synthesis of dextran polymers containing α-1,2 linked branched chains and, in the presence of maltose as acceptor and of sucrose as D-glucosyl donor, glucooligosaccharides, which contain α-1,2 glyosidic bonds [11,12]. These oligosaccharides are produced industrially (Monsan, this volume).

For leucrose, an isomer of sucrose (glucosyl- α-(1→5) fructofuranoside) a patent was granted for the production by the isolated enzyme [13,14]. By this process, leucrose became available at larger amounts (pilot-plant scale with 1000 kg leucrose per batch), so that further investigations could be carried out concerning its chemical, physical, physiological and toxicological properties. Sensory tests showed that leucrose has about 50% of the sucrose sweetening power [15] so that physiological and toxicological experiments were performed for the use of leucrose in the food sector: In vivo tests demonstrated that the α(1→5) linkage of leucrose was not cleaved by the oral microflora; so leucrose can be considered as noncariogenic. Moreover, in vitro investigations gave some indications for a small anticariogenic potential of leucrose. Metabolic studies showed that leucrose was hydrolyzed enzymatically into glucose and fructose in the jejunum and only very small amounts (0.05-0.08%) were excreted in the urine. Contrary to sugar alcohols, leucrose did not induce diarrhea and flatulence because of its rapid hydrolysis in the jejunum. Even though leucrose is not suited for diabetics because glucose is released in the digestive system. No toxic or mutagenic effects were detected in animal tests [16].

Investigations of the kinetics provided the basis for optimal concentrations of sucrose and fructose in order to obtain a high product yield (70-75%) at high leucrose productivity (3.6 g/U) [17]. These results could be realized by continuous operation in a tubular reactor of special design [18]. The most successful method for dextransucrase immobilization up to now is achieved by entrapment in alginate, first applied by Schwengers and Benecke [13]. Further investigations on this immobilization method gave a high yield of active immobilized enzyme (in the range of 80%), which could be operated up to 60 days [19].

1.2 Modelling of dextransucrase catalysis

A typical acceptor reaction is that in the presence of maltose with sucrose as substrate and glucosyl donor (Figure 1)

The scheme of reactions, that is the basis for the mathematical model, has been published by Böker et al. [17] (Figure 2). The enzyme is assumed to have three active centers: two binding sites for the glucosyl group, as discussed by Robyt et al. [21], and one acceptor site [17,22]. The reactions 1, 2 and 3 represent the growth cycle of dextran formation. A sucrose molecule S is binding at the enzyme E, fructose F is released, and the glucosyl G is transferred to the dextran chain G_i, which is always linked to the enzyme. Fructose can be linked to the acceptor site, and react with G to form leucrose FG (reactions 4 and 9). The same reactions are possible with other acceptors A (reactions 6 and 13), and acceptor products (except leucrose) may again react as acceptors (reactions 6+j and 13+2xj). Until now, these reactions are considered only for the first acceptor product AG, because kinetic tests have been only carried out at low concentrations of higher acceptor products. (The model has recently been extended to include further acceptor products [10]). Sucrose is able to bind at the acceptor site (reaction 5), but does not react to an acceptor product. So reaction 5 is responsible for inhibition by high concentrations of sucrose. 11 and 12 represent the reactions with water, which is a very weak acceptor [19]. Reactions 10 and 14 represent the chain termination and the formation of free dextran with an acceptor at one end of the chain. They are of minor quantitative importance, compared with chain growth and acceptor reactions.

Acceptors and acceptor products are assumed to be able to bind reversibly at one glucosyl binding site (reactions 19+j). This results in a competitive inhibition of the enzyme.

125

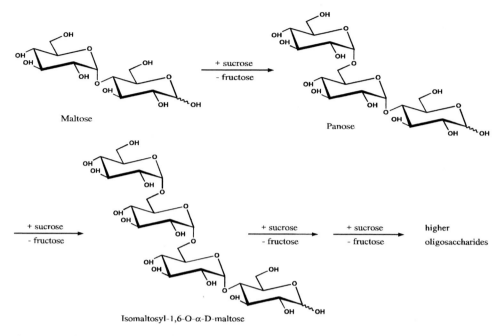

Figure 1. Maltose acceptor reaction.

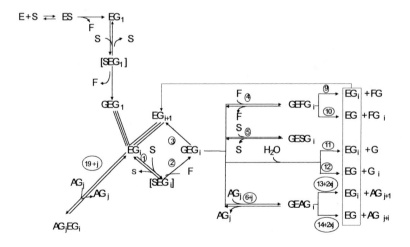

Figure 2. Scheme of reactions catalysed by dextransucrase (see text for details) [17,20].

Considering all reactions in the scheme, a system of 7 differential equations (for dS, F, FG, G, D, A, AG) with 32 kinetic parameters can be derived. In order to get reasonable results with a limited amount of experimental data, this very complex system has to be simplified: reactions 10, 14, 11, 12 (and so the differential equations for D and G) are neglected, because their contribution to product formation on a molar basis is very low. The remaining system consists of 5 differential equations with 15 independent parameters [20]. Typical results obtained by application of this model to a broad range of experimental kinetic investigations are given subsequently.

2. KINETICS AND REACTION ENGINEERING

2.1 Leucrose formation with fructose as a weak acceptor.
In the presence of fructose the only major acceptor product formed is leucrose. The overall reaction rate decreased considerably with increasing initial fructose concentration. This reflects the shift from dextran formation to the much slower acceptor reaction when fructose is added. As well as maltose, high concentrations of fructose depressed substrate inhibition [17,20,23]. The rate of leucrose formation as a function of sucrose at different concentrations of fructose as acceptor is shown in Figure 3.

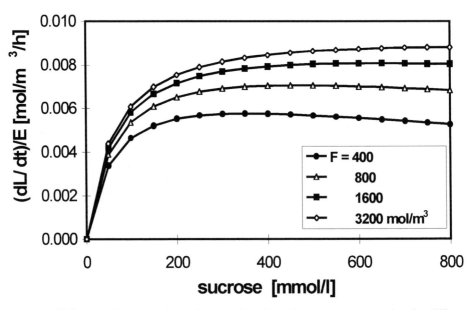

Figure 3. Initial rates of leucrose formation as a function of sucrose concentration for different levels of acceptor (fructose) concentration (model calculations) [20].

The influence of initial sucrose concentration on the yield of leucrose at fixed fructose concentration was very small. No significant increase of leucrose yield and productivity could be reached, only at low sucrose concentration (< 200 mmol/l) does the reaction rate decrease significantly. A concentration above 0.44-0.58 mol/l sucrose should be avoided because of the high viscosity of such solutions. On the contrary, initial fructose concentration influenced the kinetics and yield to a high degree (Figure 3), although the overall reaction rate decreased with increasing fructose concentration, the reaction rate of leucrose formation and yield increased up to a yield of 75% (fructose concentration 2.22-2.50 mol/l). Even higher values could be observed at more elevated fructose concentrations (up to 91% at a fructose concentration of 3.33 mol/l), but it would be critical to apply reaction solutions of such a high viscosity in technical processes.

Moreover, an influence of reaction temperature on the selectivity of dextransucrase-catalyzed reactions could be demonstrated. Measurements were carried out at 5°C, 15°C and 25°C and concentrations of 0.0584 mol/l sucrose and 2.22 mol/l fructose: The ratio of leucrose formation to the overall reaction rate increased with decreasing temperature, making it reasonable that out of all dextransucrase-catalyzed reactions, the leucrose formation became more dominant at low temperatures. Leucrose yield rose from 69% at 25°C to 75% at 5°C, whereas the byproduct ratio (isomaltulose and trehalulose relating to leucrose) decreased [17].

A strong increase of leucrose yield was observed in a temperature range of -4 C to -20 C. At these low saccharide concentrations a leucrose yield of 65% could be reached even at unfavorable concentrations, whereas only 10-15% yield was the result at 25°C under such conditions. A conceivable reason for this phenomenon might be the strong freezing-point--depression of the reaction solution due to the high concentration of dissolved substances (saccharides, dextransucrase and the buffer salts). In the range of -4°C to -20°C a relative soft ice was formed that presumably consisted of a minor liquid and a major solid part. The liquid part is assumed to have a much enhanced saccharide concentration resulting in a considerably higher leucrose yield [24].

The productivity is a measure for the amount of product related to the applied catalyst activity, which can be obtained over the useful lifecycle of the catalyst. Productivity is a key parameter with regard to the economy of an enzymatic process. So, productivity and operational stability of the batch operation process with immobilized dextransucrase were compared with continuous use of the enzyme beads. For the continuous reaction a tubular packed bed reactor (total volume 0.5 l) was constructed consisting of six compartments, each partially filled with the same amount of immobilized enzyme (25 or 10 g wet weight, 0.3 mm diameter, 2.8 U/g enzyme activity). The reaction solution with constant composition (0.292 mol/l up to 0.877 mol/l sucrose and 2.22 mol/l fructose in a 0.05 mol/l calcium alginate buffer pH 5.4) was pumped upward through the reactor at 25°C. Leucrose yield was in the range of 62-73% whereas productivity showed a wide spread of 0.7-3.6 g/U depending on the sucrose and enzyme concentrations used [18,23].

Modeling and design of continuos reactor systems was performed in order to identify optimal conditions for the production of leucrose. Four points are of major interest [20]:

1. The production rate of leucrose ($mol/m^3/h$) has to be maximized
2. The leucrose yield and concentration in the product mixture should be high

The formation of dextran and other by-products has to be minimized. The formation of dextran is detrimental, mainly because it leads to a swelling of the alginate beads and may limit the catalyst stability, which is essential for the continuous process. Other important by-products are trehalulose and isomaltulose [17].

128

3. The amount of sucrose in the product should be low. The need to separate sucrose and leucrose would increase the costs of downstream processing.

It can be seen from Figure 4, that there are optimal fructose concentrations with respect to leucrose production: $F_0=1700$ mol/m^3 for 1% residual sucrose in the product, 2100 mol/m^3 for 10%. For 10% sucrose, the maximum leucrose production rate is only 9% lower than the maximum value, achieved with the stirred vessel at high sucrose and very low leucrose concentrations. Compared with a stirred vessel with 10% sucrose output, the production rate is higher (in the range of 24% up to 125%). So it can be concluded, that a tubular flow reactor will be the best choice.

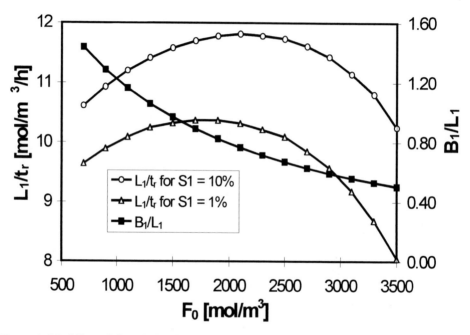

Figure 4. Modeling of the tubular flow reactor – Productivity (leucrose production rate (L_1) per residence time t_r) and formation of byproducts (B1) [20].

2.2 Acceptor products with glucose and KGPA as weak acceptors

The glycosylation of glucose (G) yields a homologous series of acceptor products with α-1,6-glycosidic bonds (isomaltose, -triose etc.) [7]. Figure 5 gives the yields of the acceptor products at elevated sucrose concentration (0.526 M) and varying G concentration. The conversion was nearly complete in these experiments (> 90% for sucrose). It is obvious that the first acceptor product (isomaltose) is obtained in rather low yield. This is due to the affinity of G as an acceptor and the very high affinity of isomaltose, isomaltotriose and tetrose as acceptors for the acceptor site of dextransucrase (see also Figure 10 and Table 1).

Table 1
Kinetic parameters of acceptor reactions (K. Demuth 1999): p7(i) * 10exp7 (L^2mmol^{-2})

	p7(Mono)	p7(Di)	p7(Tri)	p7(Tetra)	p7(Penta)
ISOMAL		**9.3**	**9.1**	6.9	
MALTOSE		5.7	**9.6**	5.7	3.9
KGPA		0.93	**8.8**	7.3	(11.8)
GLUCOSE	0.42	**9.3**	**9.1**	6.9	
GLUCAL	0.26	5.0	**8.7**	8.1	

The relative acceptor quality for a broad range of mono- and oligosaccharides and derivatives are given in Figure 10.

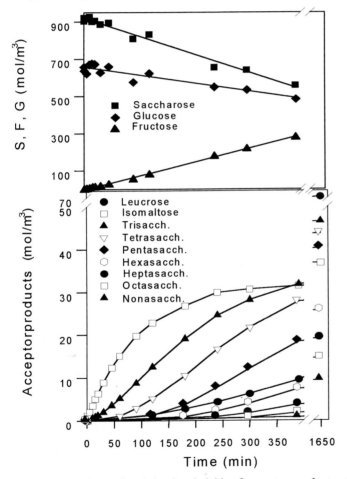

Figure 5. Experimental and simulated yields of acceptor products at different concentrations of glucose as an acceptor [10].

130

Figure 6 shows simulated results for potassium glucosylarabonate (KGPA) as a weak to medium strength acceptor (see also Figure 10 and Table 1). The first acceptor product of the homologous series, also with a new α-1,6-glycosidic bond (an isomaltosyl-α-1,5 arabonic acid), is obtained in medium yield only at low initial sucrose and high acceptor concentration (upper left). The second acceptor product is formed in maximal yield of about 30% also at rather low initial sucrose and high acceptor concentration (upper right). With increasing initial sucrose concentration the third and fourth acceptor product are formed with increasing yields (lower left and lower right).

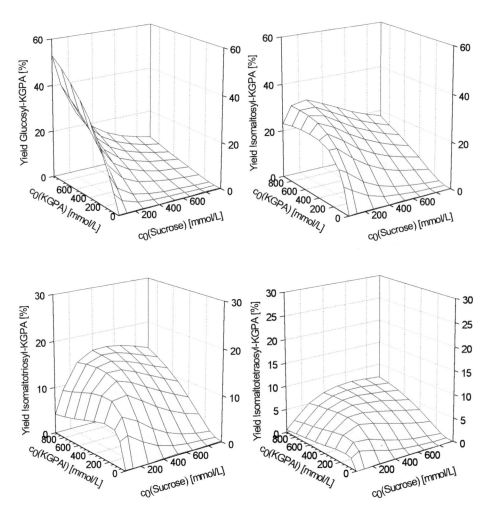

Figure 6. Simulated yields of acceptor products with glucosylarabonate (KGPA) as an acceptor, as functions of sucrose and accptor concentration (complete conversion of sucrose) [10].

2.3 Panose formation with maltose as a strong acceptor

The initial overall reaction rates for the maltose acceptor reaction calculated from the estimated parameter values as a function of the initial concentrations are shown in Figure 7. The accelerating and the inhibiting effect of the maltose acceptor on the overall reaction can be recognized, as found experimentally. The overall reaction rate (that is sucrose consumption by dextran formation and acceptor product formation) was accelerated significantly up to the factor of 2.2 by addition of maltose. This reflects the shift from dextran formation to the considerably faster acceptor reaction in the presence of maltose [25]. Furthermore, the graph shows - as expected - substrate inhibition at very low maltose concentrations. As we observed earlier in our experiments, this inhibition is eliminated when higher initial maltose concentrations are chosen. It should be mentioned that the inhibition by maltose according to this graph should be eliminated also when the initial sucrose concentrations are sufficiently high.

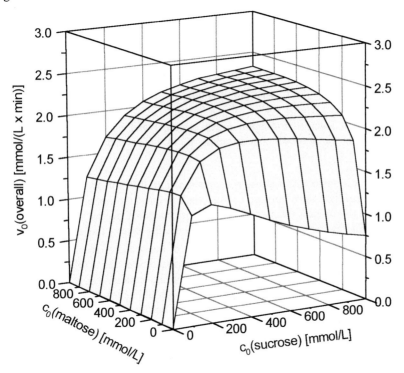

Figure 7. Calculated initial overall reaction rates for the maltose acceptor reaction (sum of dextran and acceptor product formation) [25].

Figure 8 shows that the rate of acceptor product formation dominates over dextran formation at elevated acceptor concentration: The rate of panose formation tends to reach 80% of the overall reaction rate (given by fructose formation) and the dextran formation rate to be only 20%, at the initial maltose concentration chosen.

132

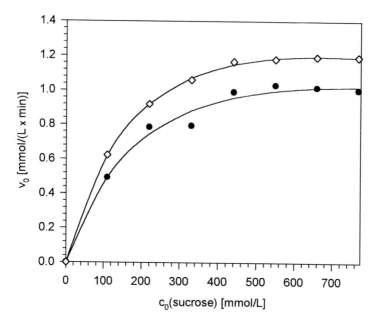

Figure 8. Initial rates of fructose formation (open symbols) and panose formation (closed symbols) as a function of sucrose concentration at an initial maltose concentration of 326 mmol/l [25].

   For technical applications of the maltose acceptor reaction, it is not only of importance to obtain high acceptor reaction rates but also to keep the side product formation rate as low as possible. Comparison of acceptor reaction rates with corresponding overall reaction rates may give valuable hints for favorable reaction conditions. For this reason the selectivity Sel_0 of acceptor product formation was defined as the quotient of the acceptor reaction rate and the overall reaction rate. Thus, according to the dextransucrase model, the selectivity of panose formation should be a function only of the maltose starting concentration. At high initial maltose concentrations the optimal value for the selectivity should be reached.

   In Figure 9 both the selectivities obtained from model calculations and those determined from the initial reaction rates from four series of experiments are shown as function of the maltose or sucrose starting concentrations. In both cases the selectivity converges to a maximum at high maltose concentrations, and no significant dependence of the experimental data on the sucrose concentration is observed. This can especially be seen from the experimental data obtained for the fourth serie of experiments, where the initial maltose concentration was kept constant and the initial sucrose concentration was varied in the range 110-769 mmol/l. Of course, this is only valid for the initial rates, since with progress of the reaction both panose and fructose react as acceptors, so that the selectivity also becomes a function of those saccharide concentrations. For technical purposes the maltose concentration should be kept above at least 400 mmol/l [25].

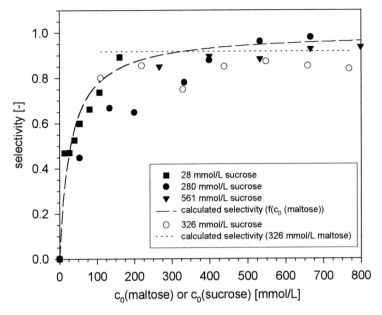

Figure 9. Selectivity of panose formation [25].

3. CONCLUSIONS

It may thus be concluded, that acceptors will compete with the substrate for the acceptor binding site.
- Strong acceptors such as maltose accelerate the overall reaction rate, up to a factor of over 2, due to a faster acceptor reaction than dextran formation rate.
- In the presence of the weak fructose acceptor, the slow acceptor reaction becomes the dominating step at high fructose concentrations resulting in a decrease of the overall reaction rate; maximal rates of leucrose formation were 20% of the reaction rate without acceptor.

Experiments were performed with free and immobilized enzyme with no significant difference in the reaction kinetics; high efficiencies of immobilized dextransucrase were obtained.

The kinetic investigations allow to give a quantitative estimate for the acceptor strength based on the kinetic parameters of the acceptor reaction (p7, Table 1) [10]. From these it may be concluded, that:
- isomalto- and maltooligosaccharide structures are optimal to fit into the acceptor binding site (in accordance with the many data published by Robyt et al. [1,2,7]),
- di- to tetrasaccharides are the best acceptors , and that
- the first acceptor product may be a functionalised sugar, a sugar alcohol or acid (such as glucosyarabonic acid).

134

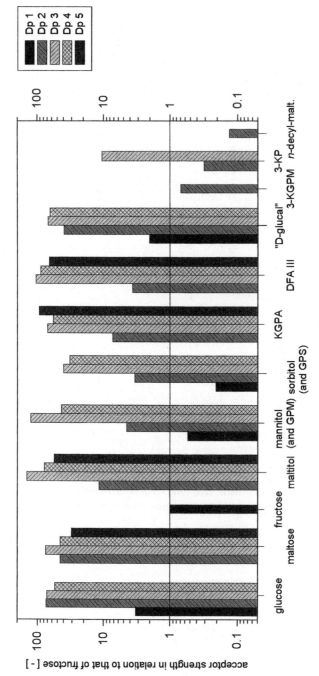

Figure 10. Acceptor strength of various mono- and disaccharides and their ensuing acceptor products in relation to the acceptor efficiency of fructose (DP1 signifies the monosaccharide, followed by the acceptor products, from di – (DP2) to pentasaccharide (DP5) (GPM: glucosyl-mannitol, GPS: glucosylsorbitol, KGPA: potassium glucosyl-arabonic acid, DFA III: difructoseanhydride III, 3-KGPM: 3-keto-glucosymannitol, 3-KP: 3-keto-palatinose, n-decyl-malt: n-decyl-maltoside) [10].

REFERENCES

1. J.F. Robyt., T.F. Walseth, Carbohydr. Res., 61(1978) 433.
2. J.F. Robyt., S.H. Eklund, Bioorg. Chem., 11 (1982) 115.
3. K.G. Ebert, G. Schenk, Z. Naturforsch., 23b (1968) 788.
4. F.H. Stodola, H.J. Koepsell, E.S. Sharpe, J. Am. Chem. Soc., 78 (1956) 2514.
5. M. Killey, R.J. Dimler, J.E. Cluskey, J. Am. Chem. Soc., 77 (1955) 3315.
6. J.F. Robyt, Adv. Carbohydr. Chem. Biochem. Vol 51 (1995) 133.
7. J.F. Robyt, S.H. Eklund, Carbohydr. Res., 121 (1983) 279.
8. D. Prinz, Ph.D. thesis. Technical University, Braunschweig, Germany, 1990
9. K. Heincke, B. Demuth, H.J. Jördening, K. Buchholz, Ann. N. Y. Acad. Sci., 864 (1998) 203.
10. K. Demuth, PhD thesis, Technical University of Braunschweig, 1999.
11. F. Paul, A. Lopez Munguia, M. Remaud, V. Pelenc, P. Monsan, US Patent 5,141,858.
12. M. Remaud-Siméon, A. Lopez Munguia, V. Pelenc, F. Paul, P. Monsan. Appl. Biochem. Biotech., 44 (1994) 101.
13. D. Schwengers, H. Behnecke, DBP-DE 34 46 380 (1984); Eur. Pat. 185 302 (1985), Pfeifer & Langen, Chem. Abstr., 105 (1986) 77815p.
14. D. Schwengers, in: Carbohydrates as Organic Raw Materials, Ed. F. W. Lichtenthaler. Verlag Chemie, Weinheim 1991, 183.
15. Pfeifer & Langen, Anwendungstechnische Informationen (1987), Pfeifer & Langen, Postfach 451080, D-50885 Köln.
16. P.S. Elias, H. Benecke, D. Schwengers: Safety Evaluation Studies of Leucrose. J. Am. College of Toxicology, 15 (1996) 205.
17. M. Böker, H.J. Jördening, K. Buchholz, Biotechnol. Bioeng., 43 (1994) 856.
18. K.D. Reh, M. Noll-Borchers, K. Buchholz:, Enzyme Microb. Technol., 19 (1996) 518.
19. A. Reischwitz, K.-D. Reh, K. Buchholz, Enzyme Microb. Technol., 17 (1996) 457.
20. B. Demuth, H.J. Jördening, K Buchholz, Biotechnol. Bioeng., 62 (1999) 583.
21. J.F. Robyt, B.K. Kimble, T.F. Walseth, Arch. Biochem. Biophys., 165 (1974) 634.
22. A. Tanriseven, J.F. Robyt, Carbohydr. Res., 225 (1992) 321.
23. K. Buchholz, M. Noll, D. Schwengers, Starch/Stärke, 50 (1998) 164.
24. B. Daum, K. Buchholz, Ann. N. Y. Acad. Sci., 864 (1998) 207.
25. K. Heincke, B. Demuth, H.J. Jördening, K. Buchholz, Enzyme Microb. Technol. 24 (1999) 523.

Food Biotechnology
S. Bielecki, J. Tramper and J. Polak (Editors)

Pyranose oxidase for the production of carbohydrate-based food ingredients

D. Haltrich, C. Leitner, B. Nidetzky and K.D. Kulbe

Division of Biochemical Engineering, Institute of Food Technology, University of Agricultural Sciences Vienna (Universität für Bodenkultur - BOKU), Muthgasse 18, 1190 Wien, Austria

Pyranose oxidase (P2O) is a tetrameric flavoenzyme found in wood degrading fungi that catalyzes the C-2 oxidation of several aldopyranoses, including glucose, galactose and some other monosaccharides commonly found in lignocellulose, to the corresponding 2-keto derivatives. These can be attractive intermediates in the production of food additives, such as fructose or tagatose, and can be easily produced in high yields. We have investigated the enzyme technological formation of these two ketoses via the aldos-2-uloses – as entirely enzymatic variations of the old Cetus process – in some detail. In this Cetus process, which was published more than 15 years ago, crystalline D-fructose is produced from D-glucose via the intermediate 2-keto-D-glucose. Whereas the first step in the traditional process is catalyzed by immobilized P2O, the subsequent reduction is performed by catalytic hydrogenation. In our enzymatic variation of this process, soluble P2O from *Trametes multicolor* was employed. D-Glucose could be converted into 2-keto-D-glucose in yields above 98%. When the biocatalyst together with the two stabilizing agents, i.e., catalase and bovine serum albumin, was separated from the product solution by ultrafiltration, it could be reutilized for several subsequent batch operation cycles. 2-Keto-D-glucose thus obtained was quantitatively reduced to D-fructose by NAD(P)-dependent aldose reductase from *Candida tenuis*. Two different enzymatic systems were successfully employed for the continuous regeneration of the coenzyme necessary in this reaction. In this way, D-fructose essentially free of D-glucose can be prepared by this simple and convenient method. In a similar way, D-galactose can be converted into D-tagatose via the intermediate 2-keto-D-galactose. Yields are, however, lower because of the slight instability of this intermediate.

1. INTRODUCTION

Nowadays enzymes are recognized as important and efficient catalysts for a number of stereospecific and regioselective reactions necessary for carbohydrate synthesis. The broad substrate specificity of some of these enzymes is often utilized to obtain a range of both synthetically useful intermediates or final products [1]. One example of such an attractive biocatalyst that can be used for the transformation of a number of carbohydrates is the enzyme pyranose 2-oxidase (P2O, glucose 2-oxidase, pyranose:oxygen 2-oxidoreductase, EC 1.1.3.10) which catalyzes the specific oxidation of several aldopyranoses at the position C-2 to yield the corresponding 2-keto products [2]. During these oxidation reactions electrons are transferred to molecular oxygen, which results in the formation of hydrogen peroxide. The existence of ample patent literature on P2O reflects the importance of this enzyme for biotechnological applications. Pyranose oxidase was employed as the key biocatalyst in the Cetus process, which has been developed more than 15 years ago for the conversion of D-glucose to crystalline D-fructose via the intermediate 2-keto-D-glucose [3–5]. In this process immobilized P2O from *Polyporus obtusus* was used for the oxidation of D-glucose, whereas 2-keto-D-glucose was reduced by catalytic hydrogenation. As an alternative to the enzymic oxidation process, the undisrupted mycelia of various P2O-producing basidiomycetes have been used in a microbial approach to convert D-glucose into 2-keto-D-glucose [6]. Despite this potential application in an industrial process, P2O has received relatively little scientific attention for biocatalytic purposes.

P2O is widely distributed among white-rot fungi, in which it is considered a constituent of the ligninolytic system supplying peroxidases with hydrogen peroxide [7]. It has been purified and characterized from several microorganisms including *Phanerochaete chrysosporium* [8,9], *Phlebiopsis (Peniophora) gigantea* [10,11], *Pleurotus ostreatus* [12], *Polyporus obtusus* [13], *Trametes (Coriolus) versicolor* [14], and the unidentified basidiomycete no. 52 [15]. The data available at present reveal some general similarities among P2Os from these different fungi. Typically, P2O is a rather large protein which contains covalently bound flavin adenine dinucleotide. P2O prefers D-glucose as its substrate which is oxidized to D-*arabino*-hexos-2-ulose (2-keto-D-glucose, D-glucosone). In addition, it also exerts significant activity on a number of other carbohydrates including D-xylose, L-sorbose or D-glucono-1,5-lactone. This property, however, varies considerably among pyranose oxidases isolated from different fungal sources.

Here we report the application of soluble pyranose oxidase from the basidiomycetous fungus *Trametes multicolor* in the conversion of two monosaccharides, i.e., D-glucose and D-galactose, into the corresponding aldos-2-uloses 2-keto-D-glucose and 2-keto-D-galactose. Both compounds are not produced on an industrial scale at present, however, they could be attractive intermediates for several proposed subsequent reactions. For 2-keto-D-glucose these include the oxidation to 2-keto-D-gluconic acid and further transformation to D-isoascorbic acid, its reduction to D-fructose, D-mannitol or D-sorbitol, as well as its reductive amination yielding 1,2-diaminosorbitol or -mannitol [16], while 2-keto-D-galactose can be reduced to D-tagatose, galactitol (dulcitol) or D-talitol. Furthermore, we describe the subsequent enzymatic reduction of 2-keto-D-glucose and 2-keto-D-galactose to D-fructose and D-tagatose, respectively, in that way establishing an entirely enzymatic process for the production of these two ketoses as an alternative to the chemo-enzymatic Cetus process. The latter reductive reaction is catalyzed by aldose (xylose) reductase (ALR) which performs the initial step in D-xylose assimilation in yeasts and mycelial fungi, i.e., the NAD(P)H-dependent reduction of this pentose to xylitol. Recently, ALR from the yeast *Candida tenuis* has been reported as a well suited biocatalyst for enzymatic, coenzyme-dependent processes [17]. An outline of the proposed reaction scheme as well as additional substrates and products – both mono- and disaccharides – that can be converted by this enzyme technological approach employing P2O and ALR are presented in Figure 1.

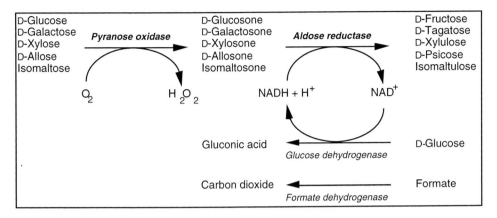

Figure 1. Reaction scheme for the production of ketoses by pyranose oxidase-catalyzed oxidation of sugars and aldose reductase-catalyzed reduction of the resulting 2-keto compounds.

Especially D-tagatose, a naturally occurring keto-hexose which is found in strongly heated milk as a degradation product of lactose [18], is of special interest for food technology as it is a low-calorie sweetener with a sweetness comparable to that of sucrose (92% relative sweetness), causes no dental cavities, and has proposed prebiotic properties, i.e., it supports growth of intestinal microorganisms beneficial for human health. Chemical syntheses of D-tagatose from D-galactose or D-fructose have been reported, albeit the relatively low yields are not satisfactory for a larger scale production. Recently, the microbiological conversion of various alditols or D-galactose to D-tagatose has been described [19]. As an alternative to these fermentative methods, a chemo-enzymatic process for the preparation of D-tagatose has been proposed using P2O to oxidize D-galactose to D-*lyxo*-hexos-2-ulose (2-keto-D-galactose, galactosone) which then is reduced by catalytic hydrogenation [20]. A disadvantage of this process, however, is that the resulting D-tagatose preparation still contains significant amounts of aldoses, mainly D-galactose.

2. MATERIALS AND METHODS

2.1. Materials and chemicals
All chemicals were of the highest purity available and obtained from Sigma (St. Louis, MO, U.S.A.) unless otherwise stated. Horseradish peroxidase (EC 1.11.1.7), grade I, was from Boehringer Mannheim (Mannheim, Germany) and glucose dehydrogenase (EC 1.1.1.47) from *Bacillus cereus* was from Amano (Milton Keynes, UK). Catalase (EC 1.11.1.6) from bovine liver was purchased from Fluka (Buchs, Switzerland). Formate dehydrogenase (EC 1.2.1.2) from *Candida boidinii* was a kind gift from Dr. Udo Kragl (KFA Jülich, Germany). Bovine serum albumin was from United States Biochemicals (Cleveland, OH, USA), material for preparative chromatography was obtained from Amersham-Pharmacia (Uppsala, Sweden), and the dye Reactive Green 19 (Procion Green H-E4BD) was from ICI (Manchester, UK).

2.2 Microbial strains and culture conditions

The wild type strain of the basidomycete fungus *Trametes multicolor* MB 49 was used as the source of pyranose oxidase. Pyranose oxidase was produced on a lactose-based medium as described elsewhere [21]. For the production of aldose reductase, the yeast *Candida tenuis* CBS 4435 (Centraalbureau voor Schimmelcultures, Baarn, The Netherlands) was grown on a medium containing 2% xylose, 1.6% yeast extract, 0.4% peptone from casein and 0.05% $MgSO_4 \cdot 7H_2O$ as previously described [22].

2.3. Enzyme assays

Pyranose oxidase activity was determined spectrophotometrically at 420 nm and 30°C by measuring the formation of H_2O_2 in a peroxidase-coupled assay using ABTS (2,2'-azinobis(3-ethylbenzthiazoline-6-sulfonic acid); ε_{420} = 43,200 $M^{-1} \cdot cm^{-1}$ [23]) as the chromogen [10]. The standard assay mixture (1 mL total) contained 1 μmol ABTS in potassium phosphate buffer (50 mM, pH 6.5), 2 U horseradish peroxidase, 100 μmol D-glucose and a suitable amount of the P2O sample. One unit (U) of P2O activity is defined as the amount of enzyme necessary for the oxidation of 2 μmol of ABTS per min under the given conditions. Catalase activity is expressed in Sigma units and was measured according to the instructions of the supplier.

Aldose reductase (ALR, EC 1.1.1.21; alditol:NAD(P)$^+$ 1-oxidoreductase) activity was assayed spectrophotometrically at 340 nm and 25°C by measuring the consumption of NADH (ε_{340} = 6,220 $M^{-1} \cdot cm^{-1}$). The standard assay mixture (1 mL) contained 707 μmol D-xylose and 0.22 μmol NADH in 300 mM phosphate buffer (pH 6.0). One unit of ALR activity refers to the amount of enzyme consuming 1 μmol of NADH per minute. Protein concentrations were determined according to the dye-binding method of Bradford [24] using bovine serum albumin as standard.

2.4. Enzyme purification

For the partial purification of pyranose oxidase washed mycelia of *T. multicolor* (2 g wet weight) were resuspended in 10 mL phosphate buffer (50 mM, pH 6.5) containing 10 mM EDTA, homogenized and then disrupted by three passages through a French pressure cell (1,380 bar, 4°C). Cell debris were removed by ultracentrifugation (30,000 g, 4°C, 20 min) and the clear supernatant loaded onto a Source 30Q column pre-equilibrated with 20 mM BisTris buffer, pH 6.5 (buffer A). Subsequently, P2O was eluted with a linear gradient of 0–100% 1 M KCl in buffer A. Fractions of the eluate were screened for P2O activity and the most active fractions pooled.

Aldose reductase was partially purified by dye-ligand pseudoaffinity chromatography [25]. Washed cells of *C. tenuis* were diluted 1:3 in 30 mM phosphate buffer (pH 7.0) and disintegrated in a continuously operated bead mill (glass bead diameter 0.24–0.5 mm) at 4°C and an average residence time of 10 min. Cell debris were removed by ultracentrifugation (110,000 g, 40 min) and the clear supernatant was loaded onto a Reactive Green 19–Sepharose 4B-CL column equilibrated with 50 mM phosphate buffer, pH 7.0. ALR was then eluted with 2 M NaCl, fractions containing ALR activity were pooled and concentrated by 70% ammonium sulfate precipitation to a concentration of approximately 100 $U \cdot mL^{-1}$.

2.5. Carbohydrate conversions

The P2O-catalyzed conversions of aldoses into the corresponding aldos-2-uloses were performed at 30°C in 500-mL conical flasks with a total working volume of 200 mL. The reaction system contained 250 mM D-glucose (100 mM D-galactose) in phosphate buffer (25 mM, pH 6.5), 1.0 U·mL^{-1} partially purified P2O (0.85 U·mL^{-1} for the conversion of D-galactose), 1,000 U·mL^{-1} catalase and 5 mg·mL^{-1} bovine serum albumin. The reaction system was continuously stirred (200 rev·min^{-1}) and oxygenated using pure oxygen (0.5 vol oxygen per vol fluid per min) which was bubbled through a porous sintered glass tube. The dissolved oxygen concentration was measured using a microoxygen electrode (Microelectrodes Inc., Londonderry, NH, U.S.A.). When necessary the pH was held constant by automatic titration with 1 M KHCO$_3$. At the times stated, samples were taken to monitor chromatographically the course of the reaction. After completion of the conversion, proteins were separated by ultrafiltration using Millipore Minitan ultrafiltration membranes, 10 kDa cut-off (Bedford, MA, U.S.A.).

For the ALR catalyzed reduction of aldos-2-uloses, which was performed at 25°C, 25 mL of the crude, ultrafiltered sugar solutions was supplemented with 1.25 U·mL^{-1} ALR and 0.25 mM NAD$^+$. For the continuous regeneration of the coenzyme two different enzymatic regenerative systems [26] were compared. When employing the system formate dehydrogenase (FDH) / formate, the reaction system contained 1.5 U·mL^{-1} FDH and 100 mM sodium formate, while the pH was maintained at 7.0 by titrating with 1 M formic acid. Alternatively, the regenerative system glucose dehydrogenase (GDH) / glucose was used (1.5 U·mL^{-1} GDH, 250 mM D-glucose, titration with Tris).

2.6. Analytical methods

Identification and quantification of D-glucose / 2-keto-D-glucose and D-galactose / 2-keto-D-galactose were done by HPLC using an Ostion LGKS 0800 Na column (250 x 8 mm; Watrex, Prague, Czech Republic) and refractive index detection. Water was used as a mobile phase at a flow rate of 0.5 mL·min^{-1} and a column temperature of 80°C. 2-Keto-D-glucose / D-fructose and 2-keto-D-galactose / D-tagatose were analyzed by HPLC employing an Aminex HPX 87C column (BioRad, Hercules, CA, U.S.A.) at 80°C with 10 mM Ca(NO$_3$)$_2$ as eluent (0.7 mL·min^{-1}) and refractive index detection.

3. RESULTS

3.1. Partial characterization of pyranose oxidase

For obtaining a P2O preparation that is suitable and stable for the transformation of sugars, the enzyme was partially purified by anion exchange chromatography. The enzyme preparation thereby obtained had a specific P2O activity of 8.8 U·mg^{-1}, corresponding to a 2.3-fold purification of the enzyme in a yield of approximately 85%. This partially purified P2O preparation was used for characterizing some catalytic properties of the enzyme. P2O from *T. multicolor* exerts activity on various carbohydrates. D-Glucose was the most readily oxidized substrate, while L-sorbose, D-glucono-1,5-lactone and D-xylose were oxidized with 99.3%, 59.5% and 56.3% activity, respectively, relative to D-glucose [27]. In contrast, D-galactose is only a poor substrate with a relative activity of 5.8%. P2O from *T. multicolor* exhibited a very broad pH optimum at 25°C and retained more than 90% of its maximum activity in the range of pH 3.5–7.0. Pyranose oxidase from *T. multicolor* was found to be very stable in the absence of its carbohydrate substrates. A partially purified enzyme preparation lost less than 5% of its initial activity when stored for one month at 30°C in 50 mM phosphate buffer, pH 6.5. This excellent storage stability of the enzyme was, however, contrasted with the stability under operational conditions, where it completely lost its catalytic activity, unless rapid destruction of hydrogen peroxide was ensured. This reactive compound is formed in stoichiometric amounts during the P2O-catalyzed oxidation of sugars and has been shown to irreversibly inactivate different enzymes including glucose oxidase and catalase [28–30]. It could be shown that the addition of 1,000 U·mL^{-1} catalase from bovine liver to the standard reaction system for the oxidation of carbohydrates significantly increased the operational stability of the biocatalyst. Further improvement of this stability could be attained when adding 5 mg·mL^{-1} bovine serum albumin. Under such reaction conditions, P2O was perfectly stable and more than 98% of the initial activity could be recovered after a single, discontinuous oxidation experiment of 100 mM D-glucose [27].

3.2. Partial characterization of aldose reductase

A single dye-ligand chromatography step was found to be sufficient for obtaining a stable ALR preparation exhibiting a specific activity of approximately 10 U·mg^{-1} protein in yields of above 85%. This enzyme preparation did not contain any contaminating enzyme activities that would interfere in the synthesis of ketoses from the corresponding aldos-2-uloses [17]. ALR from the yeast *Candida tenuis* has been extensively characterized with respect to its physico-chemical and kinetic properties [25]. In brief, ALR retains 70% of its maximum activity between pH 6.0 and 7.5 in 25–50 mM Tris/HCl and phosphate buffer and reveals notable stability at 25°C (half-life of approximately 1,500 h). ALR reduces a number of aldehyde compounds including aldoses and aldose derivatives as well as aliphatic and aromatic aldehydes. Interestingly, several 2-keto-D-aldoses are reduced by ALR at least equally well as the natural substrate D-xylose. The catalytic constant for this latter natural substrate was found to be 18.2 s^{-1}, whereas the corresponding values for 2-keto-D-glucose and 2-keto-D-galactose are 23.9 s^{-1} and 21.2 s^{-1}, respectively. The enzyme was not substrate-inhibited by aldos-2-uloses. Furthermore, the reaction product, D-fructose did not influence the ALR activity at concentrations of up to 500 mM and the reverse reaction from D-fructose to 2-keto-D-glucose was less than 0.1% of the reduction reaction. These results indicate excellent prerequisites for the enzymatic reduction of aldos-2-uloses to D-fructose or D-tagatose employing ALR as the biocatalyst.

3.3. Transformation of D-glucose into D-fructose via 2-keto-D-glucose

The pyranose oxidase-catalyzed conversion of D-glucose into 2-keto-D-glucose was carried out in a discontinuous mode of operation. One critical factor for the efficient utilization of P2O in bioconversion processes seems to be the gas-liquid mass transfer of oxygen which is only poorly soluble in aqueous solutions. Assuming that P2O is fully active and oxidizes D-glucose with its maximum catalytic capacity, 0.5 mmol of oxygen will be consumed per L and min under the selected reaction conditions. Thus, a sufficient oxygen supply is crucial for avoiding possible limitations of the reaction [31]. Furthermore, P2O from *P. gigantea* and *P. chrysosporium* have reported K_m-values for oxygen of 0.65 mM and 0.13 mM, respectively, [9,10], while the K_m-value for the *T. multicolor* enzyme is approximately 0.1 mM [Leitner and Haltrich, unpublished results]. These values are in the range of the oxygen concentration dissolved in aqueous solutions under 1 atm of air at 30°C which amounts to approximately 0.25 mM. Because of its low solubility, this P2O cosubstrate is present in the reaction system in nonsaturating concentrations and hence the enzyme-catalyzed reaction will not advance at its maximum velocity. In order to increase both the oxygen concentration and the oxygen transfer into the reaction system, the substrate solution was sparged with pure oxygen.

A typical time course of D-glucose conversion by P2O is shown in Figure 2. When adding the biocatalyst to the reaction solution, the dissolved oxygen tension (DOT) decreased very rapidly from the initial value of 100% of oxygen saturation (corresponding to approximately 1.16 mM O_2) to 15–17% and then stayed relatively constant during substrate conversion. With the depletion of D-glucose in the reaction solution the DOT rises again to its initial value. As is evident from Figure 2 the DOT is an excellent indicator for the progress of the enzymatic oxidation reaction. It was not necessary to control the pH value which was constant during the entire biotransformation reaction. Since hydrogen peroxide is continuously decomposed by catalase and thereby removed from the reaction system, the P2O-catalyzed oxidation of D-glucose is quasi-irreversible and thus substrate conversions of close to 100% are attainable (Figure 2). In these discontinuous transformation experiments using 250 mM D-glucose and pure oxygen as the substrates, volumetric and specific productivities of 9.0 g·(L·h)$^{-1}$ and 9.0 mg·(L·h·U P2O)$^{-1}$, respectively, could be obtained for the product 2-keto-D-glucose. Furthermore, no by-product formation could be detected during this oxidation reaction by both thin layer chromatography and HPLC (data not shown).

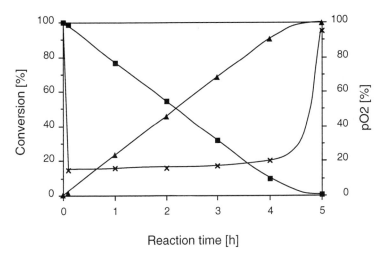

Figure 2. Discontinuous conversion of D-glucose into 2-keto-D-glucose using pyranose oxidase from *Trametes multicolor*.
Conditions: 250 mM glucose; 1.0 U·mL^{-1} P2O; 1,000 U·mL^{-1} catalase; 5 mg·mL^{-1} BSA; 25 mM phosphate buffer; pH 6.5. Symbols: –■–, glucose; –▲–, 2-ketoglucose; –×–, pO$_2$. Reprinted from Ref. 27, with permission.

After separating the enzymes and BSA from the reaction mixture by ultrafiltration (cut-off 10 kDa), the crude 2-keto-D-glucose solution could be used for the subsequent reduction, in which aldose reductase from *C. tenuis* was employed as the biocatalyst, without any further purification. P2O together with BSA could be conveniently reused for several successive batch operation cycles of substrate conversion. This excellent operational stability as well as the reusability of the biocatalyst are shown in Figure 3, showing the residual P2O activity after five consecutive 300 mL-conversion experiments. It should be noted, that the decrease in P2O activity also includes the losses during ultrafiltration of these rather small reaction volumes. In contrast, catalase from bovine liver was found to be unstable under the reaction conditions and had to be freshly added to each new bioconversion batch.

2-Keto-D-glucose was reduced to D-fructose in discontinuous batch experiments applying either formate dehydrogenase (FDH) / formate or glucose dehydrogenase (GDH) / glucose for NADH regeneration. Both of these regenerative systems can be successfully employed for the production of D-fructose with ALR and NADH as is evident from Figure 4. For both of the conjugated enzyme systems investigated in our study a conversion rate of above 99% was attained. The resulting volumetric productivities were 5.0 and 6.4 g fructose per L per h when using FDH and GDH, respectively, for coenzyme regeneration. Since ALR reduces exclusively the aldehyde group of 2-keto-D-glucose, formation of side products such as D-glucose or D-mannose, which theoretically might be obtained by reduction of the keto moiety of 2-keto-D-glucose, could not be observed by TLC and HPLC analysis of the reaction product (data not shown).

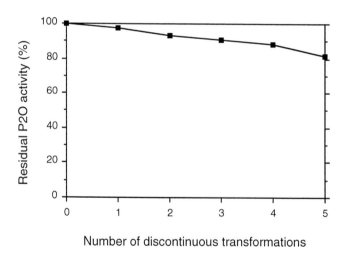

Number of discontinuous transformations

Figure 3. Residual pyranose oxidase activity after oxidation of 50 g·L^{-1} substrate in 25 mM phosphate buffer, pH 6.5, at 25°C. The biocatalyst was separated from by ultrafiltration (cut-off 10 kDa) and subsequently added to fresh substrate solution.

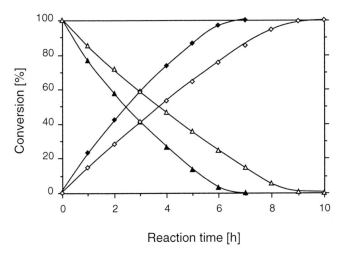

Reaction time [h]

Figure 4. Discontinuous conversion of 2-keto-D-glucose into D-fructose using aldose reductase (ALR) from *C. tenuis*. Formate dehydrogenase (1.5 U·mL⁻¹; open symbols) or glucose dehydrogenase (1.5 U·mL⁻¹; dark symbols) were used for the continuous regeneration of the coenzyme NADH.
Conditions. 250 mM ketoglucose; 1.25 U·mL⁻¹ ALR; 0.25 mM NAD⁺; 25 mM phosphate buffer, pH 7.0. Symbols: $-\triangle-$ and $-\blacktriangle-$, 2-ketoglucose;$-\Diamond-$ and $-\blacklozenge-$, fructose. Reprinted from Ref. 27, with permission.

3.4. Transformation of D-galactose into D-tagatose via 2-keto-D-galactose

The P2O-catalyzed biotransformation of D-galactose into 2-keto-D-galactose was performed in a similar, discontinuous manner as the D-glucose oxidation. Contrary to the D-glucose conversion, however, the pH value had to be controlled by automatic titration with 1 M KHCO₃. A typical time course of D-galactose conversion by P2O is shown in Figure 5. Although D-galactose was completely converted after 24 h as judged by HPLC, 2-keto-D-galactose was only obtained in yields of approximately 85%, corresponding to a volumetric productivity of 0.99 g·(L·h)⁻¹. In addition to 2-keto-D-galactose, two unidentified acidic by-products were detected in the product solution by HPLC. Presumably, these are degradation products of the aldos-2-ulose since these sugars are known to decompose to various acids, especially under alkaline conditions [32]. During one single reaction cycle of 24 h, 12% of the initially employed P2O activity was lost due to inactivation.

The ALR-catalyzed reduction of 2-keto-D-galactose was done in accordance do the reduction of 2-keto-D-glucose, using formate dehydrogenase / formate as the regenerative system. The pH of this reaction system was maintained at 7.0 by titrating with 1 M formic acid. Tween-80 (0.1%) was added since it was found to stabilize ALR to a significant extent [33]. As is shown in Figure 6, 2-keto-D-galactose was completely reduced to D-tagatose with no by-products being detectable by HLPC or TLC. The volumetric productivity of this biotransformation was calculated to be 7.2 g of D-tagatose per L per h.

146

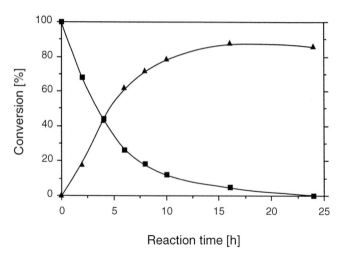

Figure 5.   Discontinuous conversion of D-galactose into 2-keto-D-galactose using pyranose oxidase from *T. multicolor*.
Conditions: 100 mM galactose; 0.85 U·mL^{-1} P2O; 1,000 U·mL^{-1} catalase; 5 mg·mL^{-1} BSA; 50 mM phosphate buffer, pH 6.5. Symbols: –■–, galactose; –▲–, 2-ketogalactose.

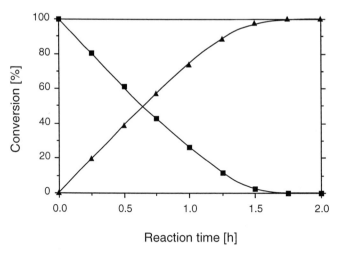

Figure 6.   Discontinuous conversion of 2-keto-D-galactose into D-tagatose using aldose reductase (ALR) from *C. tenuis* and formate dehydrogenase (FDH) for the continuous regeneration of the coenzyme NADH.
Conditions: 80 mM ketoglucose; 2.5 U·mL^{-1} ALR; 2.0 U·mL^{-1} FDH; 0.22 mM NAD$^+$; 80 mM Na formate; 0.1% Tween-80; 50 mM phosphate buffer, pH 7.0. Symbols: –■–, 2-ketogalactose; –▲–, tagatose.

4. DISCUSSION

Pyranose oxidase from *Trametes multicolor* MB 49 was found to be an excellent biocatalyst for the *C*-2 specific oxidation of pyranose sugars as has been shown in this study for the transformation of D-glucose into 2-keto-D-glucose (glucosone) and D-galactose into 2-keto-D-galactose (galactosone). In contrast to P2O from *Oudemansiella mucida, P. chrysosporium* and *T. versicolor* [34], we found that the *T. multicolor* enzyme apparently does not catalyze the over-oxidation of its reaction product 2-keto-D-glucose at position *C*-3 to give 2,3-diketo-D-glucose (D-*erythro*-hexos-2,3-diulose). This aldos-2-ulose-consuming property of P2O would certainly interfere with the quantitative conversion of D-glucose to the primary oxidation product. Furthermore, *T. multicolor* does not form the aldos-2-ulose-utilizing enzyme pyranosone dehydratase, which has been shown both in *P. chrysosporium* [35,36] and *P. obtusus* [37]. This enzyme catalyzes the conversion of 2-keto-D-glucose into cortalcerone, a β-pyrone compound. Obviously, a contaminating enzyme activity that consumes the product 2-keto-D-glucose would have negative effects on the yields that can be obtained in the transformation of sugars into the corresponding *C*-2 oxidized products. Because of this lack of pyranosone dehydratase activity in the enzyme system of *T. multicolor* only a simple chromatographic purification step is sufficient to obtain a stable P2O preparation that is suitable for the transformation of aldoses into the corresponding aldos-2-uloses.

In contrast to previous reports, in which the use of an immobilized enzyme preparation or of the fungal biomass was described for the transformation of different aldoses into the respective aldos-2-uloses [2,4,6,20,38], we employed soluble enzyme, thus avoiding possible mass transfer limitations that are frequently found in heterogeneous systems due to diffusion in particles [39,40]. Furthermore, employing a soluble P2O preparation has the advantage of utilizing the full catalytic efficiency of the biocatalyst since considerable losses of the initial enzyme activity have been reported for the immobilization of P2O [38,41].

Both 2-keto-D-glucose and 2-keto-D-galactose are attractive intermediates for subsequent biotransformations, especially for the production of food related ketoses such as D-fructose and D-tagatose. Here we could show the close to complete enzymatic conversion of both compounds into the corresponding ketoses that are essentially free of aldoses. As an alternative to the established Cetus process, in which 2-keto-D-glucose is reduced by catalytic hydrogenation [5], we employed the NAD-dependent enzyme aldose reductase from *C. tenuis*, which specifically reduces aldos-2-uloses at position *C*-1. Interestingly, several 2-keto-D-aldoses were found to be excellent substrates of ALR that were reduced by the enzyme at least equally well as the *in vivo* substrate D-xylose [33]. Both of the systems used for coenzyme regeneration, i.e., FDH/formate and GDH/glucose, were well compatible with this enzymatic reaction and could be successfully employed. Whereas the latter regenerative system has the advantage of higher productivities, a second product D-gluconic acid is formed in this reaction and has to be separated in turn. Additionally, D-glucose, which is supplemented as the substrate for this regenerative enzyme, has to be removed when not completely utilized. In contrast, the use of FDH results in no additional products and an excess of formate can be easily separated from the product D-fructose.

The novel system of enzymatic reduction of aldos-2-uloses could also be successfully employed for the production of D-tagatose which can be obtained from D-galactose in overall yields of approximately 85%. An important advantage of this entirely enzymatic process is the high regioselectivity of the biocatalyst ALR for position *C*-1 of 2-keto-D-galactose intermediate to

148

produce D-tagatose with no detectable formation of aldoses [42]. This is in contrast to a catalytic reduction step where considerable amounts of by-products, mainly D-galactose and D-talose, were formed [20]. The absence of aldoses should simplify the subsequent downstream processing in a considerable way.

5. ACKNOWLEDGEMENT

This work was supported by a grant from the Austrian Science Foundation (FWF project No. P 11459-MOB).

REFERENCES

1. Drueckhammer, D. G., Hennen, W. J., Pederson, R. L., Barbas, C. F., Gautheron, C. M., Krach, T. and Wong, C.-H. (1991) *Synthesis, 7*, 499–525.
2. Freimund, S., Huwig, A., Giffhorn, F. and Köpper, S. (1998) *Chemistry - a European Journal, 4*, 2442–2455.
3. Neidleman, S. L., Amon, W. F. and Geigert, J. (1981) *European Patent* EP 88,103.
4. Liu, T.-N. E., Wolf, B., Geigert, J., Neidleman, S. L., Chin, J. D. and Hirano, D. S. (1983) *Carbohydrate Research, 113*, 151–157.
5. Geigert, J., Neidleman, S. L. and Hirano, D. S. (1983) *Carbohydrate Research, 113*, 159–162.
6. Horwarth, R. O. and Ibrahim, O. O. (1984) *European Patent* EP 0098533.
7. Daniel, G., Volc, J. and Kubatova, E. (1994) *Applied and Environmental Microbiology, 60*, 2524–2532.
8. Volc, J. and Eriksson, K.-E. (1988) *Methods in Enzymology, 161*, 316–322.
9. Artolozaga, M. J., Kubátová, E., Volc, J. and Kalisz, H. M. (1997) *Applied Microbiology and Biotechnology, 47*, 508514.
10. Danneel, H.-J., Rössner, E., Zeeck, A. and Giffhorn, F. (1993) *European Journal of Biochemistry, 214*, 795–802.
11. Schäfer, A., Bieg, S., Huwig, A., Kohring, G.-W. and Giffhorn, F. (1996) *Applied and Environmental Microbiology, 62*, 2586–2592.
12. Shin, K. S., Youn, H. D., Han, Y. H., Kang, S. O. and Hah, Y. C. (1993) *European Journal of Biochemistry, 215*, 747–752.
13. Janssen, F. W. and Ruelius, H. W. (1968) *Biochimica et Biophysica Acta, 167*, 501–510.
14. Machida, Y. and Nakanishi, T. (1984) *Agricultural and Biological Chemistry, 48*, 2463–2470.
15. Izumi, Y., Furuya, Y. and Yamada, H. (1990) *Agricultural and Biological Chemistry, 54*, 1393–1399.
16. Röper, H. (1991) In *Carbohydrates as organic raw materials* (Ed, Lichtenthaler, F. W.) VCH, Weinheim, pp. 267–288.
17. Nidetzky, B., Neuhauser, W., Haltrich, D. and Kulbe, K. D. (1996) *Biotechnology and Bioengineering, 52*, 387–396.

18. Troyano, E., Villamiel, M., Olano, A., Sanz, J. and Martinez-Castro, I. (1996) *Journal of Agricultural and Food Chemistry*, **44**, 815–817.

19. Muniruzzaman, S., Tokunaga, H. and Izumori, K. (1994) *Journal of Fermentation and Bioengineering*, **78**, 145–148.

20. Freimund, S., Huwig, A., Giffhorn, F. and Köpper, S. (1996) *Journal of Carbohydrate Chemistry*, **15**, 115–120.

21. Leitner, C., Haltrich, D., Nidetzky, B., Prillinger, H. and Kulbe, K. D. (1998) *Applied Biochemistry and Biotechnology*, **70–72**, 237–248.

22. Kern, M., Haltrich, D., Nidetzky, B. and Kulbe, K. D. (1997) *FEMS Microbiology Letters*, **149**, 31–37.

23. Michal, G., Möllering, H. and Siedel, J. (1983) In *Methods of enzymatic analysis*, Vol. 1, Weinheim (Ed, Bergmeyer, H. U.) Verlag Chemie, pp. 197–232.

24. Bradford, M. M. (1976) *Analytical Biochemistry*, **72**, 248254.

25. Neuhauser, W., Haltrich, D., Kulbe, K. D. and Nidetzky, B. (1997) *Biochemical Journal*, **326**, 683–692.

26. Chenault, H. K. and Whitesides, G. M. (1987) *Applied Biochemistry and Biotechnology*, **14**, 147–197.

27. Leitner, C., Neuhauser, W., Volc, J., Kulbe, K. D., Nidetzky, B. and Haltrich, D. (1998) *Biocatalysis and Biotransformation*, **16**, 365–382.

28. Kleppe, K. (1966) *Biochemistry*, **5**, 139–143.

29. Altomare, R. E., Kohler, J., Greenfield, P. F. and Kittrell, J. R. (1974) *Biotechnology and Bioengineering*, **16**, 16591673.

30. Tarhan, L. and Uslan, H. (1990) *Process Biochemistry*, **25**, 14–18.

31. Li, T.-H., Su, Y.-F., Hong, C.-H. and Chen, T.-L. (1994) *Journal of Chemical Engineering of Japan*, **27**, 205–210.

32. Lindberg, B. and Theander, O. (1968) *Acta Chemica Scandinavica*, **22**, 1782–1786.

33. Neuhauser, W., Haltrich, D., Kulbe, K. D. and Nidetzky, B. (1997) *Biochemistry*, **37**, 1116–1123.

34. Volc, J., Sedmera, P., Havlícek, V., Prikrylová, V. and Daniel, G. (1995) *Carbohydrate Research*, **278**, 59–70.

35. Volc, J., Kubátová, E., Sedmera, P., Daniel, G. and Gabriel, J. (1991) *Archives of Microbiology*, **156**, 297–301.

36. Gabriel, J., Volc, J., Sedmera, P., Daniel, G. and Kubátová, E. (1993) *Archives of Microbiology*, **160**, 27–34.

37. Koths, K., Halenbeck, R. and Moreland, M. (1992) *Carbohydrate Research*, **232**, 59–75.

38. Huwig, A., Danneel, H.-J. and Giffhorn, F. (1994) *Journal of Biotechnology*, **32**, 309–315.

39. Chang, H. N. and Moo-Young, M. (1988) *Applied Microbiology and Biotechnology*, **29**, 107112.

40. Prazeres, D. M. F. and Cabral, J. M. S. (1994) *Enzyme and Microbial Technology*, **16**, 738–750.

41. Olsson, L., Mandenius, C. F., Kubatova, E. and Volc, J. (1991) *Enzyme and Microbial Technology*, **13**, 755–759.

42. Kotecha, J. A., Feather, M. S., Kubiseski, T. J. and Walton, D. J. (1996) *Carbohydrate Research*, **289**, 77–89.

Food Biotechnology
S. Bielecki, J. Tramper and J. Polak (Editors)
© 2000 Elsevier Science B.V. All rights reserved.

Biotransformation of sucrose to fructooligosaccharides: the choice of microorganisms and optimization of process conditions

A. Madlová, M. Antošová, M. Baráthová, M. Polakovič, V. Štefuca, and V. Báleš

Slovak University of Technology, Faculty of Chemical Technology, Department of Chemical and Biochemical Engineering, Radlinského 9, 812 37 Bratislava, Slovakia

The biotransformation of sucrose to fructooligosaccharides (FOSs) was investigated in this study that was aimed at the choice of microorganisms and optimization of process conditions. The best producing microorganisms were found to be two strains of *Aureobasidium pullulans* and *Aspergillus niger*. Further experiments were oriented at the effect of process conditions on the production of FOSs. The yield of FOSs was not affected by pH in the range of 4-8. The temperatures between 50-65°C affected only the first stage of the enzyme reaction and the maximal yield of FOSs was achieved after one hour of the reaction.

1. INTRODUCTION

The sweet taste and low caloric value of fructooligosaccharides (FOSs) have made them important as new alternative sweeteners in recent years. They possess a number of desirable characteristics such as non-cariogenicity, safety for diabetics, and lowering the levels of serum cholesterol, phospholipid and triglyceride. They are applied not only in the food industry, but also in other industrial sectors. Fructooligosaccharides are found as natural components in many common foods including fruit, vegetables, milk and honey [1,2].

The term fructooligosaccharides is nowadays preferably used for fructose oligomers, which contain one glucose unit and 2-4 fructose units bound together by β-2,1 glycosidic linkages. FOSs are produced by the catalytic action of fructosyltransferase (EC 2.4.1.9) on sucrose and lower oligomers when glucose is the by-product of the transfer reactions. Small amounts of glucose and fructose are usually formed by the competing hydrolytic activity of the enzyme. The sources of enzyme can be either microorganisms (*Aspergillus* sp., *Aureobasidium* sp., *Fusarium* sp., *Penicillium* sp., *Arthrobacter* sp., etc.) or plants. The production yield of FOSs using enzyme originated from plants is low and mass production of enzyme is limited by seasonal conditions. In contrast, fructosyltransferases derived from fungi provide high yields of FOSs and their mass production is not complicated.

The sucrose concentration is recommended to be in the range from 600 to 850g/l. The effect of temperature and pH depends on the origin of enzyme. The recommended optimal conditions of production of FOSs are pH of about 5.5 and temperature of 55°C for the enzyme of *Aureobasidium pullulans* [3]. The conditions for the reaction by the enzyme of the fungus *Aspergillus japonicus* are pH 5.0-6.0 and temperature 55-65°C [4, 6]. Cheng applied for the reaction by *Aspergillus japonicus* enzyme pH 4.8-5.5 and temperature 60-65°C [5]. The conditions for the production of FOSs by the *Aspergillus niger* enzyme are a pH of 5 and a temperature of 60°C [6]; the optimal pH for production by the *Aspergillus oryzae* enzyme is in the range of 5-7 [7].

This work aimed at: (i) the screening of microorganisms capable to produce the fructosyltransferase activity and the choice of the best one, (ii) the study of biotransformation of sucrose to fructooligosaccharides by the action of enzymes with transfructosylating activities originating from the selected strains, and (iii) the optimization of process conditions of FOSs production (pH and temperature).

2. MATERIALS AND METHODS

2.1. Microorganisms and cultivation conditions

The cultivation was carried out at the conditions reported by Chen and Jung [3,4]. The microorganisms *Aureobasidium pullulans* (AP) CCY 27-1-94 (AP I), CCY 27-1-93 (AP II), CCY 27-1-89 (AP III), CCY 27-1-56 (AP IV), CCY 27-1-45 (AP V) and CCY 27-1-17 (AP VI), and *Aspergillus niger* (AN I) and (AN II) from our own collection were applied for the production of fructosyltransferase. The seed culture medium of both types of microorganisms consisted of 1% (w/v) sucrose and 0.2% (w/v) yeast extract; pH was 5.5. The fermentation medium for isolation and cultivation of AP contained 20% (w/v) sucrose, 1% (w/v) yeast extract, 0.05% (w/v) $MgSO_4 \cdot 7H_2O$, 0.5% (w/v) K_2HPO_4 and 1% (w/v) $NaNO_3$; the fermentation medium of AN contained 20% sucrose, 2% (w/v) yeast extract, 0.05% (w/v) $MgSO_4 \cdot 7H_2O$, 0.5% (w/v) K_2HPO_4 and 1% (w/v) $NaNO_3$. The value of pH was set to 6.5. The cultivation ran at 28°C in 500-ml Erlenmeyer flasks for 92 h at 180 rpm. After cultivation, the cells of AP and AN were washed with physiological solution several times, then the physiological solution was removed by filtration.

2.2. Production of FOSs

The enzymatic reactions were performed in a stirred batch reactor. The concentration of sucrose in 0.1 M phosphate buffer was 700 g/L and the enzyme reaction ran for 8 hours. The mass ratio of the wet cells and sucrose solution was 1:9. The investigation of the effect of pH was realized in the range of 4-8 at the temperature of 55°C. The effect of temperature was investigated in the range from 55°C to 65°C at pH 6. The reaction was stopped by heating in boiling water for 2 minutes and the mixture was subjected to the analysis for the determination of the concentration of sugars.

2.3.Analysis of sugars

The analysis of sugars was performed by high performance liquid chromatography (HPLC). The HPLC system consisted of a pump Maxi-Star K-1000, an on-line degasser Model A1050, a refractive index (RI) detector Type 298.00, an injection valve equipped with a 10 μl loop, all from Knauer (Berlin, Germany), and a column thermostat Jetstream Plus II (Thermotechnic Products, Germany). The separation was carried out using a column EuroKat Pb (Knauer, Berlin, Germany) filled with a sulphonated SDVB copolymer in the lead form (300 x 8 mm I.D.) equipped with a guard column of the same filling. Double-distilled water filtered through a 0.2 μm membrane filter was used as the mobile phase at the flow rate of 0.8 ml/min. The column temperature was set to 80°C. The RI detector was operated at 32°C. The chromatographic data were recorded and evaluated using an EuroChrom 2000 Integration Package Software (Knauer, Berlin, Germany).

2.4. Calculation of yield of fructooligosaccharides and selectivity

The measure of enzyme activity was the total yield of fructooligosaccharides, Y_{FOS}, which was calculated from the yields of GF_2 and GF_3 as follows

$$Y_{GF_2} = \frac{2c_{GF_2}}{c_{S_0}} \qquad Y_{GF_3} = \frac{3c_{GF_3}}{c_{S_0}} \qquad Y_{FOS} = Y_{GF_2} + Y_{GF_3} \qquad (1)$$

where c_{GF_2}, c_{GF_3}, c_{FOS}, c_{S_0} are the molar concentrations of kestose, nystose, total FOSs, and initial sucrose, respectively.
The selectivity of conversion from sucrose to FOSs was calculated from the following formula

$$S_{FOS} = \frac{2c_{GF_2} + 3c_{GF_3}}{2c_{GF_2} + 3c_{GF_3} + c_F} \qquad (2)$$

where c_F is the molar concentration of fructose.

3. RESULTS AND DISCUSSION

3.1. The choice of microorganisms

The test criteria of microorganisms, which were available for the production of enzymes, were the ample growth of mycelia at the cultivation conditions and transfructosylating activities of cells. The comparison of the yield of total FOSs produced by different microorganisms is given in Figure 1. The best producing microorganisms were found to be *Aureobasidium pullulans* II and III and *Aspergillus niger* II, which provided yields of FOSs above 0.8. The enzyme activity of *Aureobasidium pullulans* IV was very low and *Aureobasidium pullulans* V and VI grew relatively slowly. All next experiments were performed using the mycelium of AP II.

3.2. Effect of pH and temperature on enzyme activity

The investigation of the effect of pH was realized in the range of pH of 4-8 and at the temperature of 55°C. The results are shown in Figure 2. In the monitored range, pH had no significant effect on the yield of FOSs. This result differs from the observations published by other authors. Jung found an optimum pH of 5.5 using the free cells of *Aureobasidium pullulans* [3]; the enzyme activity sharply decreased below pH 5 and above pH 6.5. Cheng [5] worked with immobilized mycelium of *Aspergillus japonicus*, having an optimum activity at pH 4.0-5.5; above pH 6 the activity rapidly decreased. The selectivity of conversion of sucrose to FOSs was in our case above 0.95 in the whole range of pH, as shown in Figure 3.

The effect of temperature in the range from 55 to 65°C was determined at pH 6. The increase of temperature caused the increase of the initial rate of enzyme reaction. From Figure 4 appears that the optimal temperature is between 60-65°C. However, already after 60 minutes of reaction time, the amount of total FOSs produced reached the maximum value which was independent on temperature. Jung [3] found that the dependence of enzyme activity on temperature showed a narrow maximum at 55°C, while above this temperature the activity strongly decreased. It is important to remark that in our case the time of duration of the enzyme reaction necessary to reach the final yield of FOSs was much shorter than those reported by other authors [3,8].

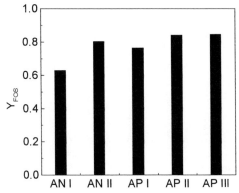

Abbreviations

AN = Aspergillus niger
AP = Aureobasidium pullulans

AP I = AP CCY 27-1-94
AP II = AP CCY 27-1-93
AP III = AP CCY 27-1-89

Figure 1. Comparison of the yield of FOSs achieved using different microorganisms.

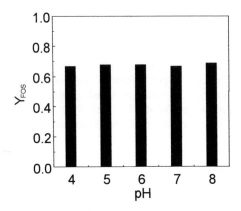

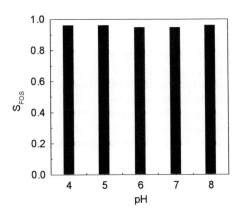

Figure 2. Effect of pH on the yield of FOSs.

Figure 3. Effect of pH on the selectivity of FOSs.

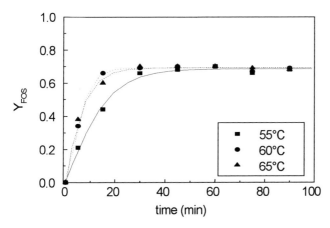

Figure 4. Effect of temperature on the intermediate yield of FOSs.

4. CONCLUSIONS

The aim of the present investigation was to optimize the conditions of the production of fructooligosaccharides (FOSs) by the action of enzyme of microbial origin. The screening of microorganisms resulted in the choice of three strains (*Aureobasidium pullulans* II and III and *Aspergillus niger* II) which were the best producers of fructosyltransferase activity. The effect of pH in the range of 4-8 on the production of FOSs was not significant and the yield of total FOSs was above 0.67 in all cases. The increase of temperature caused the increase of initial reaction rate, but, after one hour of the reaction, the yield of FOSs reached the maximum value independent of the temperature.

REFFERENCES

1. J.W. Yun, Enzyme Microb. Technol., 19 (1996) 107.
2. J.E. Spiegel, R. Rose, P. Karabell, V.H. Frankos and D.F. Schmitt, Food Technol., 48 (1994) 85.
3. K.H. Jung, J.W. Yun, K.R. Kang, J.Y. Lim and J.H. Lee, Enzyme Microbiol. Technol., 11 (1989) 491.
4. W.-C. Chen and C.-H. Liu, Enzyme Microb. Technol., 18 (1996) 153.
5. C.-Y. Cheng, K.-J. Duan, D.-C. Sheu, C.-T. Lin and S.-Y. Li, J. Chem. Tech. Biotechnol., 66 (1996) 135.
6. C.J. Chiang, W.C. Lee, D.C. Sheu and K.J. Duan, Biotechnol. Progress, 13 (1997) 577.
7. M. Kurakake, T. Onoue and T. Komaki, Appl. Microbiol. Biotechnol., 45 (1996) 236.
8. H. Hidaka, M. Hirayama and N. Sumi, Agric. Biol. Chem., 52 (1988) 1181.

Food Biotechnology
S. Bielecki, J. Tramper and J. Polak (Editors)
© 2000 Elsevier Science B.V. All rights reserved.

Enzymatic isomaltooligosaccharides production

C. Kubik, E. Galas, B. Sikora, D. Hiler

Institute of Technical Biochemistry, Technical University of Lodz,
Stefanowskiego 4/10, 90-924 Lodz, Poland

The subject of our study was the enzymatic synthesis of isomaltooligosaccharides (IMOs) from sucrose using two enzymes produced in our laboratory, i.e. dextransucrase (DS) from *Leuconostoc mesenteroides* and dextranase (D) from *Penicillium funiculosum*. Dextransucrase converts sucrose into dextran with fructose (Fru) as a by-product, whereas dextranase transforms dextran into IMOs.

The research was focused on elucidation of the effect of reaction time and enzyme and sucrose concentration, as well as the on the IMOs production.

The reactions were carried out for 24 and 48 h, at 30°C and pH 5.4, in solutions containing different amounts of sucrose (5, 10, 20 and 40% w/v).

It was shown that the composition of reaction products in particular depended on the concentration of sucrose solutions. The highest yield of isomaltose (IM), about 50% of total determined sugars, was achieved in 10 and 20% sucrose solutions. In the 10% case, glucose (Glc) predominated among the other reaction products, while it was isomaltotriose (IM$_3$) at 20%. At higher sucrose concentrations even more IM$_3$, and leucrose (Leucr), were formed.

1. INTRODUCTION

During the last decade the popularity of non-digestible oligosaccharides (NDOs) has grown rapidly. The degree of polymerization of these carbohydrates ranges from approximately 2 to 20. Since various monosaccharides can be linked in different ways, NDOs composition is very complex. They are reported to be very useful in maintaining or providing a good gastrointestinal environment. NDOs escape digestion in the upper intestinal tract and are preferentially consumed in the colon by beneficial bacteria, especially belonging to genera *Bifidobacterium* and *Lactobacillus* [1,2]

Commercially available oligosaccharides comprise: lactulose, lactosucrose, raffinose, galactooligosaccharides, fructooligosaccharides, soybean oligosaccharides, isomalto-oligosaccharides, xylooligosaccharides, cyclodextrins [3-7]. These sugars are obtained from various sources and in different ways, including (i) extraction from plants, (ii) enzymatic or acidic decomposition of natural polymers or (iii) enzymatic synthesis. The latter method provides diversity in products and comprises (1) reversion of hydrolysis, (2) transhydrolytic reactions exploiting transferase activity of hydrolases, and (2) transfer reactions catalysed by transferases [8].

Our studies were aimed at production of isomaltooligosaccharides (IMOs). This group of sugars can be obtained in the following ways:

- from starch, which is hydrolysed by α-amylase (EC 3.2.1.1) and pullulanase (EC 3.2.1.41) to di- and trimaltooligosaccharides, being substrates for α-glucosidase (EC 3.2.1.20), acting as a transferase of glucosyl residue; in this reaction mainly α-1,6 linkages are formed [9];
- from dextran, by hydrolysis with dextranase (EC 3.2.1.11) [10];
- from sucrose, using dextransucrase (EC 2.4.1.5) and glucose or low molecular weight oligosaccharides as acceptors [11-13];
- from sucrose, using the mixture of dextransucrase and dextranase. The first enzyme converts sucrose into dextran which is immediately splitted into IMOs by the dextranase[14].

In our studies the last method was used.

2. MATERIALS AND METHODS

2.1. Enzymes

Dextransucrase was synthesized by a strain of *L. mesenteroides*, applied in KZF "POLFA" (Poland) for commercial production of dextran. The enzyme activity was determined at pH 5.2 and 30°C with sucrose (75g l^{-1}) as a substrate. Reducing sugars were determined according to Miller [15] using alkaline DNS solution. One unit of activity (U) is defined as the amount of enzyme producing 1 μmol of fructose per 1 min under conditions described above.

Dextranase was produced by *P. funiculosum* 72 in a medium containing dextran as a sole source of carbon. One unit of its activity is defined as the amount of enzyme that in solutions of dextran T-110 (16.7 g/l) liberates reducing sugars [16,17] equivalent to 1 μmol of IM per 1 min at pH 5.4 and 50°C.

2.2. Isomaltooligosaccharides production

Reactions were performed at 30°C and pH 5.4 in 5, 10 or 20% (w/v) sucrose solutions, as well as in a solution in which the sucrose concentration was increased from 20 to 40% after 12 h of reaction. The synthesis was monitored by determination of reducing sugars in reaction mixtures after 24 and 48 h, and up to 72 h in case of 40% sucrose solution (as Fru, assayed with DNS). Oligosaccharides were identified by thin layer chomatography (TLC). Silica gel 60 TLC plates (Merck) were used. TLC was run using one ascent of solvent system of n-butanol/aceton/water (4/1/1 volume parts) and then, after drying of the plate, one ascent of solvent system of n-propanol/ethyl acetate/water (2/1/1 volume parts). The carbohydrates were visualized by spraying the plates with 5% (v/v) sulfuric acid in ethanol containing 0.5% naphtol, followed by heating at 100°C for 10 min. Quantitative analyses of sugars were performed by reverse-phase HPLC using Aminex HPX 87H column (BioRad) and 0.001 M H_2SO_4 as an eluent.

3. RESULTS

The studies were focused on the elucidation of effects of sucrose, DS and D concentrations and reaction time on IMOs production.

When the processes were run in 5 and 10% sucrose solutions containing 0.5-1 DS U ml^{-1} and 1-4 D U ml^{-1}, complete depletion of sucrose occurred in all the mixtures after 24 h of reaction .

In solutions containing 5% of sucrose the level of reducing sugars remained stable after 24 and 48 h of reaction (about 45 mg^{-1}) and the spots of IMOs (TLC analysis) had similar intensity.

In 10% substrate solutions the level of reducing sugars was higher after 48 h of reaction than after 24 h and slightly raised with the increase of dextranase concentration in the reaction mixtures. For example, in solutions containing 1 DS U ml^{-1} and 1 D U ml^{-1}, 70 and 73 mg of reducing sugars per 1 ml were detected after 24 and 48 h, respectively; when 2 D U ml^{-1} were used, they corresponded to 72 and 78 mg ml^{-1}, respectively, and with 3 or 4 D U ml^{-1} to about 85 and 88 mg ml^{-1}, respectively. TLC analyses showed that the smallest amounts of higher IMOs were present in the mixtures with 3 or 4 D U ml^{-1} after 48 h of reaction. In more concentrated substrate solutions (20 and 40%), the dose of 4 D U ml^{-1} was applied.

In 20% sucrose solutions containing 0.5-1 DS U ml^{-1} and 4 D U ml^{-1}, the substrate was also fully depleted after 24 h of reaction, but similarly to 10% substrate solutions, conversion of dextran to IMOs was not complete and reducing sugars raised from about 149 mg ml^{-1} after 24 h to 156 mg ml^{-1} after 48 h.

When the substrate concentration was enhanced from 20 to 40% after 12 h of the process, its full conversion occurred after 48 h in solution with 1 DS U ml^{-1} and 4 D U ml^{-1} . Reducing sugars reached the level of 330 mg ml^{-1} after 48 h, which was the same also after 72 h.

HPLC analyses showed that in 5% sucrose solutions containing 0.5-1 DS U ml^{-1} and 1-4 D U ml^{-1} the composition of products after 24 and 48 h of reaction was very similar. The same phenomenon was observed in 10 and 20% sucrose solutions, when 0.5-1 DS U ml^{-1} and 3 or 4 D U ml^{-1} (4 D U for 20% solution) were applied, and reactions lasted for 48 h. HPLC analyses of products released in 40% sucrose solutions showed the similar elution patterns of sugars in mixtures after 48 and 72 h of reaction.

Examples of HPLC analyses of reaction mixtures containing different concentrations of sucrose, 1 DS U ml^{-1} and 4 D U ml^{-1}, are presented in Fig.1, the quantitative composition of reaction products calculated from HPLC analyses in Table 1, and the percentage of their content in Table 2.

The results proved that an increase of sucrose concentration in reaction mixtures resulted in a decrease in the amount of Glc and an increase in IM and IM$_3$. In diluted sucrose solutions a large amount of Glc is released as a result of reaction with water. The content of IMOs with degree of polymerization above 3 varied from 7 to 15%, being the highest in 5% substrate solution. The amount of IM$_3$ was enhanced from about 3% in 5% sucrose solution, to about 36% in 40% solution. The highest IM content (47-50% of assayed saccharides) was observed in 10 and 20% sucrose solutions (Tables 1 and 2). When the enzymatic reaction was performed in 40% sucrose solution, IM$_3$ predominated among the reaction products and the content of IM was decreased to about 30% of all assayed saccharides.

160

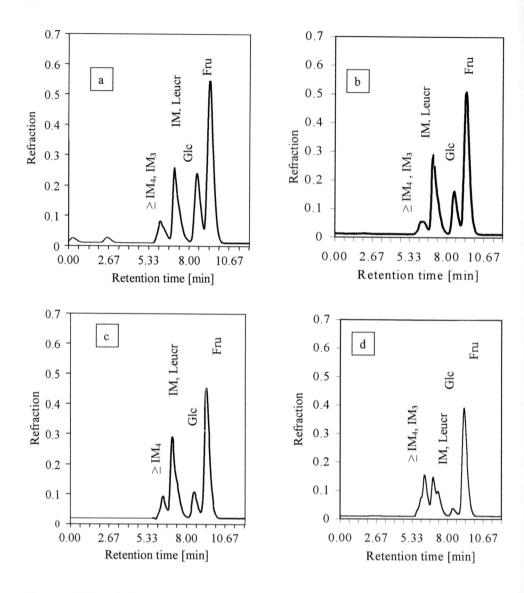

Figure 1. HPLC-elution pattern of oligosaccharides present after 48 h in reaction mixtures containing 1 DS U ml^{-1} and 4 D U ml^{-1} in sucrose solutions: (a) 5%, (b) 10%, (c) 20% and (d) 40% (enhanced from 20 to 40% after 12 h of reaction). Dilution of the mixtures is as follows: (a) none, (b) two-fold, (c) 4-fold, (d) 14-fold.

Table 1
Quantitative composition of reaction products determined by HPLC (see Fig.1)

sugars analyzed	sucrose solutions (%)			
	5	10	20	40
	mg/ml			
$\geq IM_4$	3.20	2.84	6.50	16.80
IM_3	0.65	2.60	10.40	67.00
IM	0.64	18.96	36.60	57.30
Leucr.	1.70	3.90	8.23	36.00
Glc	7.67	9.85	11.70	12.20
efficiency* of sucrose conversion	84	71	68	80

* The efficiency of conversion was calculated as a ratio of the oligosaccharides mass to the mass of sucrose introduced to the reaction mixtures, multiplied by 0.48. Leucrose was not considered in the calculations.

Table 2
Yields (%) of saccharides released in sucrose solutions containing dextransucrase (1 U ml^{-1}) and dextranase (4 U ml^{-1}) after 48 h of enzymatic reaction

sugars analyzed	sucrose solutions (%)			
	5	10	20	40
$\geq IM_4$	15	7	9	9
IM_3	3	7	14	36
IM	39	47	50	30
Leucr.	8	10	11	19
Glc	35	29	16	6

The amount of leucrose, the product of the sucrose and fructose reaction catalyzed by dextransucrase [18], raised with an increase of the concentration of sucrose in the solution. Its level reached about 8% of the examined sugars in 5% substrate solution, and 19% in 40% solution.

Our results are to some extent different from those presented by Paul et al. [14], who applied dextransucrase from *L. mesenteroides* NRRL-512F and dextranase from *Penicillium sp.* (purchased from Sigma). The authors, executing the enzymatic reaction in 5–20% sucrose solutions achieved more IM (80% in 5% substrate solution and 65% in 20% solution) among the determined sugars. Generally, the observed tendency of an increase of Glc amount in more diluted sucrose solutions, and of IM$_3$ in the concentrated ones, was the same as in our experiments.

4. CONCLUSIONS

1. Addition of 0.5 DS U ml^{-1} to 5-20% sucrose solutions ensures complete conversion of the substrate within 24 h of reaction. In 40% solution, sucrose is completely consumed after 48 h, when 1 DS U ml^{-1} is added.
2. Application of the above stated DS activity (i. e. 0.5 or 1 U ml^{-1} and the following doses of D activity: 1 U ml^{-1} in 5% solution, 3 U ml^{-1} in 10% and 4 U ml^{-1} in 20 and 40% solutions results in complete conversion of formed dextran into IMOs in 48 h of reaction.
3. The highest amount of IM (about 50%) is obtained in 10-20% sucrose solutions. In more diluted substrate solutions the amount of free Glc present in the mixture of reaction products is enhanced, whereas in more concentrated ones, more IM$_3$ and Leucr is released.

REFERENCES

1. H. Tomomatsu, Food Technol., 10 (1994) 61.
2. G.R. Gibbson, A. Willems, S. Reading and D. Collins, Proc. Nutr. Soc., 55 (1996) 889.
3. O. Tsuneyuki, Nutrition Reviews, 54 (1966) 59.
4. S. Fuji, M. Komota, Zuckerind. 116 (1991) 197.
5. Y. Haata, K. Nakajima, Y. Hosono, and M. Yamamoto, J. Japan Soc. Clin. Nutr., 11 (1989) 42.
6. K. Ajisaka, H. Nishida, and H. Fujimoto, Biotechnol. Lett., 9 (1987b) 387.
7. T. Miyake, M. Yoshida, K. Takeuchi, US Patent No 4 518 581 (1985).
8. P. Monsan, F. Paul, M. Remaud and A. Lopez, Food Biotech., 3 (1989) 11.
9. K. Imada, Lett. in Appl. Microb., 19 (1994) 247.
10. C.F. Brown and P.A. Inkerman, J. Agric. Food Chem., 40 (1992) 227.
11. D. Schwengers, US Patent No 4 649 058 (1987).
12. J.F. Robyt and S.H. Eklund, Carbohydr. Res., 121 (1983) 279.
13. F. Paul, E. Oriol, D. Auriol and P. Monsan, Carbohydr. Res., 149 (1986) 433.
14. F. Paul, P. Monsan, M. Remaud and V. Pelenc, Europ. Patent Appl. 0 252 799 (1988).
15. G.L.Miller, Anal. Chem., 31 (1959) 426.
16. M. Somogyi, J. Biol. Chem., 195 (1952) 19.
17. N. Nelson, Methods Enzymol., 8 (1957) 85.
18. M.Böker, H.J.Jördening,K. Buchholz, Biot. Bioeng., 43 (1994) 856.

Food Biotechnology
S. Bielecki, J. Tramper and J. Polak (Editors)
© 2000 Elsevier Science B.V. All rights reserved.

163

Oligosaccharide synthesis by endo-β-1,3-glucanase G_A from *Cellulomonas cellulans*

A. Buchowiecka and S. Bielecki

Institute of Technical Biochemistry, Technical University of Lodz
90-924 Lodz, ul. Stefanowskiego 4/10, Poland
e-mail: buchow@ck-sg.p.lodz.p.pl

Microbial endo-β-1,3-glucanase G_A from *Cellulomonas cellulans* displaying transglyco-sylating activity was applied for synthesis of oligosaccharide glycosides. Water-soluble β-1,3-glucan (laminaran) was used as a donor of oligomeric glycosyl residues and 4-nitro-phenyl-β-D-xylopyranoside served as their acceptor. Transglycosylation reaction carried out at 37°C with 25% acetronitrile as co-solvent resulted in 3 pairs of β-1,3-, β-1,4-regioisomers of di-, tri- and tetrasacchride glycosides with total yield 68% after 3hours process.

1. INTRODUCTION

It is now widely established that carbohydrates play an important role in virtually any biological systems [1-3]. In addition, these macromolecules have numerous medical [4] and industrial applications [5]. Thus, there is a permanent demand to search for new ways of obtaining of these substances.

Besides more efficient chemoenzymatic methods, strictly enzymatic fashions of synthesis are used implying highly specific glycosyltransferases [6-11]. However, their applications on larger scale are limited by high costs of both biocatalysts and their substrates. On the other hand, enzymatic synthesis of oligosaccharides by glycosyl hydrolases has been widely accepted because they are easily obtainable, relatively stable in reaction media and their natural substrates are not expensive. Much attention has been focused on the use of exoglycosidases in reversed hydrolysis and transglycosylation reaction leading to new products formation [12].

Publications about the application of endo-glycanohydrolases in oligosaccharide synthesis are less numerous [13] comparing to papers concerning usage of exo-glycosidases. The main reasons for it are low accessibility of sufficiently pure preparations of endo-acting enzymes as well as higher complexity of product mixtures that are often difficult to separate.

In this work we present the results of transglycosylation reaction catalysed by microbial endo-β-1,3-glucanase with the use of polymeric donor and an acceptor labeled with chromophoric aglycon.

2. MATERIALS AND METHODS

2.1. Chemicals

Laminarin from *Laminaria digitata*, 4-nitrophenyl-β-D-xylobioside, 4-nitrophenyl-β-D-cellobioside, 4-nitrophenyl-β-D-glucobioside, were from Sigma. HPTLC Kieselgel 60 F_{254} thin-layer plates were from Merck. Mixed beads ion exchanger AG 501-X8(D) resin was from Bio-Rad Lab. 4-Nitrophenyl-β-D-laminaribioside, p-nitrophenyl-β-D-laminaritrioside, p-nitrophenyl-β-D-laminaritetraoside were chemically synthesised according to the method described [14].

2.2. Enzyme

Endo-β-1,3-glucanase G_A *Cellulomanas cellulans* (formerly *Oerskovia xanthineolytica*) was isolated from a culture medium and purified to homogeneity using gel filtration chromatography followed by chromatofocusing. Endo-β-1,3-glucanase G_A is a monomeric protein of MW 27 663 Da (+/- 0.1%) by MALDI TOF MS method (Fig. 1) and of pI 5.85 by chromatofocusing. The enzyme shows optimum hydrolytic activity at pH 5.5 and temperature 55°C. Michaelis-Menten constant was determined at optimal condition for laminarin hydrolysis. Kinetic data processed with a linear-regression-analysis program resulted in $K_m=1.12\times10^{-4}$ M and catalytic constant $k_{cat}=56$ s^{-1} [15].

Enzyme assay: β-1,3-glucanase activity was determined reductometrically by monitoring the increase in reducing sugars [16-18] released in a 0.05% (w/v) solution of laminarin in 0.1 M citric acid-K_2HPO_4 buffer, pH 5.5; at 55°C. One unit of enzyme activity is defined as 1 μmol of glucose equivalents released/min. One unit corresponds to 16.67 nkat. Glucanase G_A displays transglycosylating activity, which was employed for synthetic purposes.

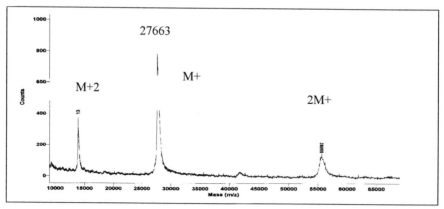

Figure 1. MALDI TOF spectrum of endo-β-1,3-glucanase G_A *Cellulomonas cellulans*. M^{+2}, M^+, $2M^+$: molecular peaks of different degrees of ionisation and aggregation

2.3. Oligosaccharide synthesis

Transglycosylation reactions were carried out in a water-organic environment containing 20%, 25% or 30% (v/v) acetonitrile and appropriate concentration of 0.1M citric acid-K_2HPO_4 buffer, pH 5.5. *Laminaria digitata* laminaran (10% w/v) was used as a donor and

4-nitrophenyl-β-D-xylopyranoside (9% w/v) as an acceptor. G_A enzyme with an approximate equimolar addition of BSA was freeze-dried and used in the ratio 0.6U/1mg of donor. The reaction mixtures were incubated 5 hours at temperatures 37°C, 45°C or 50°C. At 30 minutes intervals samples were taken, heated 5 minutes at 100°C and then analysed by TLC.

2.4. Qualitative analysis of products

Appropriately diluted samples of reaction mixtures were applied to HPTLC Kieselgel 60 F_{254} thin-layer plates, developed in ethyl acetate/acetic acid/water (2:1:0.5 by vol.) and the products were detected with α-naphthol reagent [19].

2.5 Quantitative analysis of products

UV absorption densitometry was used. Thin-layer chromatograms developed as defined above were dried and scanned at λ_{max}=295 nm using CAMAG TLC Scanner 2 with CATS version 3.0X program. Standard curve was prepared for 4-nitrophenyl-β-D-xylopyranoside in the range between 4-20 nmol giving a linear response between UV absorption and molar concentration of chromophor. Yield of synthesis was calculated as a mole percentage of the acceptor used.

2.6. Synthesis products isolation

Semi-preparative TLC method was used. Thin-layer chromatograms developed as defined above were UV viewed, then the gel areas containing synthesis products marked and subsequently scraped-off. Synthesis products were eluted with a solvent mixture consisting of acetonitrile and methanol (3:2). The solution was dried under vacuum and residue dissolved in deionized water, then desalted with AG 501-X8(D) resin. The pure isolates were filtered through 0.2μm membrane, freeze-dried and subjected to FAB MS analysis.

2.7. Molecular mass determination of synthesis products

FAB MS method was used with DTT:DTE (3:1) (dithiotreithol:dithioerytrithol) as a matrix and LiCl as a cationizing reagent [20]

2.8. Structure determination of synthesis products

The structures of synthesis products were determined on thin-layer chromatography. Rf values of the products of interest and reference substances of known structures were compared. 4-Nitrophenyl-β-D-glucobioside, 4-nitrophenyl-β-D-laminaribioside, 4-nitro-phenyl-β-D-laminaritrioside, 4-nitrophenyl-β-D-laminaritetraoside were used as the reference substances.

3. RESULTS AND DISCUSSION

The bacterial strain *C. cellulans* produces exatracellular complex of lytic enzymes containing several β-1,3-glucanases. One of them, endo-β-1,3-glucanase G_A was isolated and characterised as described above. As we reported before [21] G_A enzyme displaying transglycosylating activity was successfully applied for short-chain oligosaccharides synthesis. In this article we present qualitative and quantitative results of transglycosylation process catalysed by this enzyme.

3.1. Synthesis products determination

Preliminary studies on transglycosylation reaction were performed with 4-nitrophenyl-β-D-glucopyranoside or 4-nitrophenyl-β-D-xylopyranoside as acceptors and laminarin as a donor. The reactions were carried out for 3 hours under standard conditions assumed i.e. water-organic environment containing 30% of acetonitrile, temperature 50°C. The resulted reaction mixtures were analysed by TLC as defined above. Figure 2 presents products of both reactions compared with reference substances of known structures.

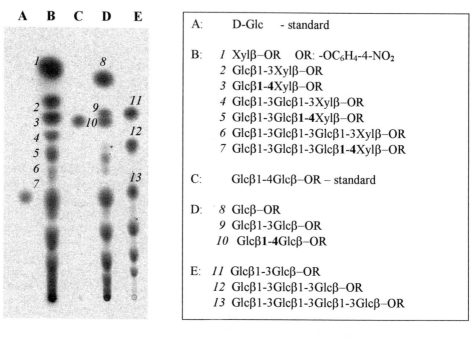

Figure 2. TLC chromatogram of the products of enzymatic glycosylation of 4-nitrophenyl–β–D-xylopyranoside (B) and 4-nitrophenyl–β–D-glucopyranoside (D) as acceptors compared to standard substances: glucose (A), 4-nitrophenyl–β–D-cellopyranoside (C), mixture of laminaribiose, -triose, -tetraose and their 4-nitrophenyl–β–D-glycosides (E).

TLC chromatograms of reaction mixtures (Figure 2) revealed the presence of:
- unused acceptors (spots with the highest Rf values on paths B and D)
- hydrolysis products (spots with Rf values lower or equal to glucose spot Rf)
- synthesis products (spots lying between acceptor and glucose spots).

Among the major products of 4-nitrophenyl-β-D-glucopyranoside glycosylation (path D), there are D9 with the Rf value corresponding to 4-nitrophenyl-β-D-laminaribioside (path E11) and D10 identified as 4-nitrophenyl-β-D-cellobioside (path C). Thus, the experiment proved that glucanase G_A synthesises β-1,3- and β-1,4-glycoside linkages; additionally, it has been shown that under separation conditions the product with β-1,3-glycosidic linkage migrates higher than its β-1,4-regioisomer.

The products of p-nitrophenyl-β-D-xylopyranoside glycosylation (B2 - B7) were isolated and their molecular masses determined by FAB MS method. The analysis proved that B2 and B3, B4 and B5, B6 and B7 are three pairs of nitrophenyl-glycosides of disaccharide, trisaccharide and tetrasaccharide respectively [21]. Their chemical formulas, shown in Figure 2, were assigned based on analogy that β-1,3-regioisomers migrate faster than corresponding β-1,4-regioisomers. In order to confirm this result larger amounts of synthesis products should be isolated and subjected to NMR analysis.

Results of investigations enabled to distinguish two types of 4-nitrophenyl-β-D-oligo-sacchrides named as β-1,3-line and β-1,4-line products based on regiochemistry of transglycosylation.

3.2. Quantitative and qualitative evaluation of synthesis progress

4-Nitrophenyl-β-D-xylopyranoside glycosylation was chosen as a model reaction to investigate effects of temperature and co-solvent concentration on synthesis results. Several combinations of reaction conditions designated with symbols A20_37, A25_37, A30_37, A20_45, A25_45, A30_45, A20_50, A25_50, A30_50 were examined. First number of the symbol indicates acetonitrile concentration and the second temperature of a process. The dynamics of a synthesis for each reaction variant was followed by densitometric evaluation of thin-layer-chromatograms performed as described above. An image of a typical UV absorption densitogram is presented in Figure 3. Figure 4 shows time course of total synthesis efficiency obtained for selected cases. Yield of synthesis was calculated as mole percentage of the acceptor used.

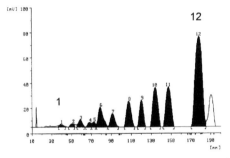

Figure 3. The typical absorption densito-gram at λ_{max} = 295 nm
Peak no. 12: unused acceptor
Peaks no. 11, 9, 7: di-, tri-, tetrasaccharide glycosides of β–1,3-line
Peaks no. 10, 8, 6: di-, tri-, tetrasaccharide glycosides of β–1,4-line
Peaks no. 5 – 1: undefined glycosides

Figure 4. The dynamics of synthesis at different reaction conditions designated with symbols: A20_37 (1), A25_37 (2), A25_45 (3), A25_50 (4), A30_45 (5) (see the explanation in text)

Higher temperature and organic solvent concentration due to fast enzyme inactivation negatively affect the efficiency of synthesis (curves 3 and 4). The best results were obtained for the reaction conducted at 37°C with 25% of acetonitrile (curve 2). After 3 hours time total yield was measured as 68%. As compared to the results achieved with 20% solvent concentration at the same temperature, an approximate 20% increase of yield was observed. This suprisingly high yield level is comparable with the results recently reported by Withers [22].

The diagrams in Figure 5 show quantitative changes in oligomeric composition of synthesis products at half-hour intervals. At both β-1,3- and β-1,4-lines of products an increase of di- and trisaccharide glycosides is observed whereas tetrasaccharide glycosides concentration remains approximately constant. Tetrasaccharide glycosides belonging to β-1,4-line accumulate on the average concentration level about 6%. Their β-1,3-regioisomers achieve half of this level (about 3%). These facts indicate that tetrasaccharide glycosides are continuously degraded by the G_A glucanase during the synthesis process. It also means that the tetrasaccharides containing β-1,4-bond are less susceptible to enzymatic hydrolysis.

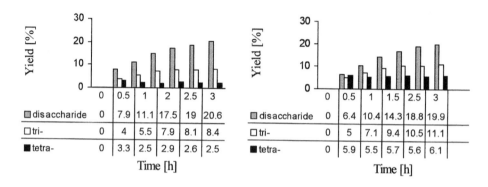

Figure 5. Time course of synthesis products belonging to β–1,3 line (left) and β–1,4 line (right).

Through the transglycosylation reaction endo-β-1,3-glucanase G_A synthesises mixture of six major oligosaccharide glycosides with polymerisation degree n=2, 3, 4. It was previously postulated [21] that these end products are formed *via* enzymatic shortening of intermediates products with longer carbohydrate chains. Densitometric analysis detected small amounts of such intermediates in the reaction mixture (Figure 3, peaks 1-5). Additionally, qualitative results concerning the efficiency of tetrasacchride glycosides also support this hypothesis.

4. SUMMARY

Microbial endo-β-1,3-glucanase G_A from *C. cellulans* apart from hydrolytic displays transglycosylating activity, which was applied for synthetic purposes. Water-soluble β-1,3-glucan laminarin was used as a donor of oligomeric glycosyl residues transferred with the enzyme onto acceptor molecules. 4-Nitrophenyl-β-D-xylopyranoside was chosen as a model acceptor for detailed investigation of transglycosylation process. Nitrophenyl label of synthesis products facilitated their separations on HPTLC and enabled sufficiently sensitive detection and qualitative evaluation by UV absorption densitometry in the course of reaction.

It was established that the enzyme synthesises β-1,3- and β-1,4-glicosidic bonds leading to the mixture of di-, tri- and tetrasaccharide glycosides with disaccharides as dominating component. The total yield of the synthesis with p-nitrophenyl-β-D-xylopyranoside as the acceptor reached 68 % (31.6% of β-1,3- and 37.1% of β-1,4-regioisomers) after 3 hours of the reaction conducted at 37°C in the presence of 25% acetonitrile.

REFERENCES

1. R.A. Dwek, Chem. Rev., 96 (1996) 683.
2. M. Mammen, S-K. Choi, G.M. Whitesides, Angew. Chem. Int. Ed. Engl., 37 (1998) 2754.
3. R.A. Laine, Pure & Appl. Chem., 69 (1997) 1867.
4. J.C. McAuliffe, O. Hindsgaul, Chemistry & Industry, 3 (1997) 170.
5. E.J. Vandamme, W. Soetaert, Biotechnologia, 2 (1996) 136.
6. C-H. Wong, R.L. Halcomb, Y. Ichikawa, T. Kajimoto, Angew. Chem. Int. Ed. Engl., 34 (1995) 412.
7. C-H. Wong, R.L. Halcomb, Y. Ichikawa, T. Kajimoto, Angew. Chem. Int. Ed. Engl., 34 (1995) 521.
8. H.J.M. Gijsen, L. Qiao, W. Fitz, C-H. Wong, Chem. Rev., 96 (1996) 443.
9. G.M. Watt, P.A.S. Lowden, S.L. Flitsch, Current Opinion In Structural Biology, 7 (1997) 652.
10. D.H.G. Cront, G. Vic, Current Opinion in Chemical Biology, 2 (1998) 98.
11. S. Kobayashi, J. Polym. Sci., (A) 37(16) (1999) 3041
12. K.G.I. Nilsson, Tibtech., 6 (1988) 256.
13. T.N. Zvyagintseva, L.A. Elyakova, Bioorg. Khim., (Russ), 20 (5) (1994) 453.
14. J-L. Viladot, V. Moreau, A. Plans, H. Driguez, J. Chem. Soc., Perkin Trans., 1 (1997) 2383.
15. A. Buchowiecka, PhD Thesis, Technical University, Lodz, Poland, (1999).
16. N.J. Nelson, J. Biol. Chem., 153 (1944) 375.
17. M. Somogyi, J. Biol. Chem., 195 (1952) 19.
18. L.G. Paleg et al., Anal. Chem., 31 (1959) 1902.
19. J.F. Robyt, R. Mukerjea, Carbohydr. Res., 251 (1994) 187.
20. Z. Zhou, S. Ogden, J.A. Leary, J. Org. Chem., 55 (1990) 5444.
21. S. Bielecki, A. Buchowiecka, Biotechnologia, 2 (1996) 184.
22. L.F. Mackenzie, Q. Wang, R.A.J. Warren, S.G. Withers, J. Am. Chem. Soc., 120 (1998) 5583.

Food Biotechnology
S. Bielecki, J. Tramper and J. Polak (Editors)
© 2000 Elsevier Science B.V. All rights reserved.

Use of native and immobilized β-galactosidase in the food industry

A. Miezeliene[a], A. Zubriene[b], S. Budriene[b], G. Dienys[b], J. Sereikaite[c]

[a] Lithuanian Food Institute, Taikos av. 92, 3031, Kaunas, Lithuania

[b] Vilnius University, Naugarduko 24, 2006, Vilnius, Lithuania

[c] Institute of Biotechnology, Graiciuno 8, 2028, Vilnius, Lithuania

Enzymatic hydrolysis of lactose at 5°C was proposed for production of low lactose milk. Stability of milk casein particles was decreased and their dispersity was increased as a results of lactose conversion. Immobilized β-galactosidase was successfully tested for lactose hydrolysis in whey.
Key words: lactose, hydrolysis, β-galactosidase, immobilization, low lactose milk.

1. INTRODUCTION

Milk and milk products are considered to be an „ideal food" because they have many valuable proteins, calcium and other valuable nutrients. Milk also is rich in the unique disaccharide lactose. Lactose, or milk sugar, is a very important source of energy for infants and newborn mammals and it promotes the intestinal absorption of calcium and phosphorus. The absorption of lactose first requires hydrolysis in the small intestine. The enzyme, responsible for this hydrolysis is β-galactosidase, commonly called lactase. A number of people have difficulty in digesting milk and dairy products, containing lactose. Symptoms of lactose intolerance are observed when there is a deficiency of the enzyme β-galactosidase in the human body or when its activity is lowered. It is estimated, that about 10-15% of Europe's population suffer from the symptoms of lactose intolerance [1] and they must avoid lactose containing dairy foods. There are two ways of dealing with this problem on a world scale. In the first case, β-galactosidase is added to dairy products before eating. The second way suggests that dairy products in which lactose is hydrolysed to monosaccharides should be manufactured in dairy plants [2].

The utilization of β-galactosidase immobilized on insoluble carriers for the hydrolysis of lactose enables multiple use of the enzyme, which is more effective economically. Besides, the enzyme that has been used for hydrolysis is then separated from the reaction products and the process can be terminated at any time. The use of immobilized β-galactosidase in dairy industry is not yet very successful because of the peculiarities of the substrate. Milk and most of the dairy products are emulsions of fat in watery medium. There is a constant danger of separation and clogging in a packed bed reactor with the immobilized enzymes [3]. Inadmissible microbial contamination is possible because there are favorable conditions for the growth of bacteria during the enzymatic treatment.

Low lactose dairy products are not produced in Lithuania at present time. It is very important to develop such products, because dairy industry is one of the main fields of economy in Lithuania. The purpose of our study was to examine the process of hydrolysis of lactose in milk by using soluble enzyme preparations and to investigate the possibility of preparing immobilized β-galactosidase suitable for application in the dairy industry.

2. MATERIALS AND METHODS

Raw milk was obtained from a cheese-making plant in Kaunas. Titrable acidity (17°Th), fat content (3.0%), density (1.027 g/cm^3), initial lactose content (4.75%) of milk were measured. Raw milk was pasteurized at 78°C for 15 min and cooled to hydrolysis temperature.

A liquid β-galactosidase preparation, Lactozym 3000L, with an activity of 3000 lactase activity units (LAU) per ml from the yeast *Kluyveromyces fragilis* produced by Novo Nordisk, Denmark and Maxilact L×5000L with an activity of 5000 LAU per ml from the yeast *Kluyveromyces marxianus var. lactis* produced by Gist Brocades, the Netherlands, were used for the hydrolysis of lactose in milk. Lactozym 3000L and β-galactosidase from the *Penicillium canescens* by "Biosinteze", Lithuania, were used for immobilization.

The reaction of lactose hydrolysis was carried out for 24h at 5°C, and for 4h at 30°C and 40°C. The dosages of enzymes were established in model experiments. They were 1.2 ml/100 ml of milk for Lactozym and 0.7 ml/100 ml for Maxilact in the first case, 1.5 ml/100 ml of milk for Lactozym and 0.9 ml/100 ml for Maxilact in the second case and 1.0 ml/100ml of milk for Lactozym and 0.6 ml/100 ml milk for Maxilact in the last case. These conditions gave an activity of 3700, 4500 and 3000 LAU, respectively. The mixtures of milk and enzyme were kept in a thermostate at temperatures of 5, 30 and 40°C. After completion of the previously determined time intervals, enzymes were inactivated by heating at 85°C for 2 min and the mixtures were cooled down immediately.

In order to estimate changes of milk proteins the hydrolysed milk, after inactivation of enzymes, was cooled to 37°C and 5% of a selected strain of *Lactobacterium acidophilum* was added.

The degree of lactose hydrolysis was measured by a cryoscopic method using ANM-1M Cryoscope. The main principle of the method is that the hyhrolysis of disaccharide to monosaccharides causes the depression of the freezing point of the solution [4].

The stability of the casein particles was determined by the method of Inichov and Brio [5]. The mean diameter of the casein particles was found by the method of light scattering [6].

3. RESULTS AND DISCUSSION

A desirable degree of hydrolysis can be obtained by choosing the right dosage of enzymes, reaction time and temperature. The effect of hydrolysis time at 5°C (A), 30°C (B) and 40°C (C) on the conversion of lactose is presented in Figure 1. The highest activity of the enzymes was found at 35-45°C and pH 6.0-7.0. These conditions are favorable for promoting the activity of dairy microflora, but the problem of milk spoilage due to harmful bacterial action may occur. Therefore the conversion at 5°C temperature is most convenient for producing low lactose dairy products.

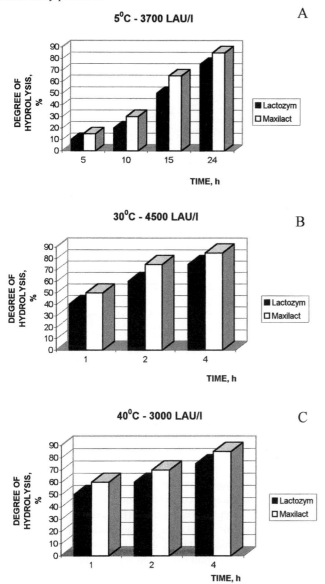

Figure 1. The impact of the reaction conditions and dosage of enzymes on the degree of lactose conversion in milk. A: reaction temperature 5°C, B: reaction temperature 30°C, C: reaction temperature 40°C.

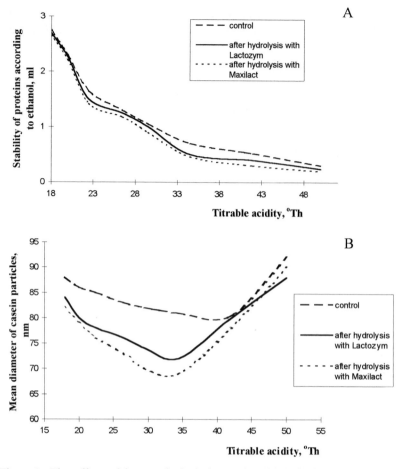

Figure 2. The effect of lactose hydrolysis on the physical-chemical characteristics of milk proteins: A: Influence on the stability of the casein particles, B: Influence on the dispersity of the casein particles.

Changes of milk proteins in hydrolysed milk samples were investigated. Influence of monosaccharides which are formed during lactose conversion on the stability of proteins according to the ethanol (A) and dispersity of the casein particles (B) is presented in Figure 2. Results suggest that the conversion of lactose affects the characteristics of milk proteins. Destruction of internal structure of casein particles started when the acidity of milk samples began to rise. The process of disaggregation continued, while the acidity of milk reached 34-35°Th in hydrolysed milk samples and 40°Th in control milk samples. The difference of mean diameter of casein particles amongst control and hydrolysed milk samples was about 14 and 9 nm. The stability of casein particles in hydrolysed milk samples was lower too. Thus, conversion of lactose in milk by using β-galactosidase preparations increased the dispersity and decreased the stability of milk casein particles. That phenomenon must be taken into account in technologies of sour dairy products.

As a result of the research, technology was developed for the manufacture of low lactose milk and fruit-flavored low lactose milk. The amount of lactose in these products was reduced to 20-30%. This is sufficient for people suffering from lactose intolerance [7]. The technology is very simple and can be realised by using ordinary dairy equipment.

Immobilization of β-galactosidase from *Kluyveromyces fragilis* by covalent binding to carriers containing amino groups using glutaraldehyde as crosslinking agent [3] resulted in high inactivation. The yield of immobilization was in the range of only 4-9% from the initial activity of the enzyme used for immobilization. These results are in agreement with literature data that yeast β-galactosidases are sensitive to glutaraldehyde action [3]. Attempts to immobilize the enzyme by entrapment into 2-hydroxyethylmetacrylate gel [8] were also unsuccessful, because complete inactivation occurred. β-Galactosidase from *Kluyveromyces fragilis* was adsorbed satisfactory on anion exchanger DEAE-Sephadex. The yield of immobilized activity was 30-35%, but it was gradually lost due to desorption.

Mold β-galactosidases are more stable than yeast β-galactosidases to chemical agents and to low pH values. Therefore effective immobilization by adsorption on cation exchangers may be expected. Indeed, β-galactosidase from *Penicillium canescens* was successfully immobilized by sorption on CM-cellulose and CM-Sephadex (yield of immobilization 40-45%). Immobilized preparations had good stability. They were tested for lactose hydrolysis in whey. The obtained results were positive and research in that area will be continued.

REFERENCES

1. H. W. Modler, A. Gelda, M. Yaguchi, S. Gelda, Bulletin of the IDF. 289 (1993) 357.
2. F. L. Suarez, D.A. Savaiano, M.D. Levit, New Eng. J. Med., 333 (1995) 1.
3. W. Hantmeier, Immobilized Biocatalysts, Springer, Berlin etc. (1986).
4. J. P. Ramet, G. Novak, P. A. Evers, H. H. Nijpels, Le Lait 1-2 (581-582) 46-53.
5. G. Inichov, N. Brio, Methods of analysis of milk and dairy products. Moscow, Pischevaja promislenost, (1971).
6. P. Djacenko, I. Vlodavec, Colloidal J., 16 (1954) 294.
7. J.A. Hourigan, Australian J. Dairy Technology, 39 (1984) 114.
8. M. Cantarella, F. Alfani, A. Gallifuoco, Methods in Biotech., 1 (1997) 67.

Food Biotechnology
S. Bielecki, J. Tramper and J. Polak (Editors)
© 2000 Elsevier Science B.V. All rights reserved.

Protein hydrolysis by immobilised *Bacillus subtilis* cells

M. Szczęsna and E. Galas

Institute of Technical Biochemistry, Technical University of Łódź
90-924 Łódź, Stefanowskiego 4/10
E-mail: mirszcz@ck-sg.p.lodz.pl

Bacillus subtilis cells immobilised in polyvinyl alcohol (PVA) cryogel beads and in triacetylcellulose (TAC) fibres secreted proteolytic enzymes (metalloproteinase & subtilisin) and were successfully applied for hydrolysis of casein. The porosity of both carriers allows diffusion of proteins in the supports. The productivity of the biocatalysts, i.e. TAC-fibres in a fixed bed column and PVA-cryogel in a fluidised bed reactor, reached the levels of 3 and 8 mg g^{-1} h^{-1}, respectively. Up to approximately 25% of peptide bonds in casein were hydrolysed. The application of the described biocatalysts enables tailoring of the peptide composition in protein hydrolysates.

1. INTRODUCTION

Proteinases of various origins are recently more commonly applied for protein hydrolysis and modification [1-5]. However, soluble proteolytic preparations, which remain in protein hydrolysates, often negatively affect their sensory properties. On the other hand, the application of immobilised proteinases is expensive, since prices of their preparations comprise the cost of these enzymes' isolation and immobilisation. Besides, the preparations display low operational stability [6]. The advantage of protein hydrolysis with immobilised whole microbial cells producing proteinases is that the costs of the process are reduced, and the biocatalyst containing alive cells may be regenerated, i.e. it is possible to regenerate its proteolytic activity.

This application of immobilised cells gained our interest when we found that *B. subtilis* IBTC-3 cells entrapped in polymeric carriers (PVA cryogel or TAC fibres precipitated in toluene) and activated in a growth medium, containing casein, could hydrolyse soluble proteins.

2. MATERIALS AND METHODS

2.1. Chemicals

Polyvinyl alcohol (MW 72kDa, DP 1600, DH 97,5-99,5 Fluka), triacetylcellulose, plant oil, haemoglobin, isoelectric casein, azocasein, crystal violet.

2.2. Microorganisms

B. subtilis IBTC-3, the serine proteinase producer, was obtained from the microbial culture collection at the Institute of Technical Biochemistry of Technical University of Lodz. Bacterial biomass was collected after cultivation of the bacteria in a liquid medium containing starch, lactose, casein, corn steep liquor and salts, for 72 or 54 h at 30°C and shocking (for 15 min at 80°C) before immobilisation.

2.3. Immobilisation methods

Entrapment of the cells in PVA cryogel was done by the "freezing-thawing" (*f-t*) method [7-8]. Spherical particles of the biocatalyst (Figure 1a) were obtained by carrying out the cryogel formation in a hydrophobic phase. The biomass mixed with aqueous solutions of polyvinyl alcohol was dropped into the oil and the mixture was frozen (for at least 10 h) and thawed (at a rate of about 0.5°C per minute) several times. After the polymer had solidified, the beads were collected and washed with 0.14 M NaCl.

In order to immobilise the *B. subtilis* IBTC-3 cells in TAC fibres, their suspension was homogenised with 10 g of triacetylcellulose, 106 ml of methylene chloride and 4.5 ml of glycerol [9]. The fibres (Figure 1b) were formed by means of the polymer precipitation in toluene at a rate of 8 m min^{-1}. The average fibre length was about 1.2 m.

2.4. Hydrolysis of proteins

Enzymatic hydrolysis of proteins using the *B. subtilis* cells immobilised in TAC fibres was performed using a system consisting of a column (20x250 mm, Lösungsmittelfest-Germany), thermostat, peristaltic pump, pH-meter and magnetic stirrer. The column was filled with 5 g TAC fibres containing entrapped *B. subtilis* cells and connected to the thermostat, and to vessel containing substrate (100 ml of 0.25 or 0.5 % casein solution) and products released during hydrolysis. Hydrolysis was executed at 40°C with the liquid phase of reaction mixture recycling (25 ml min^{-1}) through the column. pH of the reaction mixture was all the time adjusted to 8.5 using 0.06 N NaOH.

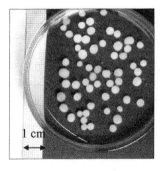 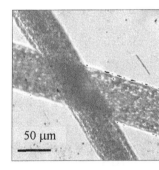

Figure 1. The biocatalyst preparations:
(a) PVA-cryogel beads (Ø 1-3 mm)
(b) TAC fibres

1 cm

50 µm   (b)

Hydrolysis of proteins with *B. subtilis* cells entrapped in the PVA-cryogel beads was conducted using a system of three identical, parallel working, fluidised bed reactors (volume 30 ml) filled with this biocatalyst (1.2 g of the beads per 100 ml of the substrate). Casein solutions were recycled through the reactors at the velocity of 20 ml min^{-1} and at 40°C. They were supplied with 100 or 200 mM NaCl and/or glucose (0.5%) and their pH was between 6.0 and 10.0. For induction of proteolytic properties (activation after the cells immobilisation or regeneration after each hydrolysis process) the whole-cell PVA-biocatalyst was incubated in a liquid nutrient medium (18 – 20 h, 30°C) and washed with 70% ethanol and 0.8% NaCl solution. Only activated or regenerated biocatalyst was applied in a hydrolysis process.

2.5. Analytical methods

Aromatic amino acids content was determined with the Folin reagent. Free NH_2- groups before and after precipitation of the protein with trichloroacetic acid were assayed with ninhydrin solution [10]. Protein concentration in the hydrolysates was determined according to Lowry et al. [11]. Proteolytic activity of biocatalysts was estimated by modified Anson method. The biocatalyst specific activity [mjA] was expressed as an amount of μmoles of tyrosine released from the substrate (haemoglobin) for 1 min by 1 g of beads of biocatalyst. Free B. subtilis cells present in the hydrolysates were detected as CFU. Samples of substrate and its hydrolysates were collected, stored at -10°C and analysed by ultrafiltration and HPLC-RP techniques as described below.

2.6. Ultrafiltration and HPLC-RP analysis

Samples of protein hydrolysates were prepared according to the HPLC manual. Peptides in selected samples were fractionated by membranes: 10000 Da (PM10) and 500 Da (YC05) placed in an ultrafiltration apparatus (Sartorius). HPLC-RP analysis was carried out in a column packed with the type C_8 hydrophobic carrier. Volumes of the samples were 20 or 100 μl and their pH ranged from 6.0 to 8.0. Peptides were eluted with 0.1% TFA in water, supplemented with acetonitrile (from 0% to 100%) at a rate of 1ml min^{-1}, and detected spectrophotometrically at 220 nm.

3. RESULTS & DISCUSSION

Viable B. subtilis cells immobilised in PVA-cryogel beads and in TAC fibres according to the method described in paragraph 2.3 secreted proteolytic enzymes, i.e. both metalloproteinase and serine proteinase (subtilisin) [9]. Their application for hydrolysis of soluble proteins was studied.

3.1 Casein hydrolysis with immobilised bacteria.

For the first time, application of viable, entrapped cells of Serratia marcescens and Humicola lutea for casein hydrolysis was reported by Vuillemard [12] and Grozeva [13, 14], respectively. In both cases, casein was one of the components of the complex growth medium, and under such conditions cells growth, biosynthesis of proteinases and protein hydrolysis occurred simultaneously. The hydrolysates contained the proteolytic enzymes, undigested components of the growth medium, as well as products of the cell metabolism and of the protein digestion. Because of the growth of these bacteria in gels, and their release from these carriers, free cells were also present in the reaction mixture [14]. In contrast to the above-presented experiments, we applied solutions of casein in water, enriched in some cases with Na^+ or Ca^{+2} ions or glucose.

The immobilised B. subtilis cells were applied for casein hydrolysis as described in the paragraph 2.4. TAC fibres were applied in a fixed bed column and PVA-cryogel beads in a fluidised bed reactor, in both cases with substrate (0.25% casein solution) recycling.

The dynamics of proteolysis within several successive runs performed with the same portion of the biocatalysts are presented in Figures 2 and 3.

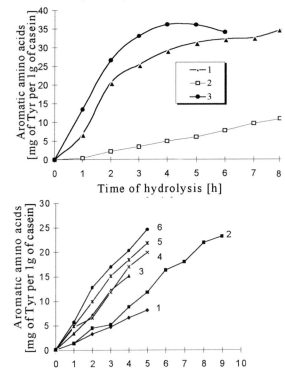

Figure 2. The dynamics of casein hydrolysis using PVA-biocatalyst containing viable *B. subtilis* cells. 1- directly after immobilisation and activation of bacteria in the nutrient medium; 2- after its 24 h exploitation for casein hydrolysis; 3- after its 40 h exploitation and regeneration (incubation of beads in the nutrient medium and treatment with 70% ethanol)

Figure 3. The dynamics of casein hydrolysis in six successive runs carried out using TAC-fibres biocatalyst (containing 10^9 *B. subtilis* spores per g) in a column reactor

Better dynamics and proteolysis efficiency gave the PVA-biocatalyst, but after a few dozen hours of exploitation its activity was markedly decreased. To avoid this, the simple method of biocatalyst regeneration was applied. It comprised of washing the beads with physiological saline solution, treatment with 70% ethanol (permeabilisation of cells), followed by 18-24 hours incubation in the nutrient medium containing 1% casein, and repeated washing with physiological saline solution. This procedure resulted in complete regeneration of the PVA-biocatalyst activity (Figure 2). Such processing was not necessary in the case of TAC-fibres biocatalyst because its activity didn't decrease but rose during protein hydrolysis (Figure 3). In both cases neither free cells nor soluble proteolytic enzymes were detected in the solutions, even after ten hours of casein hydrolysis.

The maximal productivity (expressed as the amount of soluble peptides containing aromatic amino acids released from casein) of the TAC-fibre biocatalyst and the PVA-cryogel biocatalyst, reached the levels of 3 and 8 mg $g^{-1} h^{-1}$, respectively. Casein digestion is markedly influenced by the pH of reaction mixture (Table 1).

Supply of the casein solution with glucose and/or NaCl provided an increase of operational stability of the PVA-biocatalyst. The addition of NaCl (100 mM) and glucose (0.5%) to the reaction medium increased the productivity of the PVA-cryogel catalyst (efficiency: 11 mg Tyr g^{-1} casein h^{-1}) and the degree of casein hydrolysis (Table 2).

Table 1.
Effect of pH on the PVA-biocatalyst productivity and casein hydrolysis efficiency.

| pH | Productivity [mmol Tyr g^{-1} biocatalyst h^{-1}] | | Efficiency |
	after 1 h	after 4 h	[mg Tyr g^{-1} casein]*)
initial 6.0	18.0	11.0	41
" 8.0	15.5	10.0	38
" 10.0	8.0	5.3	28
constant 6.0	21.0	17.6	42
" 8.0	23.0	18.7	53
" 10.0	5.5	6.0	25

*) efficiency after 6 hours hydrolysis

Table 2
Effect of different hydrolysis conditions on concentration of free NH$_2$ groups, aromatic amino acids and protein in protein hydrolysates obtained using the PVA-biocatalyst

No of sample	Hydrolysis conditions	Free NH$_2$ [μg Leu ml^{-1}]	Aromatic amino acids [μmol Tyr ml^{-1}]	Protein [mg ml^{-1}]
1	0.25% isoelectric casein (substrate)	114.6	1.39	0.994
	hydrolysates of isoelectric casein obtained under different conditions:			
2	pH$_0$ 6.0, E/S = 1.69	890.5	2.81	0.765
3	pH$_0$ 8.0, E/S = 1.02	597.8	2.03	0.842
4	pH$_0$10.0, E/S = 1.26	218.1	2.34	0.982
5	pH 8.0, E/S = 1.54, 100 mM NaCl, 0.5% glucose	1141.8	3.01	1.103
6	pH 8.0, E/S = 1.50, 200 mM NaCl, 0.5% glucose	799.4	1.59	0.848
7	0.25% technical casein (substrate)	84.2	1.43	1.454
8	hydrolysate of technical casein, pH 8.0, E/S= 0.3	613.0	1.94	1.091
9	0.25% solution of skim milk (substrate)	16.6	0.50	0.551
10	hydrolysate of skim milk proteins, pH 8.0, E/S = 0.3	215.4	0.68	0.406

E/S – the ratio of initial proteolytic activity of the PVA-biocatalyst [μjA] to the volume of substrate solution [ml]; time of hydrolysis: 5-6 h; temperature: 40°C

The concentration of free amino groups in these hydrolysates was twice as high as in digests of isoelectric casein (see sample 3 and 5, Table 2). Under these conditions up to approximately 25% of peptide bonds of casein were hydrolysed after 5 hours of this process. It might result from the enhanced secretion of proteinases by the immobilised viable B. subtilis cells (mono- and bivalent cations can stimulate the serine proteinase secretion [15]), as well as structural changes of the substrate molecule in the presence of Na$^+$ and glucose. It was noticed that an increase in NaCl concentration up to 200 mM in the reaction mixture, decreased the content of free NH$_2$– groups and peptides containing aromatic amino acids detectable with the Folin reagent (sample 5 and 6, Table 2).

Table 3 Composition of casein hydrolysates obtained using the PVA biocatalyst containing whole, viable *B. subtilis* cells.

Sample	HPLC-RP analysis					Ultrafiltration		
	Time of retention [min] (acetonitrile concentration [%])					Percentage [1]) of fractions containing peptides with molecular weight:		
	0-9 (0)	10-19 (0-37)	20-29 (38-60)	30-39 (61-84)	40-50 (85-100)	(kDa)		
	Percentage of fraction					>10	from 0.5 to 10	<0.5
Isoelectric casein	8.3	49.9	27.8	6.7	7.4	99,7	0,3	0
Isoelectric casein hydrolysates obtained after 6 h, under different conditions:								
pH_0 6.0, 0.54 mM Ca^{+2} *)	2.9	82.7	7.7	3.8	3.0	(0) 0	(39.3) 37.5	(60.7) 62.5
pH_0 8.0, 0.54 mM Ca^{+2}	3.5	82.7	6.3	5.1	2.4	(9.6) 22.6	(34.3) 33.7	(56.2) 43.6
pH_0 10.0, 0.54 mM Ca^{+2}	46.4	18.0	32.3	3.3	0	(43.8) 49.3	(21.5) 25.5	(34.7) 25.2
pH 8.0, 100 mM NaCl, 0.5% glucose	6.4	83.3	5.6	1.3	3.4	(0) 6.5	(32.4) 45.2	(67.6) 48.4
pH 8.0, 200 mM NaCl, 0.5% glucose	6.4	85.5	4.2	1.0	2.9	(41.8) 53.9	(18.8) 3.8	(39.4) 42.3
pH 8.0, without any additives	7.4	83.3	3.9	1.5	3.9	(17.6) 3.5	(38.9) 54.5	(43.8) 42.0
Technical casein	39.0	2.3	5.1	53.3	0	-	-	-
Technical casein hydrolysate, pH 8.0	22.8	61.5	1.3	14.3	0	-	-	-

[1]) The percentages of individual fractions in the hydrolysates was determined on the basis of per cent concentrations of aromatic amino acids (values in brackets) and free amino groups in these fractions.

*) Ca^{+2} ions were added to stabilise subtilisin.

The above presented results prove that both glucose and NaCl, as well as H^+ ions, change proteolysis profile, and in this way influence peptide composition of the hydrolysates (determined by ultrafiltration and HPLC-RP, and presented in Table 3). Casein hydrolysates obtained in the presence of 100 mM Na^+ ions and 0.5% glucose (at pH 8.0), as well as at pH 6.0 without any additives, contained mainly short chain peptides (MW<500Da). The content of hydrophobic peptide products rose with increasing concentration of H^+ ions in the milieu of hydrolysis. It is known that hydrophobic fractions in casein digests (especially β–casein) are the rich source of peptides displaying unique properties, such as casomorphin, casokinin, ACE inhibitor and others [16]. The proteinases produced by viable, immobilised *B. subtilis* cells may be the perfect tool to produce such peptides from casein.

3.2 Assessment of diffusive properties of the biocatalysts.
The studies performed with the image analysis computer system proved that the porosity of the carriers enabled diffusion of high-molecular weight compounds. Results of studies on the PVA-cryogel biocatalyst are presented in Table 4.

Table 4 Diffusion of various substances into the PVA cryogel*)

Compound	Molecular weight, Da	Distance of penetration into gel after 30 min, d[μm]
crystal violet	372	724
azocasein	21000	161
azoalbumin	45000	91
haemoglobin	68000	64

*) Diffusion of various compounds into the cryogel (sliced beads) was monitored and measured *on-line* using an image analysis system (IMAL-Poland).

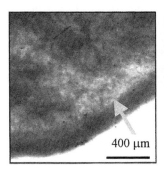 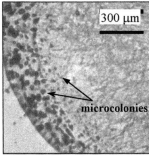

Figure 4. (left) Diffusion of azocasein (2% solution in water) into the PVA-cryogel beads; the arrow shows the front of penetrating substance after 15 min diffusion
Figure 5. (right) The micrograph of a sliced bead of activated PVA-biocatalyst showing microcolonies of *B. subtilis*

The rate of coloured substances penetration into samples of cryogel, prepared from 12% PVA solution containing 6×10^8 *B. subtilis* spores, was measured. A correlation between the rate of molecules penetration into the gel and their size was found. Low-molecular weight compounds (such as crystal violet, which molecular weight is equal to molecular weights of di- to tetrapeptides) can easily penetrate the anisotropic structure of the PVA cryogel. However, the distance of protein molecules penetration into the gel is limited and depends strongly on their molecular weights.

Azocasein penetrates into the gel for a distance of about 160 μm (Figure 4), while the bacterial microcolonies in the biocatalyst, after its activation in the nutrient medium (Difco) containing 1% of casein, are observed in a more than 300 μm deep surface zone (Figure 5). The bacteria present in deeper zones of biocatalyst beads can thus utilise for growth mainly lower–molecular weight compounds, including peptides, diffusing to internal parts of the beads. It is one of the reason of heterogeneous growth of bacteria entrapped in the gel (clearly seen differences in diameters of microcolonies).

4. SUMMARY

Viable *B. subtilis* cells immobilised either in TAC fibres (in the packed bed column reactor) or in the PVA-cryogel beads (in the fluidised bed reactor) are suitable for protein hydrolysis. The biocatalysts can be applied several times if they are washed with sterile physiological saline solution in between the proteolysis runs, and in the case of the PVA-cryogel biocatalyst, also with 70% ethanol for 3-5 minutes. Ethanol and some other solvents (e.g. chloroform, butanol) increase the permeability of the cytoplasmic membrane [17], and probably in this way improve secretion of extracellular proteinases. Bacterial cells immobilised in TAC fibres initially shows a lower productivity (~ 3 mg of Tyr per 1g of casein per h), but they display a higher operational stability and are activated in the presence of the protein. Cells entrapped in the cryogel require incubation in the nutrient medium to regain their proteolytic activity. Processed in this way, they reach a productivity of 8 mg of Tyr per 1g of casein per h, and in the presence of 100 mM Na^+ as much as 11 mg of Tyr per 1g of casein per hour. It has not been elucidated if the positive effect of Na^+ ions on the protein digestion results from the enhanced secretion of proteinases from *B. subtilis* cells, or from conformational changes of casein molecules in their presence, making them more susceptible to proteolysis. The simultaneous occurrence of both phenomena is also possible. The important advantage of this procedure is that, even after 24 hours of the process, neither soluble proteinases nor free *B. subtilis* cells were detected in the hydrolysates. Moreover, casein hydrolysis was not accompanied by the growth of bacterial microcolonies in the PVA-cryogel, and it is well known that the growth of microbes in carriers applied for their immobilisation, is the basic reason for their release into reaction mixtures.

The comparison of casein hydrolysis by *Bacillus* cells immobilised either in PVA-cryogel beads or TAC fibres is presented in Table 5.

Table 5
Comparison of basic properties and hydrolysis conditions of the two biocatalysts containing viable *B. subtilis* cells immobilised in (1) PVA-cryogel beads (2) TAC fibres

		(1) PVA cryogel.	(2) triacetylcelullose
1	Carrier		
2	Type of carrier	beads, $\varnothing$ 2-3mm	fibres, $\varnothing$ ~20μm.
3	Type of reactor	fluidised bed reactor (FBR)	packed bed column (PBCR)
4	Method of activation of biocatalyst	incubation of beads in nutrient medium (30°C, 18-24 h)	the flow of casein solution through the fibre-bed
5	Substrate concentration	0.25-0.5%	0.25-0.5%
6	Amount of biocatalyst per 100 ml of substrate	2 g	5 g
7	Maximal productivity	7 - 8 mg Tyr g^{-1} casein h^{-1}	3 mg Tyr g^{-1} casein h^{-1}
8	Mechanical stability of carrier	very high	high
9	Stability of biocatalyst during hydrolysis process	low *regeneration of biocatalyst in nutrient medium - indispensable*	very high *activation of biocatalyst during hydrolysis process*
10	Proteolytic activity in hydrolysates	0	0
11	Free *B. subtilis* cells in hydrolysates	not detected	not detected
12	Bacterial growth in biocatalyst during hydrolysis	not detected	—

The porosity of both carriers (PVA cryogel and TAC fibres) provides diffusion of proteins. The application of the described above biocatalysts facilitates execution of protein hydrolysis and enables tailoring of peptide composition of the hydrolysates.

REFERENCES

1. J. Lash, R. Koelsch and K. Kretschmer, Acta Biotechnol., 7 (1987) 227.
2. R. Lopez-Fandino, M. Ramos et all., J. Dairy Res., 58 (1991) 461.
3. B.D. Rebeca., M.T. Pena-Vera and M. Diaz-Castaneda, J. Food Sci., 56 (1991) 309.
4. I.W. Glusniev, I.S. Golubiev and B.W. Moskvitchiev, Biotechnologiya (Russ), 4 (1991) 58.
5. S.J. Ge and L.X. Zhang, Acta Biotechnol., 13 (1993) 151.
6. M. Angelova, E. Petrichova and L. Slokoska, J. Biotechnol., 39 (1995) 41.
7. O. Ariga, T. Yamagawa, H. Fuijmatsu and Y. Sano, J. Ferment. Bioeng., 68 (1989) 295.
8. V.I. Lozinski., A.L. Zubov and E.F. Titova, Enzyme Microbiol. Technol.,18 (1996) 561.
9. M. Szczesna, PhD Thesis, Technical University of Lodz, Lodz, Poland, 1998.
10. H. Rosen, Arch. Biochem. Biophys., 67 (1957) 10.
11. O.H. Lowry, N.J. Rosebrough, A.L. Farr, R.I. Randall, J. Biol. Chem., 19385 (1951) 265.
12. J.C. Vuillemard, J. Goulet, J. Amiot, Enzyme Microb. Technol., 10 (1988) 1.
13. L. Grozeva, B. Tchorbanov, P. Aleksjeva, Appl. Microbiol. Biotechnol., 39 (1993) 512.
14. P. Aleksjeva, E. Petricheva, et al., Acta Biotechnol., 11 (1991) 255.
15. A.V. Artemov, R.M. Barski, et al., Biochimiya (Russ), 51 (1986) 830.
16. A.M. Fiat and P. Jolles, Molecular and Cellular Biochemistry, 87 (1989) 5.
17. M.V. Flores, C.E. Voget and R.J.J. Ertola, Enz. Microb. Technol., 16 (1994) 340.

Food Biotechnology
S. Bielecki, J. Tramper and J. Polak (Editors)
© 2000 Elsevier Science B.V. All rights reserved.

Degradation of raw potato starch by an amylolytic strain of *Lactobacillus plantarum* C

K.J. Zielińska, K.M. Stecka, A.H. Miecznikowski., A.M. Suterska

Institute of Agricultural and Food Biotechnology, Warsaw, Poland.

A new microorganism belonging to the lactic acid bacteria has been isolated from naturally fermented alfalfa. It is identified as the amylolytic strain *Lactobacillus plantarum* C (No KKP/788/p culture collection of IAFB).
Raw potato starch treated by this amylase was partially dextrinized and degraded. The breaking down of starch granules was accompanied by an increase of production of lactic acid and reducing sugars. Light microscopic evaluation indicated that starch granules retained their morphological integrity, although they were substantially degraded. It also became apparent from the micrographs, that *Lactobacillus plantarum* C abundantly colonizes starch granules. The starch granules were transformed into hollow, pitted and fragmented forms, which is a typical result of alpha-amylase activity.

1. INTRODUCTION

The synthesis of starch in plants occurs in a highly biochemically organized environment, and the structure of starch is connected strongly to its origin. Every kind of plant produces its own specific type of starch. They differ in the amylose and amylopectin content and in the structure, size, and shape of the granules.

There is a clear correlation between the sensitivity of raw starch to enzyme activity and its origin, and the surface structure of the granules. Raw potato starch shows the lowest sensitivity to enzyme activity because of the smooth surface of the granules. In this case amylolytic enzymes need more time to penetrate inside the granules and degrade its structure [1].

During the grueling process, in the temperature range of 50°C to 70°C, starch swells, its structure changes, chains of amylose and amylopectin loosen up, and then it is most sensitive to amylolytic enzymes.

Depending on the characteristics of the amylolytic enzymes starch is decomposed to different final products. For example: potato or rye starch treated with bacterial alpha-amylase is decomposed to amylose, amylopectin, and partially to glucose, while treated with plant alpha-amylase starch is decomposed to dextrins of different molecular weight and to maltose.

In recent years amylolytic enzymes biosynthesized by lactic acid bacteria have been discovered. They are able to decompose raw, not glued potato starch. A variety of microorganisms producing amylolytic enzymes have been described, but only few of them have the ability to decompose raw starch. For example, among 80 strains from the genus

Bacillus only two species, *Bacillus stearothermophilus* and *Bacillus amylolyticus,* effectively degrade potato starch [2]. The ability to decompose raw rye starch was found in some lactic acid bacterial strains existing in the digestive tracts of animals [3].

Isolated from the fermenting roots of cassava, strain *Lactobacillus plantarum* A6 produces an extracellular alpha-amylase, which hydrolyzes raw starch in cassava [4,5]. The research results concerning *Lactobacillus plantarum* A6 amylase suggest that this strain produces mainly alpha-amylase, although it may have many forms, and we can not exclude the activity of beta-amylase and glucoamylase [6].

Currently, in research centers all over the world the following subjects are studied in this field:
- degradation of rye, corn and potato starch with amylases from lactic acid bacteria,
- determination of the characteristics of pure amylolytic enzymes from lactic acid bacteria.

On the basis of these results Xiaodong and Xuan suggest that many possibilities of using amylolytic lactic acid bacteria exist in the following types of industries: starch, food, feed, pharmaceutical and chemical (for producing biodegradable plastics) [7].

Research on the selectivity and on finding strains of lactic acid bacteria with the ability to degrade potato starch have been conducted by the authors of this paper since 1995. We have selected from the natural environment a couple of lactic acid bacteria strains able to use raw potato starch for lactic acid production. One of them was isolated from alfalfa (*Lactobacillus plantarum* C, strain KKP/788/p - culture collection of IAFB). It is characterized by the ability to hydrolyze raw and soluble potato starch and to convert to lactic acid (L+ and D-) as degradation product, because of biosynthesis of amylolytic enzymes released into the culture medium. This strain is characterized by a rapid growth of biomass and production of lactic acid from either soluble or raw potato starch, in the growth temperature of 30°C and 37°C. Paralel to the growth of biomass, lactic acid is intensively produced from starch in the first 24 hours of the culture. After 72 hours of growth the amount of lactic acid does not increase anymore and equals 1.95 g for soluble starch and 1.85 g for raw starch per 100 ml of medium [8].

The aim of the research was to determine the degree of potato starch degradation by extracellular alpha-amylase of the *Lactobacillus plantarum* C bacteria strain, grown on soluble or raw potato starch, which was the only source of carbon.

2. MATERIALS AND METHODS

Microorganism: *Lactobacillus plantarum* C (KKP/788/p strain culture collection of IAFB). The strain was cultured 24-96 hours at 37°C under anaerobic conditions in a fermentation tank (Bioflo III type). Culture medium MRS (Difco) and modified medium MRS, in which glucose was substituted by soluble (2%) or raw (5% sterilized UV) potato starch. At the end of the culture, the remaining starch in the medium was centrifuged (16 000 rpm).

The activity of amylolytic enzymes was measured in the supernatant with the application of methods based on:
a) measurement of time of starch hydrolysis to dextrins [8],
b) the amount of glucose released from starch was evaluated by the enzymatic method (with glucose oxidase and peroxidase) with the application of the Sigma Aldrich test.

The amylolytic activity was given in AS and GA units. One unit of the activity of alpha-amylase (AS) is defined as the amount of enzyme, which within 1 hour at 30°C, hydrolyzes 1 g of soluble potato starch to dextrins. The unit activity for glucoamylase GA is defined as: the amount of enzyme which releases from soluble potato starch 1 mg of glucose in one hour at 30°C.

The starch recovered from the culture medium, not used thus by the bacteria for biomass and lactic acid production, was washed with water and left to dry at 40°C.

The reducing sugar content was evaluated by the colorimetric method with the 3,5-dinitric salicylic acid (DNS) [9].

Microscopic observations of raw starch treated and not treated by amylolytic bacteria were done with an optic-light microscope Nikon Optiphot II. After fixing the starch samples with glutaric aldehyde, dehydrating them with alcohol and dusting them on a thin carbon and silver film (duster JEOL), the granules of starch treated and not treated by bacteria were observed under a scanning electron microscope JSM-35 JEOL.

3. RESULTS

We determined the dynamics of the amylases release into the medium during growth of bacteria on MRS, containing soluble or raw potato starch as the only source of carbon at 37°C, optimal for growth. We observed that the amylase activity in the liquid, expressed in AS and GA units, was greater when soluble starch rather than raw starch was applied. When soluble starch was the only source of carbon, alpha-amylase activity in the liquid after 72 hours was 10 UAS/100ml and glucoamylase 25 UGA/100 ml. When raw starch was the only source of carbon, the alpha-amylase activity was 5 UAS/100 ml and glucoamylase 19 UAG/100 ml. The amylase activity was increasing until 72 hours and then stayed on a constant level (Table 1).

Table 1
Amylase activity produced by *Lactobacillus plantarum* C (37°C; time period 48-96 h)

MRS –potato starch	Time (h)	Amylase activity units/100 ml	
		AS	GA
Soluble	48	8	15
	72	10	25
	96	10	25
Raw	48	4	12
	72	5	19
	96	5	19

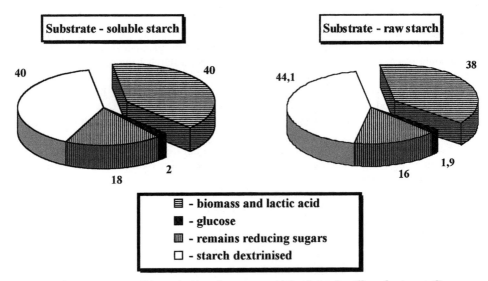

Figure 1. The percentage of degradation of potato starch by *Lactobacillus plantarum* C

In order to examine the degree of potato starch degradation caused by the investigated bacterial strain, mass decrement of starch used for production of biomass and bacterial metabolites have been determined. In the case of using soluble potato starch as a source of carbon, mass decrement of starch was 40%. The remaining starch has been degraded to water soluble reducing sugars – 20% (including 2% of glucose) and insoluble sediment – 40%, characterized by 10% content of reducing substances. In the case of using raw potato starch as a source of carbon, mass decrement was 38%. The remaining starch has been degraded to water soluble reducing sugars – 17.9% (including 1.9% of glucose) and insoluble deposit – 44.1%, characterized by 10% content of reducing substances (Figure 1).

Granules of potato starch stained with iodine; light microscopy micrographs (400 x)

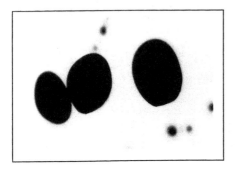

Figure 2. Raw starch

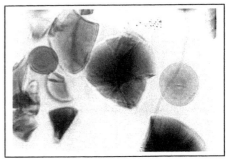

Figure 3. Starch after treatment with bacterial amylases

Examination of the morphological and chemical changes in granules of raw potato starch, either treated or not by bacterial amylase, was carried out with the usage of a light microscope with the magnification of 400 times. The starch samples were stained with iodine (see pictures).

Figure 2 shows raw potato starch granules with the blue color of the iodine complex. Figure 3 shows a micrograph of raw potato starch granules, morphologically and chemically changed by amylolytic bacteria. The color of the iodine starch complex changes from violet to rouge, which proves the decomposition of starch to dextrins. In the microscopic picture we can see starch granules with fuzzy edges and small pieces formed from their decomposition.

Surface, inner, and outer morphological structure changes in raw potato starch granules, treated or not treated by bacterial amylase, were also examined with an electron scanning microscope. Magnifications of 4000 times were applied.

Granules of potato starch; electron scanning microscopy micographs (4000 x)

   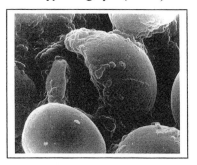

Figure 4. Raw starch

Figure 5. Starch after treatment with bacterial amylases, granules hollow, pitted and broken down

Surface of potato granule; electron scanning microscopy micrograph (15 000 x)

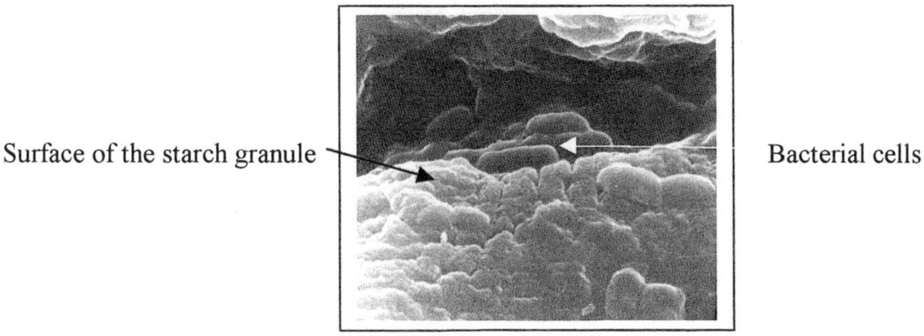

Surface of the starch granule   Bacterial cells

Figure 6. The transformed surface of the starch granule, abundantly colonized by bacterial cells

Raw potato starch granules have a very firm structure, smooth surfaces without pits or even scratches (Figure 4).

A microscopic picture of starch treated by amylolytic bacteria shows morphological changes in the surface structure of granules – one can see holes and pits reaching inside the granule after attack by amylolytic enzymes (Figure 5).

When applying a 15 000 magnification, one can see the changed surface of the granules of starch colonized by bacterial cells, which use starch as a food source (Figure 6).

Based on our results we conclude that in the supernatant of *Lactobacillus plantarum* C cultures beside alpha-amylase activity also glucoamylase activity can be found.

The raw potato starch treated by amylase *Lactobacillus plantarum* C was partially dextrinized and degraded. The breakdown of starch granules was assumed as the increase of lactic acid and reducing sugars including glucose.

Comparing the changes of degraded potato starch treated with *Lactobacillus plantarum* C amylase and corn starch treated with *Lactobacillus amylovorus* alpha-amylase [10] and cassava starch treated with *Lactobacillus plantarum* A₆ [6], we concluded from the analyzed microscopic pictures (scanning microscopy) that the morphologically altered granules of raw starch are very similar.

On the basis of our observations we conclude that the amylases synthesized by the lactic acid bacteria *Lactobacillus amylovorus* and *Lactobacillus plantarum* C and *Lactobacillus plantarum* A₆, have similar properties and that their activities are characteristic for alpha-amylases.

The use of above mentioned strain can be envisaged in several fields of application. The high biomass yield and the high growth rate on raw starch mean that the microorganism could be used in starter cultures to improve the conservation and stability of starch-based fermented foods and feeds.

REFERENCES

1. H. Fuwa and M. Nakaima, Cereal Chem., 54 (1977) 230
2. B.G. Dettori- Campus, F.G. Priest, and J.R. Stark, Process Biochem., 27 (1992) 17
3. M. Champ, O. Szylit, P. Raibaud and N. Ait-Abdelkalder, J. of Appl. Bacteriol., 55 (1983) 487
4. E.L. Giraud, A. Brauman and S. Keleke, Appl. Microbiol. Biotechnol., 36 (1991) 379
5. E.L. Giraud, L. Gosselin, B. Marin, J.L. Parada and M. Raimbault, J. of Appl. Bacteriol., 75 (1993) 276
6. E.L. Giraud, A. Champailler and M. Raimbault, Appl. and Environ. Microbiol., 12 (1994) 4319
7. W. Xiaodong, G. Xuan and S.K. Rakshit, Biotechnol. Lett., 9 (1997) 841
8. K.J. Zielińska, R. Sawicka-Żukowska, K.M. Stecka and B. Jędychowska VII Intern. Starch Convent Poland Cracow (1996)
9. G.L. Miller, Anal. Chem., 31 (1959) 426
10. S.H. Imam, A. Burgess-Cassler, S.L. Cote and S.H. Gordon, Curr. Microbiol., 22 (1991) 365

Food Biotechnology
S. Bielecki, J. Tramper and J. Polak (Editors)
© 2000 Elsevier Science B.V. All rights reserved.

Removal of raffinose galactooligosaccharides from lentil (*lens culinaris* med.) by the *Mortierella vinacea* IBT-3 α-galactosidase

H. Miszkiewicz and E. Galas

Institute of Technical Biochemistry, Technical University of Lodz
Stefanowskiego 4/10, 90-924 Lodz, Poland

The efficiency of the α-galactosidase from *Mortierella vinacea* IBT-3 in reducing raffinose and stachyose content in lentil flour was studied and compared with traditional processing. The enzyme was most active at 50°C and pH 4.0-5.0. Its specific activity was 20.5 U mg^{-1}. The α-galactosidase was stable at neutral and alkaline pH. It was stable for 1 hour at 50°C but it lost 56% of the initial activity after 1 hour at 60°C. The optimal conditions for removal of raffinose and stachyose were: pH 5.0, 50°C, 1 hour, 15% (w/v) suspension of lentil flour and 0.85 U g^{-1} of lentil flour.

The enzyme almost completely removed raffinose family sugars from lentil flour, whereas its cooking for 1 hour at 100°C diminished raffinose and stachyose contents by 78 and 85%, respectively. Concentrations of these sugars were also decreased by autoclaving (121°C, 30 min.) by 15 and 29%, respectively.

1. INTRODUCTION

Edible lentil (*Lens culinaris* Medic.) is one of the leguminous, protein containing plants cultivated since ancient ages because of high nutritious and sensory properties, little demand for fertilizers and resistance to drought. Lentil seeds contain balanced protein (28.3-34.5%), valuable fibre, macro- and microelements, vitamins B, C and E, and therefore they belong to extremely attractive and nutritious foods [1]. Edible lentil has gained even more attention for the last 25 years. The dynamic increase of its production was accompanied by studies performed in many important research institutes in the USA, Canada and other countries, which revealed that nutritious and tasteful lentil seeds displayed also some medicinal properties. Because of such advantages, lentil was rated among healthy food products, and seems to be the most perfect leguminous plant.

The increased interest in lentil, observed also in Poland, prompted our research into this crop aimed at development of its processing and gaining more information on this unique source of nutritive compounds. However, occurence of some undesirable components (antinutritional factors and α-galactosides) present in lentil seeds, limits their wider use. Most of the antinutritional factors, including phytic acid, tannic acid, amylase and proteinase inhibitors, can be removed easily by traditional cooking and soaking methods [2,3].

α-Galactosides include raffinose, stachyose and verbascose. Because of lack of an α-galactosidase (EC.3.2.1.22) activity in the small intestine these sugars cannot be hydrolized and absorbed. Microorganisms present in the large intestine utilize them and lead to flatulence formation [4]. There are reports on the application of microbial α-galactosidases for degradation of oligosaccharides present in soymilk and other legumes [3,5,6,7].

Our studies were focused on an application of the α-galactosidase from *M. vinacea* IBT-3 for degradation of sugars related to raffinose (antinutritive compounds), present in lentil seeds, as well as comparison of our method with cooking or sterilisation, which also, to some extent, reduce content of these sugars.

2. MATERIALS AND METHODS

2.1. Materials
Edible lentil seeds var. Anita were applied for the studies. Legume seeds were purchased at a local supermarket and were stored at 4°C until used.

2.2. Preparation of α-galactosidase
α-Galactosidase preparation was obtained by protein precipitation with $(NH_4)_2SO_4$ (60-100% of saturation) from *M. vinacea* IBT-3 mycelium extract.

2.3. Assay of α-galactosidase activity
The reaction mixture contained 0.5 ml of 0.5% (w/v) raffinose solution and 0.5 ml of enzyme solution in 0.05 M acetate buffer pH 5.0. The mixture was incubated at 50°C for 1 hour and the reaction was terminated at 100°C for 10 min. The amount of galactose liberated from raffinose was determined by the Somogyi [8] and Nelson method [9]. One unit of α-galactosidase activity is defined as the amount of enzyme that liberates 1 μmol of galactose under the assay condition.

2.4. Properties of α-galactosidase
The effect pH and temperature on the rate of raffinose hydrolysis by the α-galactosidase was determined at pH from 2.0-8.0 and at temperature from 20 to 80°C. Thermostability of enzyme preparation was estimated at temperatures from 10 to 80°C (1 hour, optimal pH) and their pH-stability at pH 3.0 to 10.0 (McIlvaine buffer solution, 24 hours, 4°C).

2.5. Estimation of protein content
The protein content of the partially purified enzyme was determined by the Lowry method, using bovine serum albumin as a standard [10].

2.6. Estimation of reducing sugars
The amount of reducing sugars was determined according to the Somogyi [8] and Nelson method [9].

2.7. Determination of oligosaccharides content

Lentil seeds were crushed into flour and put through a 200 μm sieve. Ground seeds were extracted with 80% ethanol at 60°C, twice for 45 min. The ratio of ground seeds to the solvent was 1:10 (m/v). Extracts were centrifuged (15 min, 8000 rpm) and the supernatant was decanted. Combined supernatants were dried under vacuum at 35°C and the residue was dissolved in 1 ml of deionized water and transferred to Dowex 50 W×8 (200-400 mesh) column (10×0.7 cm). The column was washed with deionized water (3×3 ml). Concentration of selected oligosaccharides (sucrose, raffinose, stachyose) in the eluate (volume adjusted to 4 ml with deionized water before HPLC analysis) was determined using HPX-87H column (sample volume 10 μl, 60% CH_3CN in water, flow rate 0.6 ml min^{-1}, 20°C) calibrated with these sugars.

2.8. Processing of lentil seeds

Cooking
Whole lentil seeds were soaked in distilled water (1:10 w/v) for 18 hours at ambient temperature, cooked in distilled water (1:3 w/v) for 15, 30, 45 or 60 min, rinsed, dried and crushed. The oligosaccharides content was determined as above.

Autoclaving
Whole soaked lentil seeds were autoclaved in distilled water (1:2 w/v) for 25 min at 121°C, rinsed, dried at 60°C, milled into flour, and analyzed as described above for oligosaccharides content.

Enzyme treatment
Lentil samples were ground and sieved through 850 μm aperture. 15% (w/v) suspension of lentil flour was digested with the α-galactosidase (0.85 U per 1 g of the substrate). The hydrolysis was carried out on an orbital shaker (120 rpm) at pH 5.0 and 50°C for 1 hour. After 1 hour, the hydrolysates were centrifuged. The precipitate was dried and crushed and the oligosaccharides were assayed as described above.

3. RESULTS AND DISCUSSION

3.1. Properties of α-galactosidase

The specific activity of the partially purified α-galactosidase was 20.5 U per 1 mg of protein. The maximal rate of raffinose hydrolysis was observed at 50°C and pH 4.0-5.0. Stability studies showed that the enzyme was stable at 30-50°C for 1 hour, at neutral or alkaline pH.

3.2. Effect of cooking

Lengthening of cooking time of lentil seeds diminished the oligosaccharides content, i.e. this process increased their solubility. Cooking performed for 1 hour decreased sucrose, raffinose and stachyose contents by 70, 78 and 85%, respectively (Table 1, Figure 1).

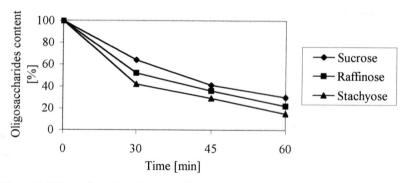

Figure 1. Effect of cooking time on oligosaccharides content in lentil flour.

Table 1
Content of oligosaccharides in raw, cooked, autoclaved and enzyme digested lentil flour (g per 100 g of dry mass)[*]

Sample	Galactose		Sucrose		Raffinose		Stachyose	
	[g/100g]	[%]	[g/100g]	[%]	[g/100g]	[%]	[g/100g]	[%]
Raw	0.286	100	0.743	100	0.799	100	0.894	100
Cooked [60 min]	N.D.	N.D	0.223	30	0.176	22	0.134	15
Autoclaved [25 min, 121°C]	N.D.	N.D.	0.773	104	0.679	85	0.635	71
Digested with enzyme	0.482	168.5	0.122	16.4	0.080	10.05	0.042	4.74

*Each value is the average of triplicate determination

Somiari and Balogh [11] reported that cooking of cowpea for 50 min reduced the raffinose content to 44% and stachyose to 28.6%. Rao and Belavady [12] found that the levels of the raffinose family sugars increased during cooking. Mulimani [3] observed a decrease of concentration of the raffinose family sugars (raffinose by 87%, stachyose by 85% and verbascose by 25%) after cooking of soybean for 60 min.

3.3. Effect of autoclaving

Autoclaving (25 min, 121°C) of lentil seeds reduced raffinose and stachyose content by 15 and 29%, respectively, and increased sucrose content up to 104% (Table 1, Figure 2).

Ruperez [13] claimed that cooking of lentil seeds (116°C, 172 kPa) for 10 min reducd by 13.1% stachyose and by 24.9% sucrose. Troszynska [1] repoted that sterilization of lentil seeds (117°C, for 2×30 min) resulted in lowering of the galactosugars content by: 15% stachyose, 20% verbascose, and almost complete decomposition raffinose. This was thought to be due to the release of bound oligosaccharides from other seed components and partial hydrolysis of high molecular weight α-galactosides.

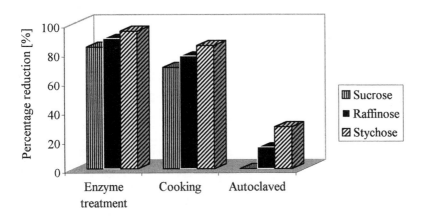

Figure 2. Reduction of oligossacharides content by different techniques of lentil flour treatment.

3.4. Effect of enzyme treatment

The optimal conditions for α-galactosides hydrolysis using this *M. vinacea* IBT-3 enzyme were determined considering the increase of reducing sugars concentration.

Optimal conditions for decomposition of the raffinose family sugars present in lentil seeds were: 15% w/v suspension of lentil flour, 0.85 U g^{-1} of the substrate, 1 hour, 50°C, pH 5.0 (Figures 3-5).

The partially purified α-galactosidase from *M. vinacea* IBT-3 almost completely decomposed galactooligosaccharides in lentil flour (stachyose by 95% and raffinose by 90%, Table 1, Figure 2).

Somiari and Balogh [11] using crude preparations of α-galactosidase from *A. niger* (2 hours, 50°C) for the removal of raffinose and stachyose present in cowpea flour, decreased content of these sugars by 93% and 82%, respectively. Mansour [14] effectively reduced (by almost 100%) the raffinose and stachyose content of chickpea flour using crude α-galactosidase extracts from *Cl. cladosporides* (40°C, pH 5.0), *A. oryzae* (50°C, pH 4.5) and *A. niger* (50°C, pH 5.0). Cruz [15] and Cruz and Park [16] reported that raffinose and stachyose, present in soymilk, were totally hydrolysed by α-galactosidases form *Cl. cladosporides* and *A. oryzae*.

The results prove that enzymatic digestion of lentil seeds with this enzyme is the most effective method of removal of raffinose oligosaccharides which induce flatulent phenomena in humans after ingestion of legumes. Therefore reduction of their content enhances digestibility of legume seeds. The *Mortierella vinacea* IBT-3 α-galactosidase is potentially useful for processing of food products containing such sugars.

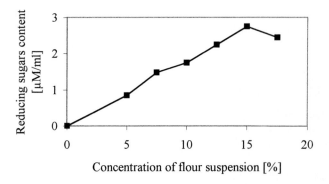

Figure 3. The effect of lentil flour concentration on the reducing sugars content.

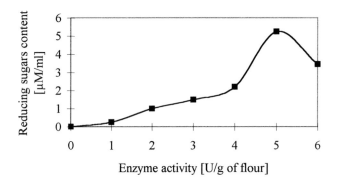

Figure 4. The influence of enzyme activity on the reducing sugars content.

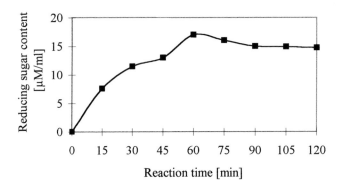

Figure 5. The influence of reaction time on the reducing sugars content.

REFERENCES

1. A. Troszyńska, J. Honke, M. Milczak, H. Kozłowska, Pol. J. Food Nutr. Sci., 2/43 (1993) 49.
2. B. Jacórzyński, Przemysł Spożywczy, 11 (1988) 323.
3. V.H. Mulimani, S. Devendra, Food Chem., 64(4) (1998) 475.
4. A.C. Olson, G.M. Gray, M.R. Gumbmann, C.R. Sell and J.R. Wagner, Food and Nutrition Press Inc., Westport, CT 1981.
5. H. Sugimoto, J.P. van Buren, J. Food Sci. 35 (1971) 655.
6. D. Thananunkul, M. Tanaka, C.O. Chichester, Tung-Ching Lee, J. Food Sci., 41 (1976) 173.
7. R. Somiari, E. Balogh, J. Sci. Food Agr., 61 (1995) 339.
8. M. Somogyi, J. Biol. Chem., 195 (1952) 19.
9. N. Nelson Methods in Enzymol., S.P. Colowick, N.O. Kaplan, Academic Press, N.Y., 8, 85, 1957.
10. O.H. Lowry, N.J. Rosenbrough, A.L. Farr, R.J. Randal, J. Biol. Chem., 193 (1951) 165.
11. R. Somiari, E. Balogh, World J. of Microbiol. and Biotechnol. 8 (1993) 564.
12. P.V. Rao and B. Belavady, J. Agric. Food Chem., 26 (1978) 316.
13. P. Ruperez, Z Lebesin Unters Forsch A 206 (1998) 130.
14. E.H. Mansour, A.H. Khalil, J. Sci. Food Agic. 78 (1998) 175.
15. R. Cruz, J.C. Bastistela, G. Wosiacki, J. Food Sci., 46 (1981) 1196.
16. R. Cruz, Y.K. Park, J. Food Sci., 47 (1982) 1973.

Food Biotechnology
S. Bielecki, J. Tramper and J. Polak (Editors)
© 2000 Elsevier Science B.V. All rights reserved.

Optimisation of physical and chemical properties of wheat starch hydrolyzates

E. Nebesny [a], J. Rosicka [a] and M. Tkaczyk [b]

[a] Technical University of Łódź, Department of Chemical Food Technology,
ul. Stefanowskiego 4/10, 90-924 Łódź, Poland, e-mail: justynar@snack.p.lodz.pl

[b] University of Łódź, Department of Physical Chemistry,
ul. Pomorska 165, 90-236 Łódź, Poland

Wheat starch shows physical and chemical properties different from potato starch. It contains higher amounts of fat – mainly lysophospholipids - forming gelatinous complexes with amylose, proteins and arabinoxylans, negatively affecting physical and chemical properties of its hydrolyzates, i.e. their colour, transparency and aroma, and decreasing the productivity of the hydrolysis process. This necessitates application of enzymes decomposing the mentioned above compounds, i.e. lysophospholipase and xylanolytic enzymes, for enzymatic hydrolysis of wheat starch, together with traditionally used enzymatic preparations (α-amylase for liquefaction and glucoamylase for saccharification).
The best physical and chemical properties of hydrolyzates, i.e. their colour, transparency, filtration rate and the highest degree of saccharification, are obtained using the xylanase preparation Shearzyme 500L together with Spezyme GA 300W, containing both glucoamylase and lysophospholipase.

1. INTRODUCTION

Starch hydrolyzates are products of acidic or enzymatic depolymerization of starch, comprising maltodextrins, starch syrups and glucose. They are mainly produced from cereal starches. However, in the past, in Poland, the potato starch was the predominant raw material. At present, cereal starches are also used for this purpose, for economic reasons.

The origin of starch influences its chemical composition, including content of proteins, fat and non-starch polysaccharides, technology of production of starch hydrolyzates, as well as their physical and chemical properties [1,2]. Wheat starch shows physical and chemical properties different from potato starch [3]. It contains higher amounts of fat – mainly lysophospholipids - forming gelatinous complexes with amylose, proteins and arabinoxylans. This has a disadvantageous effect on physical and chemical properties of the obtained hydrolyzates, i.e. their colour, transparency and aroma, and also decreases the productivity of the hydrolysis process. Wheat starch hydrolyzates are characterised by a much lower rate of filtration, in comparison to maize starch hydrolyzates [4-8]. There are many theories trying to explain what factors influence the rate of filtration of wheat and maize starch hydrolyzates. After enzymatic hydrolysis, maize starch hydrolyzates contain insoluble particles of 0.2 to

1.0 μm in size. According to Brumm [9], these particles consist of nongelatinized starch or according to Hebeda [10], of amylose-lipid complexes that may deteriorate the filtration characteristics. These components may also effect the filtration rate of wheat starch hydrolyzates [4]. Konieczny-Janda and Richter [7] suppose that the lysophospholipids present in wheat starch negatively influence the filtration rate because they form complexes with amylose or proteins in the hydrolyzates. The presence of these complexes reduces the power of water-binding as well as the power of swelling and dissolution of starch [11-13].

Unlike potato and maize starch, wheat starch is contaminated with pentosans, which are composed of 1,4-β-linked xylopyranose units with some β-1,3-linked L-arabinofuranose branchings [2,7]. These arabinoxylans display a high water-absorbing capacity and can increase the viscosity of wheat starch hydrolyzates as they form highly viscous gels. In this way, they can influence the filtration rate of wheat starch hydrolyzates. Pentosans also form complexes with proteins, which cannot be eliminated by means of physical treatment only. They hinder the process of filtration of enzymatic hydrolyzates. It is presumed that the residues of nongelatinized granules of starch, pentosans, amylose-lipid complexes or complexes between lysophospholipids and proteins affect the rate of filtration of wheat starch hydrolyzates. It is still unknown which of these components has the largest effect on the filtration characteristics.

That is why it seems reasonable to conduct research on the enzymatic hydrolysis of wheat starch, aimed at production of wheat starch hydrolyzates displaying optimal physical and chemical properties, i.e. easy to filter, transparent and colourless. This purpose was achieved by means of the application of lysophospholipase, xylanase, and cellulolytic enzymes for enzymatic hydrolysis of wheat starch, which providing the hydrolysis of fats and arabinoxylans. The stability of complexes of lysophospholipids and amylose during enzymatic hydrolysis of wheat starch into glucose, was tested simultaneously.

2. MATERIALS AND METHODS

Wheat starch containing 88.2% of dry matter and having pH 6.4, the acidity of 1.3°N, the alkalinity of 1.1°N, fat content 1.0% of d.m., protein content 1.1% of d.m., and ash content 0.2% of d.m., was the substrate.

The following enzymatic preparations were used in the experiments:
- α-amylase Termamyl 120L produced by Novo Nordisk A/S, with an activity of 240 KNUg^{-1},
- glucoamylase AMG 300L produced by Novo Nordisk A/S, with an activity of 300 AGUml^{-1},
- Spezyme GA 300W Y553, an enzymatic preparation containing glucoamylase and lysophospholipase, produced by Gamma Chemie GmbH, with an activity of 300 SGUml^{-1},
- xylanase-cellulase Gammazym CX 4000L Y552 produced by Gamma Chemie GmbH,
- xylanase Shearzyme 500L produced by Novo Nordisk A/S, with an activity of 500 FXU g^{-1}.

2.1 Analytical methods

The hydrolyzates were assayed for:

- the content of reducing sugars, according to the Lane-Eynon method [4], after previous removal of proteins;
- the filtration rate;

  The filtration was executed at 60°C. 90-ml samples of hydrolyzates were taken at determined time intervals and the concentration of each sample was brought to 37°Bx. The samples were filtered through a fluted filter paper of a diameter of 205 mm, constant for all of the tests, and an area of 3.3011 mm². The filtration area was equal to 3.1420 mm². The filtrate volume, obtained after 5, 10, 15, 20 or 25 minutes was measured.
- colour index [15];

  The filtrate absorbance was measured at 400 and 720 nm then calculated per 1g of product, at pH 5.6-5.7 and the concentration of hydrolyzates equal to 30°Bx.
- transparency index [15];

  The filtrates absorbance in the same conditions as above was measured at 720 nm.
- the properties of amylose-lipide complexes, using the DSC method;

  The micro DSC III calorimeter produced by Setaram was used for research, at temperatures from 20 to 120°C;
- fat content using the Soxleth method [14].

2.2 The course of the hydrolysis process

Enzymatic hydrolysis of wheat starch to glucose comprised two stages, i.e. its liquefaction and saccharification. The liquefaction of 33% suspension of starch was executed at pH 6.5, and 95°C, using the Termamyl 120L α-amylase (0.36 KNU g^{-1} of starch d.m.).

For saccharification (at 60°C and pH 4.5), the enzymatic preparation Spezyme GA 300W Y553 (0.90 SGU g^{-1} of starch d.m.) was used. It contains both glucoamylase and lysophospholipase. For comparison the glucoamylase preparation AMG 300L (0.90 AGU g^{-1} of starch d.m.) was applied. Xylanolytic enzymes, i.e. Gammazym CX 4000L containing xylanase and cellulase (0.06% per g starch d.m.) and Shearzyme 500L containing only xylanase (0.30 FXU g^{-1} of starch d.m.), were used together with the mentioned above preparations.

3. RESULTS AND DISCUSSION

Results of analysis of products released from wheat starch, previously liquefied with the α-amylase, by the Spezyme GA 300W, containing glucoamylase and lysophospholipase, and by xylanolytic enzymes are presented in this work. For comparison, the glucoamylase AMG 300L preparation was used instead of Spezyme GA 300W. The results are presented in Figures 1-6.

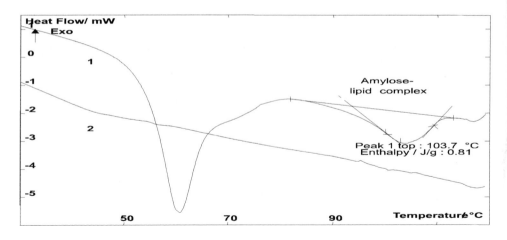

Figure 1. Calorimetric thermograms of wheat starch suspension and of the nonsoluble residue present in glucose syrup obtained using the Spezyme GA 300W, after 72 hours .
1-endotherm of wheat starch suspension, 2-endotherm of the nonsoluble residue from glucose syrup

Calorimetric analysis executed using the DSC method (Fig. 1) of a suspension of wheat starch and of a fraction isolated from its hydrolyzate resulting from saccharification, using the Spezyme GA 300W preparation, after 72 hours, shows distinct differences between them. The double endotherm was registered for the suspension of wheat starch. The first peak is characteristic for starch gelatinization, whereas the second one is characteristic for the presence of an amylose-lipid complex in a native wheat starch. For the residue isolated from the starch hydrolyzate processed with the Spezyme GA 300W preparation, we did not record the endotherm that is characteristic for the occurrence of the amylose-lipid complex at temperatures from 80 to 108°C. This confirms the complex decomposition by the lysophospholipase present in the Spezyme GA 300W preparation.

Determination of fat content in the wheat starch hydrolyzates (Fig. 2) revealed that hydrolyzates obtained using the Spezyme GA 300W, (glucoamylase and lysophospholipase), contain considerably less fat (0.31% of d.m.) than those obtained using the glucoamylase AMG 300L for saccharification (0.60% of d.m.). This also confirmed that the Spezyme GA 300W preparation decomposed the amylose-lipid complex.

The application of the Spezyme GA 300W (glucoamylase and lysophospholipase) for saccharification, together with different xylanolytic preparations, resulted in considerably higher degree of saccharification (from 98.6 to 99.7 DE) in comparison to the application of the glucoamylase AMG 300L for saccharification instead of the Spezyme. The glucoamylase AMG 300L provided the degree of saccharification of only 96.9 to 97.6 DE.

The lysophospholipase, due to the decomposition of amylose – lipid complex, and xylanases, due to the destruction of cell walls constituents of starch grains, facilitate the access of amylolytic enzymes to starch. As a result, higher values of glucose equivalents are obtained when the Spezyme preparation together with Gammazym CX 4000L and Shearzyme 500L are used. The dependence of saccharification degree of wheat starch hydrolyzates on the composition of enzyme mixtures is shown in Fig. 3.

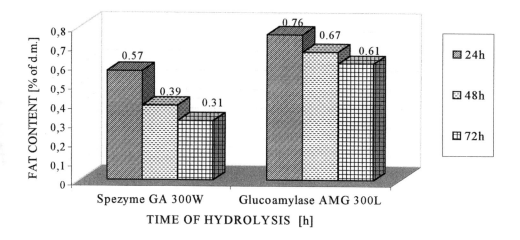

Figure 2. The influence of the Spezyme GA 300W (glucoamylase and lysophospholipase) on fat content in wheat starch hydrolyzates after saccharification.
The initial fat content in the starch – 1.0% of d.m., the fat content after its processing with the α-amylase–0.9% of d.m.

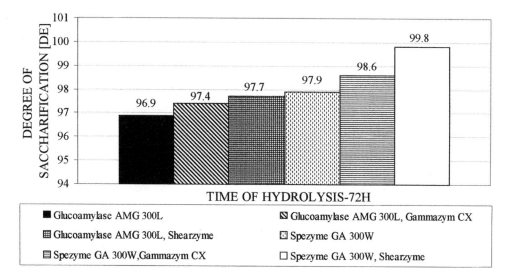

Figure 3. Degree of saccharification of wheat starch hydrolyzates obtained using either Spezyme GA 300W or glucoamylase AMG 300L and combined with xylanolytic preparations.

206

Because lysophospholipids contain single fatty acid residues in their molecules, they may form micelles. These in turn may clog pores of filters applied for filtration of wheat starch hydrolyzates, thus reducing the filtration rate. These ingredients impair the filtration characteristics also by forming amylose-lipid complexes. Lysophospholipase present in the Spezyme GA 300W preparation enhances the filtration rate of wheat starch hydrolyzates because of decomposition of amylose-lipid complexes. Hydrolyzates obtained using the Spezyme GA 300W preparation together with the Shearzyme 500L xylanase for saccharification, were characterised by the highest filtration rate, equal to 1.74 ml m^{-2}s^{-1}. The xylanase, converting non-soluble pentosans into the soluble ones, changes the character of their remains from gel-like into solid, thus increasing the filtration rate. The filtration rate of hydrolyzates obtained using the glucoamylase AMG 300L was lower and equal to only 1.65 ml m^{-2}s^{-1}. The filtration rate of the wheat starch hydrolyzates is shown in Fig. 4.

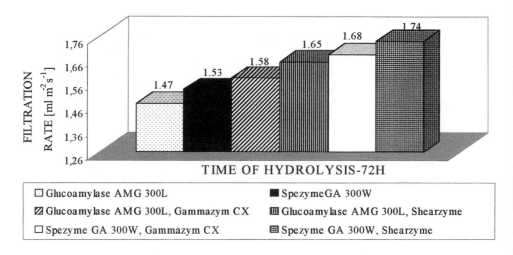

Figure 4. The comparison of filtration rates of wheat starch hydrolyzates obtained using either the Spezyme GA 300W or the glucoamylase AMG 300L and combined with xylanolytic preparations.

Wheat starch hydrolyzates resulting from saccharification, executed in presence of xylanases, are less coloured and more transparent than the ones obtained without the latter enzymes. The best one was obtained using the glucoamylase and lysophospholipase (the Spezyme GA 300W) and the xylanase (Shearzyme 500L). The colour index obtained using this enzyme mixture was equal to 337, and the transparency index to 249. In the case of hydrolyzates obtained with the participation of the glucoamylase AMG 300L these factors were lower (371 and 130, respectively). The colour and transparency factors of wheat starch hydrolyzates are shown in Figures 5 and 6, respectively.

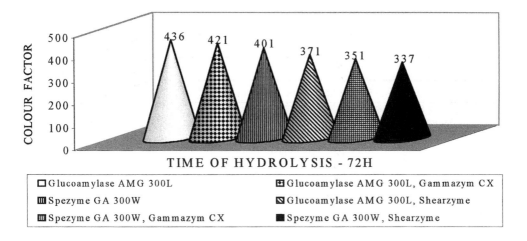

Figure 5. The comparison of the colour factor of wheat starch hydrolyzates obtained using either the Spezyme GA 300W or the glucoamylase AMG 300L and combined with xylanolytic preparations.

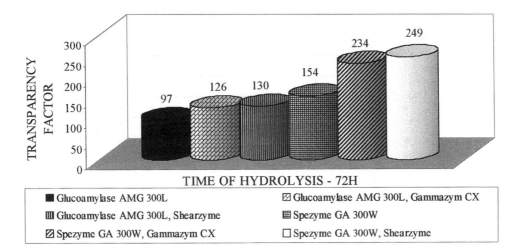

Figure 6. The comparison of the transparency factor of wheat starch hydrolyzates obtained using either the Spezyme GA 300W or the glucoamylase AMG 300L and commbined with xylanolytic preparations.

208

4. CONCLUSIONS

The application of xylanases together with glucoamylase preparations for saccharification of wheat starch, results in an additional increase of saccharification degree of its hydrolyzates.

Higher filtration rate is observed for wheat starch hydrolyzates obtained using glucoamylase and lysophospholipase (the Spezyme GA 300W) and xylanases, in comparison to the ones obtained using the glucoamylase AMG 300L and xylanases, only.

The application of the Spezyme GA 300W preparation and xylanases for saccharification of wheat starch, makes its hydrolyzates less coloured and more transparent, than when using the xylanases together with the glucoamylase AMG 300L.

The best physical and chemical properties of wheat starch hydrolyzates, i.e. their colour, transparency, filtration rate and the highest degree of saccharification, are obtained using the Shearzyme 500L xylanase preparation together with the Spezyme GA 300W for saccharification.

REFERENCES

1. L. Słomińska, Food Industry (in Polish), 3 (1997) 9.
2. A.M. Matser and P.A M. Steeneken, Cereal Chem., 2 (1998) 241.
3. J.J.M. Swinkels, Starch/Stärke, 1 (1985) 1.
4. P. Bowler and P.J. Towersey, Starch/Stärke, 10 (1985) 351.
5. F.G. H. Derez, Carbohydrate rafining process and enzyme compositions suitable for use therein, EP Patent No. 0 219 269 (1987).
6. P. Ducroo, Improvements relating to the production of glucose syrups and purified starches from wheat and other cereal starches containing pentosanes, EP Patent No. 0 228 732 (1987).
7. G. Konieczny-Janda and G. Richter, Starch/Stärke, 8 (1991) 308.
8. A. M. Matser and P.A.M. Steeneken, Cereal Chem., 3 (1998) 289.
9. P.J. Brumm and R.E. Hebeda and W.M. Teague, Starch/Stärke, 41 (1989) 343.
10. R.E. Hebeda and H.W. Leach, Cereal Chem., 51 (1974) 272.
11. D.P. Atkins, J.F. Kennedy, Starch/Stärke, 12 (1985) 421.
12. M. Kugimiya, and J.W. Donovan and R.Y.Wang, Starch/Stärke, 8 (1980) 265.
13. J. Szejtli and E. Banky-Elöt, Starch/Stärke, 11 (1975) 368.
14. Cz. Świechowski, Technical Analysis in the Food Industry (in Polish) WPLiS, Warsaw, 1968, 359.
15. PN 78/A – 74701 Polish Standard, Starch Hydrolyzates, Testing Methods.

Food Biotechnology
S. Bielecki, J. Tramper and J. Polak (Editors)
© 2000 Elsevier Science B.V. All rights reserved.

Kinetics of olive oil hydrolysis by *Candida cylindracea* lipase

I. Sokolovská, M. Polakovič and V.Báleš

Slovak University of Technology, Faculty of Chemical Technology, Department of Chemical and Biochemical Engineering, Radlinského 9, 812 37 Bratislava, Slovakia

In this study, we investigated the kinetics of olive oil hydrolysis in an oil-in-water emulsion in the absence of emulsifier. As this reaction is catalyzed by the lipase acting at the interface of oil and aqueous phase, the heterogeneous character of the process was taken into account during the formulation of the model of enzymatic hydrolysis of olive oil. Five kinetic equations were derived from a general mechanism. Their parameters were optimized using a multiresponse method considering all experimental data obtained at different enzyme concentrations. A model that incorporated the distribution of enzyme between the aqueous phase and oil-water interface was chosen as the best one.

1. INTRODUCTION

The catalytic action of lipases is strictly dependent upon the presence of a lipid in the reaction system. This phenomenon is known as the "lipase activation by interface" and consists in the conformation rearrangement of the lipase molecules resulting in the revelation of the active site. This can occur only in the presence of a hydrophobic interface in the reaction mixture. The reactions catalyzed by lipases were described by several mechanisms including the conventional form of Michaelis-Menten equation [1-3], as well as a mechanism that considers the activation of lipase by interface and consecutive adsorption of activated lipase at the substrate molecules present at the surface of oil droplets [4-6]. Malcata et al. (1992) [7] used a complex, Ping-Pong Bi Bi mechanism to describe the kinetics of reactions catalyzed by the lipase.

2. EXPERIMENTAL

The kinetics of olive oil hydrolysis were studied in an oil-in-water emulsion in the absence of emulsifier. The reactions were done in a batch reactor (BIOSTAT 2, Braun, Germany) under the conditions: working volume 2 dm^3, stirring rate 500 rpm, temperature 30°C, aeration by an air flow of 84 dm^3/h. The aqueous phase was composed of KH_2PO_4 (6.0 g/dm^3), $MgSO_4 \cdot 7H_2O$ (1.0 g/dm^3), urea (4.0 g/dm^3), and micronutrients: $FeCl_3 \cdot 6H_2O$ (10.0 mg/dm^3), inositol (0.4 mg/dm^3), biotin (0.8 mg/dm^3), and thiamin HCl (0.2 mg/dm^3).

Lipase (EC 3.1.1.3) type VII produced by the yeast *Candida cylindracea* was purchased from Sigma (St. Luis, USA) and it was dissolved in the aqueous phase. Its specific activity was determined as 13.4 U/mg powder using the back titration method. 1 unit (U) of lipase is defined as the amount of lipase producing 1 µmol of fatty acids per minute under assay conditions (emulsion of olive oil (30 % v/v) and phosphate buffer 0.1 M, pH 7, 37°C, stirring rate 300 rpm).

Olive oil of commercial grade (Borges, Spain) with a density of 912.7 kg/m³ and a content of free fatty acid of 2.4 mg/g of oil was dispersed in the aqueous phase. The droplet diameter was determined to be 2.6×10^{-5} m from photographs of fermentation medium taken during the yeast cultivation. Gas chromatography analysis of olive oil showed that oleic acid represents the main fraction (80%) of the fatty acids bound through ester bonds in triacylglycerols (TAG).

Several experiments were performed differing in the total lipase concentration (0.1-20 U/cm³). The olive oil fraction was maintained at 0.01 (v/v) in all hydrolysis runs. The concentration of fatty acids liberated within the reactions was determined by titration (0.2-0.4 M KOH) using phenolphthalein as indicator. The product concentration values were transformed into the mass fraction values of the substrate using the mass balance and stoichiometry of the reaction.

3. MODEL FORMULATION

A general mechanism of the enzymatic hydrolysis of olive oil is suggested (Eq. 1). This mechanism assumes that free lipase from the aqueous phase can be reversibly adsorbed at the oil-water interface through its active center on substrate or product molecules present at the interface.

$$E(aq) + S(oil) \underset{k_{-1}}{\overset{k_1}{\rightleftharpoons}} ES(oil/aq) \overset{k}{\longrightarrow} E(aq) + P(oil)$$

$$E(aq) + P(oil) \underset{k_{-2}}{\overset{k_2}{\rightleftharpoons}} EP(oil/aq)$$

$$(1)$$

E represents the free enzyme present in the aqueous phase, S and P are the free substrate, resp. fatty acid molecules at the interface oil-aqueous phase, ES represents the complex formed after the enzyme binds to substrate, EP characterizes the complex of enzyme and product.

The first step of the process is the lipase diffusion from the bulk of the aqueous phase towards the interface. The rate of diffusion depends on the total lipase concentration and hydrodynamic conditions, mainly on the intensity of agitation. Then the lipase adsorption at the substrate, resp. product molecules present at the interface follows. The "quality" of interface (electric potential, surface pressure) and the physical properties of the aqueous phase (pH, temperature, and ionic strength) affect the adsorption rate [4,5]. Generally it is assumed that the lipase adsorption at the interface is reversible and it is described by the Langmuir adsorption isotherm [1,7]. The rate constants k_1, k_{-1}, k_2, k_{-2} are the constants of adsorption and desorption processes. The equilibrium of the lipase adsorption at the hydrophobic interface can be characterized by the dissociation constants K_s and K_p (Eqs. 2-3). It was supposed that lipase adsorption at the interface induces very fast structural rearrangement in the enzyme molecule and consequently the formation of the complex ES and/or EP. The quantitative characterization of the amount of individual species at the interface was made through activities, which were expressed in a relative form to a maximum binding capacity of the unit area of interface.

$$K_s = \frac{1}{K_1} = \frac{k_{-1}}{k_1} = \frac{c_e a_s}{a_{es}} \quad (2)$$

$$K_p = \frac{1}{K_2} = \frac{k_{-2}}{k_2} = \frac{c_e a_p}{a_{ep}} \quad (3)$$

a_{es} is the activity of complex ES, a_s (a_p) is the activity of free substrate (free product) present at the interface of oil and aqueous phase, a_{ep} is the activity of EP complex.

The activity coefficients, γ, characterize the relationship between the total composition of oil fraction and representation of the individual lipid structures at the interface (Eqs. 5-6) while the mass balance of the interface must satisfy the condition expressed by Eq.(4).

$$a_{es} + a_s + a_{ep} + a_p = 1 \quad (4) \qquad a_{es} + a_s = \gamma_s w_s \quad (5) \qquad a_{ep} + a_p = \gamma_p w_p \quad (6)$$

w_s and w_p represent the mass fraction of substrate and product in the oil phase.

Lipase acts at the substrate through the ester bonds and, as it is a non-specific enzyme towards the position of ester bond in the TAG molecule [8], it gradually accomplishes the total olive oil hydrolysis (Eq. 7):

$$TAG + 3\,H_2O \underset{}{\overset{lipase}{\rightleftharpoons}} 3\,RCOOH + C_3H_8O_3 \qquad (7)$$

The fatty acids liberated during the reaction remain in the oil phase because of their very low solubility in water, while glycerol molecules diffuse to the aqueous phase. The lipid fraction composition changes in the course of the reaction and the mass balance of oil fraction is expressed by the equations (Eqs. 8-9)

$$m_o = m_s + m_p \quad (8) \qquad w_s = \frac{m_s}{m_o} \qquad w_p = \frac{m_p}{m_o} \qquad (9)$$

m_o, m_p, m_s are the mass of total oil fraction, fatty acids and substrate respectively.

Since the affinity of lipase towards the interface is high, it was assumed that the reaction rate was controlled by the rate of decomposition of the complex ES (Eq. 10). The hydrolysis rate was related to the unit area of this interface.

$$r = \frac{dm_s}{A\,dt} = -k a_{es} \qquad (10)$$

r is the reaction rate (g/m^2/min), rate constant k has the same dimension as the reaction rate, A is the area of interface of oil and aqueous phase (m^2).

From the general scheme represented by Eqs. (2-10) several models were derived. The heterogeneous character of the enzymatic olive oil hydrolysis was considered during the model formulations as the reaction occurs at the lipid interface. The surface area of dispersed lipid phase was calculated from Eq. (11), assuming spherical shape of the oil droplets, and supposing it to be constant during the hydrolysis. Keeping constant experimental conditions allowed us to assume the constant oil droplet diameter d_o. The density of oil fraction ρ_o did not change in the course of the reaction because of the density of olive oil and oleic acid is about the same.

$$A = 6m_o /(d_o \rho_o)$$ (11)

The concentration of active enzyme adsorbed at the interface was determined through the mass balance of total enzyme present in the reaction system. An overview of the variables of the kinetic models is shown in Table 1.

Models I and II were derived under the assumption that lipase forms the active complex only with the substrate present at the interface. Moreover, Model I considers that a negligible fraction of total enzyme is adsorbed at the interface. On the other hand, Model II considers that the distribution of the enzyme between the aqueous phase and the interface can result in a significant decrease of the free enzyme concentration in the aqueous phase. The concentration of active enzyme adsorbed at the interface was calculated from Eq. 12

$$c_{eads} = q_{emax} a_{es} A / V_{aq} = c_{e0} - c_e$$ (12)

c_e, c_{e0} is the concentration of free, resp. total enzyme (kg/m^3), q_{emax} represents the maximal amount of the enzyme adsorbed at the interface (kg/m^2), V_{aq} is the volume of aqueous phase.

The remaining models include the lipase capability to form also the complex with the fatty acid molecules present at the interface. Model V is the most complex one since it considers both significant lipase distribution between the phases and its uneven affinity to product and substrate. Both Models III and IV represent simplified forms of Model V. Model III assumes an equal affinity of lipase to product and substrate molecules present at the interface, whereas Model IV neglects the distribution of lipase between the phases. The values of activity coefficients of substrate γ_s and product γ_p were calculated under the assumption that the ratio of the free substrate and product present at the interface is equal to the ratio of total amount of these components in the oil fraction.

Table 1
Overview of relationships for the variables characterizing individual models.

No.	γ_s	γ_p	c_e	a_{es}
I	$\gamma_s=1$	-	$c_e = c_{e0}$	$a_{es} = \dfrac{w_s c_e}{K_s + c_e}$
II	$\gamma_s=1$	-	$c_e = c_{e0} - a_{es} q_{emax} \dfrac{A}{V_{aq}}$	$a_{es} = \dfrac{w_s c_e}{K_s + c_e}$
III	$\gamma_s=1$	$\gamma_p=1$	$c_e = c_{e0} - a_{es} q_{emax} \dfrac{A}{w_s V_{aq}}$	$a_{es} = \dfrac{w_s c_e}{K_s + c_e}$
IV	$\gamma_s \neq 1$	$\gamma_p \neq 1$	$c_e = c_{e0}$	$a_{es} = \dfrac{c_e K_p w_s}{c_e \left(K_p w_s + K_s w_p\right) + K_s K_p \left(w_s + w_p\right)}$

4. RESULTS AND DISCUSSION

The parameters of kinetic equations were optimized from all experimental progress curves using a multiresponse method of evaluation. The statistics of the fitting procedures are shown in the Table 2. The best fit of experimental data was achieved using Model II (Figure 1). This model fitted well also the data of the hydrolysis carried out at low lipase concentration (0.1-2.5 U/cm^3). The results showed that the consideration of enzyme distribution between the interface and aqueous phase is not negligible in the process description (Figure 1).

The models II and III differ in the assumption of the lipase capability to form the complex also with the product molecules present at the interface. Since the sum of squared deviations using the models II and III for the experiments performed at the low lipase concentrations are considerably different (Table 3), we can assume that lipase does not bind to the fatty acid molecules occurring at the interface.

Assessing the courses of the substrate conversion (Figure 1), we can conclude that the concentration of lipase sufficient for the saturation of interface oil-water was 7.5 U/cm^3.

Table 2
The statistical evaluation of the experimental data fitting by the individual models.

MODEL	I	II	III	IV	V
Sum of squared deviation	0.47	0.428	0.455	0.458	0.442
kA (g/min)	0.387	0.356	0.364	0.385	0.349
k (g/m^2/min)	0.074	0.068	0.07	0.074	0.067
δk (%)	3.6	4.0	4.5	8.5	8.0
K_s (g/dm^3)	0.1105	0.0504	0.064	0.1071	0.060
δK_s (%)	8.0	28.8	32.0	11.3	31.7
K_p (g/dm^3)	-	-	$K_s=K_p$	0.1076	0.067
δK_p (%)	-	-	-	15	33.1
q_{emax} (kg/m^2)	-	3.2×10^{-5}	1.8×10^{-5}	-	1.7×10^{-5}
δq_{emax} (%)	-	24.6	44.6	-	43.4

δk, δK_s, δK_p, δq_{emax} - average errors of estimated parameters.

Table 3
Sum of squared deviations for the experiments made at low enzyme concentrations (H1: c_{e0}=0.1 U/cm^3; H2: c_{e0}=0.3 U/cm^3) and olive oil concentration of 10 g/dm^3, evaluated using the individual models.

MODEL	I	II	III	IV	V
Sum of squared deviations H1	0.046	0.018	0.033	0.040	0.026
Sum of squared deviations H2	0.034	0.026	0.038	0.030	0.030

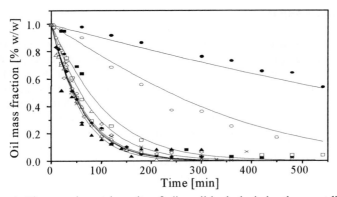

Figure 1. The experimental results of olive oil hydrolysis by the yeast lipase and their fit with Model II. Olive oil fraction 1% (v/v) and total enzyme concentration (U/cm^3): (●) 0.1, (○) 0.3, (■) 1.5, (□) 2.5, (▲) 7.5, (Δ) 10.0, (♦) 15.0, (◊) 20.0; (×) olive oil fraction 6.6% (v/v), lipase 1.5 U/cm^3.

5. CONCLUSIONS

The olive oil hydrolysis by the lipase under our experimental conditions can be described by the kinetic equation of the model II, so we can make following conclusions:

• The first step of the olive oil hydrolysis is the lipase adsorption at the interface described by the Langmuir adsorption isotherm.
• The decomposition of the complex ES is the rate limiting step.
• It is important to consider the lipase distribution between the interface and the aqueous phase.
• Lipase binds only to the substrate molecules resulting in the formation of the complex ES.

Finally, it is important to stress that, unlike in most previous studies, the heterogeneous character of the process has been considered in the formulation of models. The results indicate that this heterogeneous nature and controlling role of the interface area should not be neglected in the modelling of liquid-liquid systems.

REFERENCES

1. G. Benzonana and P. Desnuelle, Biochim. Biophys. Acta, 105 (1965) 121.
2. Y. Kosugi and H. Suzuki, J. Ferm. Technol., 61 (1983) 287.
3. T. Tatara, T. Fuji, T. Kawase and M. Minagawa, JAOCS, 62 (1985) 1053.
4. R. Verger, M.C.E. Mieras and G.H. de Haas, J. Biol. Chem., 248 (1973) 4023.
5. Y.J. Wang, F.F. Wang, J.Y. Sheu, Y.C. Tsai and J.F. Shaw, Biotechnol. Bioeng., 39 (1992) 1128.
6. H.I. Ekiz and M.A. Caglar, Chem. Eng. J., 38 (1988) B7.
7. F.X. Malcata, H.R. Reyes, H.S. Garcia, C.G. Hill and C.H. Amundson, Enzyme Microb. Technol., 14 (1992) 426.
8. S. Benjamin and A. Pandey, Yeast, 14 (1998) 1069.

Food Biotechnology
S. Bielecki, J. Tramper and J. Polak (Editors)
© 2000 Elsevier Science B.V. All rights reserved.

Enzymatic resolution of some racemic alcohols and diols using commercial lipases

J. Kamińska, J. Góra

Institute of General Food Chemistry, Technical University of Lodz
ul. Stefanowskiego 4/10, 90-924 Lodz, Poland

Application of different commercial lipases as chiral catalysts for kinetic resolution of several racemic alcohols in multigram scale by enantioselective acetylation with vinyl acetate in organic solvents, is described. The most important factors affecting these transformations are briefly discussed from a preparative viewpoint. Regioselectivity of investigated lipases in acetylation of the primary hydroxyl group in diols containing a primary and secondary hydroxyl group was also observed.

1. INTRODUCTION

The ability of hydrolytic enzymes, eg. lipases, to act as stereo- and regioselective catalysts has now been well recognised for several years [1,2]. At present the number of commercially available lipases, isolated from different sources, is still increasing. Although frequently they are not pure enzymes, they have proved to be quite suitable for synthetic purposes [3].

Organic chemists are interested in preparation of enantiomerically pure compounds as building blocks for pharmaceuticals, agrochemicals, food additives, and other bioactive substances. Application of enzymes as chiral catalysts has become very attractive since the discovery that they work not only in water, but in organic solvents as well [4].

The enzymatic resolution of racemic alcohols or carboxylic acids and regioselective reactions can be accomplished *via* hydrolysis, esterification or transesterification. Actually resolution of racemic alcohols by means of lipases, as a source of chirality, seems to be a very elegant route to homochiral products. The wide spectrum of structurally different alcohols resolved by lipases allowed to formulate experimental rules for prediction which enantiomer is favoured by a lipase [5]. However, the knowledge of active site architecture and reaction mechanism in nonaqueous medium is still insufficient to predict whether a given alcohol can be resolved by a given lipase or not. Moreover, the activity and selectivity of an enzyme depends strongly on the solvent's properties [6,7]. Therefore, for every substrate, experimental determinatiom of biotransformation conditions is necessary.

In this report we present some of our results on enantioselective and regioselective acetylation of alcohols and diols with vinyl acetate in organic solvents by several commercial lipases.

2. RESULTS

2.1. Resolution of substituted furan derivatives

Various furan derivatives substituted at position 2 are versatile starting materials for synthesis of natural products such as carbohydrates, macrolides, pheromones, alkaloids. 1-(2-Furyl)ethanol - the object of our first investigation is a convenient precursor for preparation of 6-deoxyhexoses (eg. daunosamine, ristosamine). Depending on the configuration of the chiral carbon it will provide D- or L- hexoses. The goal of our research was to find a convenient preparative procedure to both enantiomers of high purity, in multigram quantities.

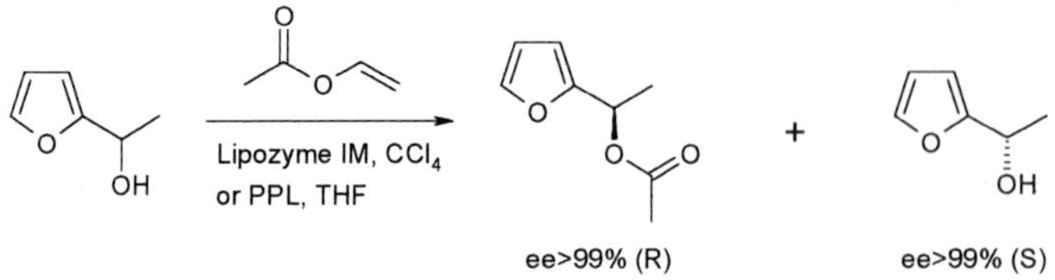

Scheme 1. Resolution of 1-(2-furyl)ethanol.

Table 1
The influence of solvent on 1-(2-furyl)ethanol resolution

Lipozyme IM				Porcine pancreas lipase		
E	(S)-alcohol %ee	(R)-acetate %ee	Solvent	E	(S)-alcohol %ee	(R)-acetate %ee
82	**59**	**96**	**hexane**	86	8	70
69	71	94	isooctane	-	-	-
57	79	92	cyclohexane	5	6	66
70	92	90	benzene	-	-	-
50	71	92	toluene	39	18	94
94	22	97	dichloromethane	214	8	>99
106	48	97	chloroform	225	12	>99
64	**>99**	**87**	**tetrachloromethane**	227	14	>99
83	46	96	**tetrahydrofuran**	**266**	**30**	>99
83	35	97	dioxane	258	26	>99

We have chosen several commercial lipases of relatively low price as potential catalysts - namely microbial lipases from *Mucor miehei* (Lipozyme IM® - Novo-Nordisk), *Candida antarctica* (Novozym 435® - Novo-Nordisk), *Pseudomonas* (PS – Amano), *Aspergillus* (A – Amano), *Candida rugosa* (AY – Amano), *Candida cylindracea* (CCL - Sigma), and porcine pancreas lipase (PPL - Sigma). In the first screening we looked for a highly active and enantioselective enzyme for resolution of 1-(2-furyl)ethanol, Lipozyme IM and PPL lipase were found to meet these requirements. Then for both lipases the further search for the proper solvent system was undertaken. The results are shown in Table 1.

Enantioselectivity factor E which is a measure of enzyme preference for one enantiomer over another, was calculated according to the equation given by Chen [8]:

$$E=\ln[1-c(1+ee_P)]/\ln[1-c(1-ee_P)] \qquad c=ee_S/(ee_S+ee_P)$$

The highest E values were observed in chloroform in the case of Lipozyme IM, and in tetrahydrofuran for porcine pancreas lipase. However, the transesterification rate in chloroform was low, while in hexane or tetrachloromethane very high. Therefore, for preparation of homochiral (S)-alcohol the reaction was carried out in tetrachloromethane in the presence of Lipozyme IM. Enantiomerically pure (R)-acetate was obtained by transesterification in the presence of PPL in tetrahydrofuran or in hexane using Lipozyme IM [9,10].

In spite of our expectation, resolution of structurally similar 2-(2-furyl)-2-hydroxyethyl acetate (Scheme 2) was not effective under the above conditions. Enantioselectivity of Lipozyme IM was found insufficient for this furan derivative. In lipase screening for this substrate the highest enantioselectivity was found with lipase PS. Then the solvent screening was done and results are shown in Table 2.

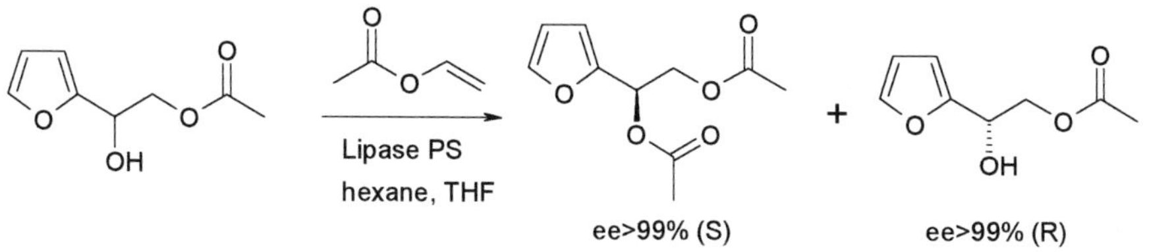

Scheme 2. Resolution of 2-(2-furyl)-2-hydroxyethyl acetate.

The best enantioselectivity value of 74, reached in a hexane – THF mixture, did not allow the obtention of both pure enantiomers in a single transformation. Therefore, the first transesterification was continued to ca. 55% conversion yielding pure (R)-substrate, and (S)-enriched diacetate, which was then deacetylated to diol and treated again with vinyl acetate and PS lipase in the same solvent mixture. After the second transesterification pure (S)-diacetate was obtained [11].

Table 2
The influence of solvent on 2-(2-furyl)-2-hydroxyethyl acetate resolution

Solvent	(R)-substrate %ee	(S)-product %ee	E
hexane	82	85	33
isooctane	93	82	35
cyclohexane	90	85	38
benzene	73	85	26
toluene	69	86	27
dichloromethane	42	82	15
chloroform	64	88	30
tetrachloromethane	95	86	53
1,2-dichloroethane	28	83	15
diisopropyl ether	51	88	26
vinyl acetate	40	83	15
carbon tetrachloride:hexane 3:1	85	88	41
hexane : tetrahydrofuran 8:2	**99**	**87**	**74**
hexane : tetrahydrofuran 7:3	65	83	21

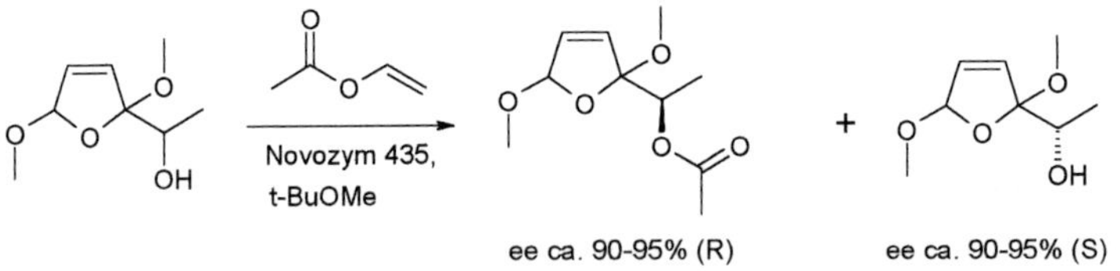

Scheme 3. Resolution of 1-(2,5-dimethoxy-2,5-dihydro-2-furyl)ethanol.

Resolution of 1-(2,5-dimethoxy-2,5-dihydro-2-furyl)ethanol is now under investigation. The most active lipase for this substrate resolution was found to be Novozym 435. A high reaction rate is attained in *tert*-butyl methyl ether. However at the moment we are not able to determine the exact value of enantioselectivity, because in every case there is a mixture of four or more diastereomers, and we cannot yet reach full separation of all of them on our chiral GC columns.

2.2. Resolution of aliphatic alcohols and diols

Concerning aliphatic alcohols and diols we were interested in resolution of racemic 6-methyl-5-hepten-2-ol, 2-ethyl-1,3-hexanediol, and 1,2-pentanediol. The first two compounds are known to possess biological activity - (S)-6-methyl-5-hepten-1-ol is an attractant for *Gnatotrichus* sp., 2-ethyl-1,3-hexanediol is active against mosquitoes. For both aliphatic diols none of the commercial lipases tested was effective. The trials of resolution *via* mono- or diacetylated derivatives resulted in very poor enantiomeric excesses. Moreover, after acetylation of the primary hydroxyl group, the acetylation of secondary hydroxyl groups proceeds with great difficulty. However, it is noteworthy that these substrates can be acetylated with vinyl acetate exclusively at the primary hydroxyl group in the presence of a lipase catalyst. The reaction is very fast at room temperature and 100% selective when using a stoichiometric quantity of diol and acylating agent.

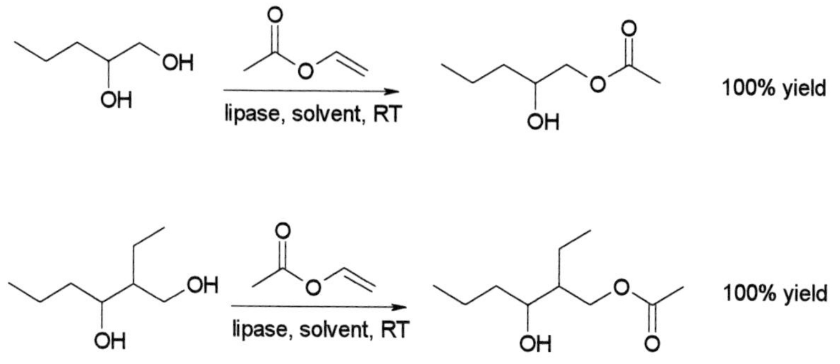

Scheme 4. Regioselective acetylation of acyclic diols catalysed by lipases.

The other compound we succeeded to resolve in a single transesterification step is 6-methyl-5-hepten-2-ol. From lipase screening, Novozym 435 was chosen as the most active catalyst.

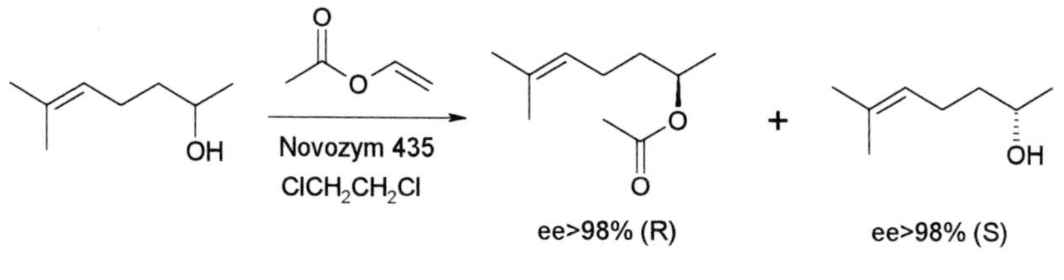

Scheme 5. Resolution of 6-methyl-5-hepten-2-ol.

Trying out several solvents for this substrate – enzyme system allowed to find conditions with an enantioselectivity ratio high enough to get both pure enantiomers in one step (see Table 3). Varying reagents and catalyst ratio in 1,2-dichloroethane, yielded at 50% conversion both enantiomers of high optical purity, i.e. >98%.

Table 3
The influence of solvent on 6-methyl-5-hepten-2-ol resolution.

Solvent	(S)-alcohol %ee	(R)-acetate %ee	E
hexane	85	21	4
cyclohexane	50	27	3
toluene	>99	35	9
tetrahydrofuran	87	90	55
diisopropyl ether	98	17	4
tert-butyl methyl ether	74	39	5
dichloromethane	14	>99	236
tetrachloromethane	>99	64	22
1,2-dichloroethane	**>99**	**92**	**124**

3. CONCLUSIONS

In summary, we found conditions for kinetic resolution of several furan derivatives and 6-methyl-5-hepten-2-ol - chiral precursors in organic synthesis – using commercial lipases as a source of chirality.

REFERENCES

1. A.M. Klibanov, Acc. Chem. Res., 23, (1990), 114.
2. R.J. Kazlauskas, A.N.E. Weissfloch, A.T. Rappaport, L.A. Cuccia, J. Org. Chem. 56, (1991), 2656
3. K. Faber, Biotransformations in Organic Chemistry, 3-rd ed. Springer-Verlag, Berlin, 1997
4. J. Tramper ed. Biocatalysis in Non-Conventional Media, Elsevier Science Publisher B.V., 1992
5. R.J. Kazlauskas, Trends Biotechnol., 12, (1994), 464.
6. G. Carrera, G. Ottolina, S. Riva, Trends Biotechnol., 13, (1995), 63.
7. D.A. MacManus, E.N. Vulfson, Enzyme Microb. Technol., 20, (1997), 225.
8. C.-H. Chen, Y. Fujimoto, G. Girdaukas, C.J. Sih, J. Am. Chem. Soc. 104, (1982), 7294.
9. J. Kamińska, M. Sikora, J. Góra, O. Achmatowicz, B. Szechner, Polish Patent Application, P313607, 1996
10. J. Kamińska, I. Górnicka, M. Sikora, J. Góra, Tetrahedron: Asymmetry, 7, (1996), 907.
11. J. Kamińska, D. Łobodzińska, J. Góra, Polish Patent Application, 1998

Food Biotechnology
S. Bielecki, J. Tramper and J. Polak (Editors)
© 2000 Elsevier Science B.V. All rights reserved.

Activity of immobilised *in situ* intracellular lipases from *Mucor circinelloides* and *Mucor racemosus* in the synthesis of sucrose esters*

T. Antczak, D. Hiler, A. Krystynowicz, M. Szczęsna, S. Bielecki, E. Galas

Institute of Technical Biochemistry, Technical University of Lodz,
Stefanowskiego 4/10, 90-924 Lodz, Poland

The activity of intracellular, immobilised *in situ* lipases from *Mucor circinelloides* and *Mucor racemosus* can be changed by means of chemical modifications of the reaction milieu, using some substances isolated from *Mucor* cells. The substances act ambivalently (as activators or inhibitors) on the lipases. The yield of sucrose monocaprylate synthesis and the time to reach the reaction equilibrium state were determined in mono- and biphasic systems. The investigations proved that in a milieu of di-n-pentyl ether saturated with water, 92% of sucrose was esterificated, and the location of the lipase on the interface between the phases, markedly diminished the time equilibrium to reach.

1. INTRODUCTION

Esters of saccharides and fatty acids belong to a class of natural, "green", nonionic surfactants. They are synthesised from natural components and show prebiotic properties, enabling their application in food production as dietetic fats, emulsifiers and plasticizers [1,2]. Several lipases in organic solvents can catalyse enantio- and regioselective synthesis of saccharides esters [3,4,5,6].

Intracellular lipases from *M. circinelloides* (the former name *M. javanicus*) and *M. racemosus* were not reported so far as catalysts of sucrose-esters synthesis. When the lipases are bound to the mycelium, they display high activity in hydrolysis and synthesis of esters [7,8]. They are very stable in a milieu of organic solvents at 100°C [9].

Known methods of enhancement of enzymatic activity in the organic milieu are based on regulation of pH of an essential water layer or modification of this layer using substances able to form hydrogen bonds [10].

The activity of the intracellular *Mucor* lipases in synthesis of butyl oleate may be ambivalently changed using di- or triethanolamine [9,11].

Our studies were focused on the activity of the intracellular lipases from *M. circinelloides* and *M. racemosus* in the medium containing di-n-pentyl ether and water, or in the medium of di-n-pentyl ether modified with some substances, not described so far, acting dependently on their concentrations in reaction mixtures, as activators or inhibitors of sucrose-esters synthesis.

---

* Grant of the State Comittee for Scientific Research (KBN) nr 6PO4B00311

2. MATERIALS AND METHODS

2.1. Preparations of lipases and ambivalently acting substances

M. circinelloides and *M. racemosus* were cultivated in 14 l Chemap C-2000 fermenters in a medium containing 5.4% (v/v) of corn steep liquor and 2% (v/v) of sunflower oil at 30°C for 48 hrs. The mycelium was separated by filtration, washed twice with water and dried by threefold washing with acetone [12]. After drying, the immobilised *in situ* lipase was ground in a vibrating mill to particles 1.5-5.0 μm in diameter. The residual, after distillation of acetone applied for dehydration of the mycelium, was a source of substances applied for regulation of the *Mucor* lipases. These substances are further named the Ambivalent factors I and II for *M.circinelloides* and *M. racemosus*, respectively. Astaxanthin in a form of mono-, di- or triesters was extracted from shrimp carapaces according to the method described in ref. [13].

2.2. Activity assays

One unit (U) of synthetic activity of lipase was defined as the amount of enzyme that synthesised 1 mole of an ester equivalent per 1 second under the following conditions: 1 mmole of caprylic acid, 1 mmole of sucrose, 0.05 g of lipase preparation and 5 ml of di-n-pentyl ether (saturated with water at 50°C), were incubated for 15 minutes at 50°C on a shaker (220 rev min^{-1}). Free fatty acids were titrated with 0.05 M NaOH up to pH 10.0. The yield of synthesis (mol%) was calculated from the amount of the acid consumed in the synthesis reaction. Reaction conditions different from described above are stated under figures.

2.3. Synthesis of esters in microreactors

Synthesis of esters was carried out in 25 ml hermetically closed microreactors. Di-n-pentyl ether was applied as a solvent. Four different procedures were applied. In the Monophase I procedure, the ester synthesis was executed in the solvent saturated with water (water concentration 4.372 mg/ml). In this procedure, regulation of lipases activity by chemical substances was carried out. The Monophase II procedure was run in the solvent dehydrated with molecular sieve 4Å (water concentration 0.335 mg/ml). The Biphase I procedure was run in the biphasic system di-n-pentyl ether: water. The lipase was placed on the interface between the two phases. The Biphase II procedure was executed in the biphasic system di-n-pentyl ether: water and the lipase were dispersed in the organic phase.

2.4. Other analytical methods

Sucrose and sucrose caprylate were determined using the HPLC Gold Beckman system equipped with refractometer and ODS-IP column. A mixture of methanol, acetonitrile and water (6:1:3). Water concentration in organic solvents was determined according to the Karl Fisher, titration method, using Mettler DL18 apparatus.

3. RESULTS AND DISCUSSION

The highest yield of sucrose caprylate synthesis using the lipases from *M. circinelloides* and *M. racemosus* was obtained in the milieu of di-n-pentyl ether (Table 1). In the milieu of this organic solvent both lipases synthesise monoesters. This reaction does not occur in the medium of aliphatic hydrocarbons C_5-C_{18}, 2-pirolidone, pyridine, DMSO or other solvents applied for sugar-esters synthesis. Because of hydrolysis in water, the reaction in esters like pentyl acetate and butyl acetate occurs only in the presence of substances binding water (Table 1).

The dynamics of sucrose caprylate synthesis in the milieu of di-n-pentyl ether are presented in Fig. 1. In this solvent saturated with water (Monophase I), the time of the attainment of the reaction equilibrium was many times shorter than in the dehydrated solvent (Monophase II).

Table 1
Synthesis of sucrose caprylate in various solvents

Solvent	Activity ($U \times 10^{-6}$)	
	M. circinelloides	*M. racemosus*
di-n-pentyl ether[a]	6.12	5.35
pentyl acetate[b]	4.65	4.02
di-n-butyl ether[a]	2.34	2.25
butyl acetate[b]	2.21	2.47
butyl ethyl ether[a]	1.98	1.45
1,4-dioxane[b]	0.39	0.37
3-pentanol[b]	0.30	0
acetone[b]	0.22	0

[a] saturated with water at 50°C
[b] dehydrated with molecular sieve 4Å

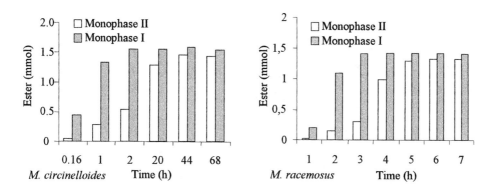

Figure 1. Dynamics of accumulation of sucrose caprylate synthesised according to the methods: Monophase I and II. Reaction conditions: 2 mmol of sucrose, 2 mmol of caprylic acid, di-n-pentyl ether 10 ml, 0.100 g of lipase, temp. 50°C, 240 rpm.

For the *M. circinelloides* lipase it took 2 hours to obtain 78% conversion of the substrates in the Monophase I procedure, and 44 hours for 73% conversion in the Monophase II one. During the synthesis run in the solvent saturated with water, the water synthesised in the reaction, formed the second phase, which was favourable for the process. Sucrose incubation at 50°C (Monophase II) resulted in its partial conversion into colourful substances, which decreased the ester yield in the equilibrium state.

The preferred organic acids for synthesis of sucrose esters with the *Mucor* lipases, contain at least eight carbon atoms in a molecule (Fig. 3). The highest yield was observed in the synthesis of sucrose caprylate, mirystate and stearate, and very low in the case of sucrose propionate synthesis. The degree of conversion of the selected substrates depends mainly on molar ratio of sucrose and an acid, which was demonstrated for caprylic acid (Fig. 4).

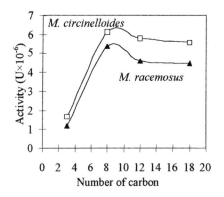 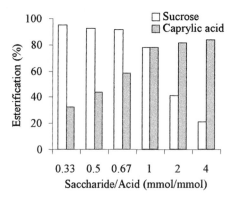

Figure 3. Influence of organic acid chain length on the lipase activity in the synthesis sucrose esters. Monophase I system.

Figure 4. Influence of molar ratio of caprylic acid and sucrose on the yield of esterification. System Monophase I. Reaction time 20 h.

For example, during the synthesis done according to the Monophase I procedure, and the substrates molar ratio equal to 1:1, about 78% of them combined and produced the ester. The optimal molar ratio of sucrose and the acid was found to be 1:1.5 and the corresponding degree of the sucrose conversion was 92%.

Substances such as di- and triethanolamine, pyridine or DMF (dimethyl formamide), which activated the *Mucor* lipases during the synthesis of butyl oleate or 16-hydroxyhexadecanoic acid lactone [8,9,11] had no effect on the enzymes during sucrose-esters synthesis.

Recently, we have found other substances activating the *Mucor* lipases in the latter reaction (Table 2). These substances displayed features of detergents and were as follows: cetylpyridinium bromide, astaxanthin (a mixture of mono-, di- and triesters) and some native compounds isolated from *Mucor* cells, able to activate or inhibit the *Mucor* lipases and called the ambivalent factors.

The substances acted ambivalently and their influence depended on their concentration. For example, the addition of 0.8 mg of the ambivalent factor I enhanced 2.4 times the

M. circinelloides activity. The doses above 60 mg inhibited the enzyme; 120 mg of this factor halved the lipase activity. Twenty mg of the ambivalent factor II increased the *M. racemosus* lipase activity by a factor 2.3 but doses exceeding 100 mg inhibited the enzyme (Fig. 5).

Synthesis of sucrose esters was also carried out in biphasic di-n-pentyl ether: water systems, in two variants different from each other in the lipase location. In the system Biphase I, the enzyme was situated at the interface between the two phases, and in the system Biphase II, this hydrolase was placed in the organic phase. In both cases, the maximal reaction yield was observed between the limiting values of the phase volume ratios A ($A = V_{organic}/V_{water}$) (Fig. 6 and 7).

Table 2
Influence of selected substances on the synthetic activity of the *Mucor* lipases

Substance	Amount ($mmol^a$, mg^b)	Activity ($U \times 10^{-6}$)	
		M. circinelloides	*M. racemosus*
None	0	6.10	5.35
Cetylpyridynium bromide	0.042^a	7.65	7.87
Astaxanthin	14^b	11.29	---
	80^b	---	10.28
Ambiwalent factor I from *M. circinelloides*	0.8^b	14.89	---
Ambiwalent factor II from *M. racemosus*	20.0^b	---	12.45

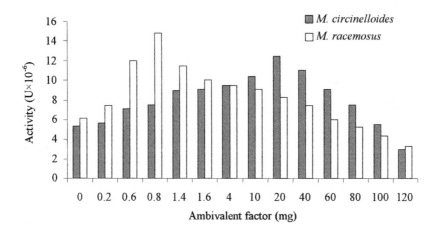

Figure 5. Influence of native substances (Ambivalent factors) accumulated in *Mucor* cells on the synthetic activity of the lipases during the sucrose-monocaprylate synthesis, executed according to the system Monophase I procedure. Reaction time 20 h.

The reaction velocity in the biphasic system is significantly affected by the values of substrates and products partition coefficients (P) between the organic and water phase. In the case of sucrose-caprylate synthesis, the value of the partition coefficient $P_{Sucrose} \ll 1$ is unfavourable, because it corresponds to low sucrose concentration in the organic phase. A favourable effect of the partition coefficients for ester with $P_{Ester} \ll 1$ and the acid with $P_{Acid} \gg 1$ was found.

When the reaction was executed according to the Biphase I procedure, the equilibrium state corresponding to 70% yield was attained within 5 minutes. In this system, the lipase was situated at the interface between the two phases assuring a high concentration of both substrates. In the system Biphase II, the reaction achieved the equilibrium state after 180 minutes and only 70% of each substrate interacted. The factor limiting the synthesis velocity was the low sucrose concentration in the organic phase.

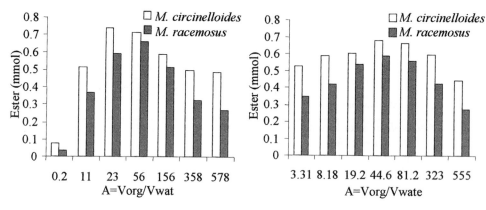

Figure 6. Influence of phase volume coefficient (A) on yield of synthesis of sucrose caprylate by lipase in biphasic system, according to the Biphase I procedure.

Figure 7. Iinfluence of phase volume coefficient (A) on yield of synthesis of sucrose caprylate by lipase in biphasic system, according to the Biphase II procedure.

Sucrose-esters synthesis yield and the time of reaching the reaction equilibrium state may be regulated by means of changes of physicochemical factors of reaction mixtures [14]. The enzyme location in biphasic systems, or water concentration in monophasic systems are the factors affecting the *Mucor* lipases activity. The lipase activity in sucrose-ester synthesis may be regulated using selected substances, such as cetylpyridine bromide or astaxanthin, activating the lipases in non-water milieu. The activity of *Mucor* lipases may be also regulated with native substances isolated from *Mucor* cells, which act ambivalently.

REFERENCES

1. O.P. Ward, J. Fang, Z. Li, Enzyme. Microb. Technl. 20 (1997) 52.
2. A.M. Klibanov, Biotech. Bioeng., 31 (1988) 208.

3. Y.M. Sin, K.W. Cho, T.H. Lee, Biotech. Lett. 20 (1998) 91.
4. F.J. Plou, M.A. Cruces, M. Bernabe, M. Martin-Lomas, J.L. Parra, A. Ballesteros, Annals of the New York Academy of Sciences, 750 (1995) 332.
5. J.A. Arcos, M. Bernabe, C. Otero, Biotech. Bioeng., 57 (1998) 505.
6. G. Liunger, P. Adlercreutz, B. Mattiasson, Biotechnol. Lett., 16 (1994) 1167.
7. T. Antczak, A. Krystynowicz, E. Galas, Biotechnologia PI, 2 (1995) 82.
8. U. Antczak, J. Góra, T. Antczak, E. Galas, Enzyme. Microb. Technol. 13 (1991) 589.
9. T. Antczak, J. Morowiec-Białoń, S. Bielecki, A.B. Jarzębski, J.J Malinowski, A.I. Lachowski, E. Galas, Biotechn. Techn., 11 (1997) 9.
10. A. Zaks, A.J. Russell, J. Biotechnol., 8 (1988) 259.
11. T. Antczak, D Hiler, E. Galas, Biotechnologia PI, 1 (1993) 59.
12. T. Antczak, A. Krystynowicz, E. Galas, Polish patent No.150601 (1991).
13. K.L. Simpson, T. Katayama, C.O. Chichester (eds.), In carotenoid as colorants and vitamin accelerators, Accademic Press Inc., 1982.
14. T. Antczak, A. Krystynowicz, D. Hiler, S. Bielecki, E. Galas, Biotechnologia PI, 1 (1999) 153.

Food Biotechnology
S. Bielecki, J. Tramper and J. Polak (Editors)
© 2000 Elsevier Science B.V. All rights reserved.

229

Properties and yield of synthesis of mannosylerythritol lipids by *Candida antarctica*

M. Adamczak, W. Bednarski

Institute of Food Biotechnology, Olsztyn University of Agriculture and Technology, Heweliusza 1, 10-718 Olsztyn, Poland

1. INTRODUCTION

Biosurfactants are amphipathic molecules that are capable of reducing surface and interfacial tensions, and of forming different types of water in oil and oil in water microemulsions [1]. The worldwide market for surfactants is expected to increase at the rate of 35% towards the end of the century. Almost all currently used surfactants are produced by chemical processes from petroleum. However, the interest in utilization of microbial surfactants has been increasing in recent years due to diversity, biodegradability and biocompatibility of surfactants as well as their application for oil recovery enhancement, pollution removal etc. [2]. Biosurfactants may play an important role as additives in production of sauces, mayonnaises or margarines.

Biosurfactants are produced by variety of microorganisms, which are able to grow mainly on water-immiscible substrates. They are either extracellular or cell-bound [1]. Glycolipids, i.e. rhamnolipids, trehalolipids and sophorolipids are the best known group of microbial biosurfactants [3].

There are just a few reports about the synthesis of mannosylerythritol lipids (MELs) [4,5]. Recently, Kitamoto et al. [6,7] had discussed the conditions for submerged shake method of MELs synthesis using *C. antarctica*. Also, some of the MELs properties were described.

The cultivation of microorganisms by submerged method in fermentor is the most appropriate method for synthesis of bioproducts on commercial scale. Medium composition, aeration requirements and foaming increase due to biosurfactants release to the medium should be considered when selecting the fermentation conditions for *C. antarctica* in fermentor.

The objective of study was to determine the growth conditions for *C. antarctica* in the fermentor as well as their effect on the yield of biosynthesis and properties of MELs.

2. METHODS

2.1. Microorganism

Candida antarctica ATCC 20509 from American Type Culture Collection, USA, was used in the experiments. The strain was grown on wort agar slants at 30°C for 4 days and was used for studies either immediately or after storing at 4°C for 1 month.

230

2.2. Medium and inoculum preparation

The yeast inoculum was prepared by suspending yeast colonies from agar slant in 10 ml of physiologic saline solution followed by subsequent transfer into 500 ml Erlenmeyer flask containing 100 ml of the seed medium. The seed medium used for inoculum preparation had the following composition (per litre): glucose (4 g), $NaNO_3$ (2 g), $MgSO_4 \times 7H_2O$ (0.2 g), KH_2PO_4 (0.2 g) and yeast extract(1 g). The whole inoculum was transferred to basal fermentation medium in the fermentor after 48 h cultivation in Incubator Shaker series 25 (New Brunswick, USA) at 30°C/200 rpm.

2.3. Yeast cultivation conditions

C. antarctica was grown at 30°C for 144 h in BioFlo III fermentor (New Brunswick, USA) containing 2 l of the medium of the following composition (per litre): carbon source (30 g), $NaNO_3$ (2 g), $MgSO_4 \times 7H_2O$ (0.2 g), and yeast extract (1 g).

The yield of glycolipids synthesis by yeast was determined in relation to:

A/ oxygen saturation; for all cultures oxygen level in the medium was maintained at 50% saturation, which gave the highest yield of biosurfactants (data unpublished). The oxygen concentration in the medium during cultivation was controlled by so-called active control, which depending on the oxygen content, changed the revolutions of a mixer from 100 to 500 per minute at a constant air flow rate 1 or 2 vvm;

B/ carbon source and feeding procedure; *C. antarctica* was cultivated for 48h in medium with initial glucose content 40 g/l followed by feeding the medium with 80 ml of soybean oil after 48 h and 96 h of cultivation, each time;

C/ type of carbon source, i.e. soybean oil, poultry fat, glucose or lactose added in the amount of 8% v/v or w/v.

During cultivation of *C. antarctica*, samples taken every 48 h were assayed for glycolipids and biomass content as well as lipolytic activity [4,5]. Lipolytic activity was measured at 37°C on 5% w/v olive oil emulsion (pH 8,0) stabilized by addition 2% w/v arabic gum. pH-stat titration method was used. Twenty ml of the emulsified solution, 470 ml of $CaCl_2$ solution (22% w/v), and culture supernatant were mixed, and liberated fatty acids were titrated automatically with 0,01 M NaOH to maintain the pH constant at pH 8.0. One unit of lipases activity was defined as the amount of enzymes that liberates 1 μmol fatty acid/min under assay conditions. The changes in surface tension of post-cultivation liquid were also determined using a stalagmometer.

The biosurfactants were characterized by determining the micelle critical concentration lipid elutibility from sand, and the effect of lipids on the growth of microorganisms. The emulsification yield using the Swift`s method [1] were also determined.

3. RESULTS AND DISCUSSION

3.1 Synthesis of biosurfactants
Submerged cultivation of yeast in a fermentor requires, among others, identification of proper aeration conditions. During the synthesis of surface active substances, oxygenation is of great importance because it influences the growth conditions of microorganisms hence yield of surfactant synthesis. Moreover, intensive aeration of cultivation medium, which has a low surface tension, causes an intensive foaming of the medium. It is important problem in the process.
Based on our studies, we found that the rate of air flow through the medium was a critical variable that influenced the intensity of foaming of the medium and yield of biosurfactants synthesis (Fig. 1A,B). However, the same intensity of medium oxygenation (50%) could be achieved, by maintaining the air flow rate at the level 1 or 2 vvm, using the so-called active regulation of stirrer revolutions. It was found that the greatest yields of glycolipids synthesis could be obtained at the air flow rate of 1 vvm independent from carbon source, although, at the air flow rate 2 vvm, relatively slightly greater yield of biosurfactants was obtained in the medium with glucose (Fig. 1A,B).

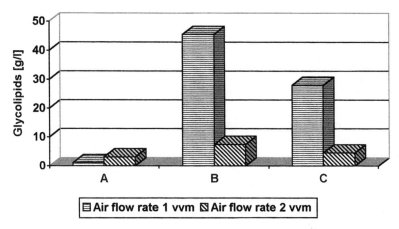

Figure 1. Yield of glycolipids synthesis by *C. antarctica* depending on the intensity of air flow through the medium. Cultivation time 144h. Medium with glucose (A), soybean oil (B), two-stage cultivation with glucose and soybean oil (C).

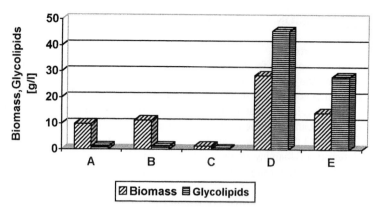

Figure 2. Biomass increase and yield of glycolipids synthesis by *C. antarctica* depending on carbon source in medium. Culturing time 144 h, aeration 50%, air flow rate 1 vvm. Medium with glucose (A), lactose (B), poultry fat (C), soybean oil (D), two-stage cultivation (E).

An increase in the air flow rate to 2 vvm intensified the foam formation of which breaking was impossible to achieve the usage of the available methods. After 24 – 48h of cultivation the foaming was already so intensive that the foam poured out of the fermentor. After foaming out, there was not any more foam in the fermentor. The reason for this could be that the surface active substances and substrate for their synthesis were removed from the cultivation medium together with the foam or too intensive aeration could have a detrimental effect on yeast. The foam that poured out of the fermentor was aseptically collected and assayed for the presence of surface active substances and lipids. The foam was proved to contain lipids (carbon source from the medium) and biosurfactants. The content of surface active substances in the foam was proportional to their content in the cultivation liquid, varying from 0.87 g/l (medium with glucose) to 4.23 g/l (medium with soybean oil).

Further experiments aimed at lowering the medium foaming and increasing the biomass concentration were involved. In the two-stage cultivation of *C. antarctica*, yeast were initially grown in the medium with 4% w/v glucose and after 48 h and 96 h, 80 ml soybean oil was added each time. Soybean oil was the carbon source, the substrate for biosurfactants synthesis and the foam breaking agent. Under such conditions, it was possible to prevent foaming out but the yield of glycolipids synthesis was low. In the two stage cultivation similarly to that in one stage cultivation, the highest yield of biosurfactants synthesis was obtained at air flow rate 1vvm (27.97 g/l) not at 2vvm (4.49 g/l) (Fig. 1C).

Cultivation of *C.antarctica* was performed at air flow rate 1 vvm in the next experiments. The final yield of glycolipids synthesis by *C. antarctica* in the medium with different carbon sources ranged from 0.21 to 45.49 g/l after its cultivation in the fermentor for 144 h. The yield of yeast biomass and glycolipids synthesis was dependent on the carbon source in the medium (Fig. 2). The highest yield of glycolipids was obtained after cultivation of *C. antarctica* in the medium with soybean oil. The obtained yield (45.49 g/l) was comparable with that reported by Kitamoto et al. [6], who achieved about 40 g of glycolipids per 1 litre of medium, growing *C. antarctica* in the medium with soybean oil by submerged shake method. It has to be pointed out that glycolipids were synthesised in the media with glucose and lactose, too (Fig. 2). Moreover, it was found out that there was a correlation (r=0.89) between the biomass growth and the amount of synthesised biosurfactants. The content of glycolipids in the medium increased up to 144 h of yeast cultivation, while the greatest productivity (0.36 g×l/h) occurred between 24 and 72 h of cultivation (Fig. 3).

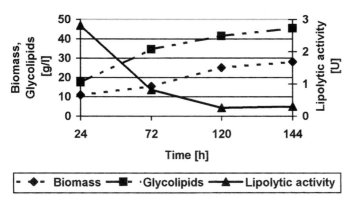

Figure 3. Effect of cultivation time on yeast growth, activity of extracellular lipases and yield of glycolipids synthesis by *C. antarctica*. Medium with 8% v/v soybean oil, medium level of oxygen 50% at 1 vvm air flow rate.

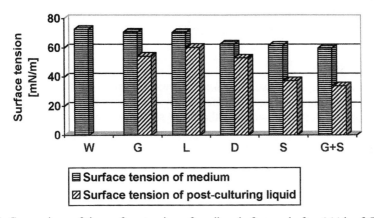

Figure 4. Comparison of the surface tension of medium before and after 144 h of *C. antarctica* cultivation in different media. W-water, medium with glucose (G), lactose (L), poultry fat (D), soybean oil (S), two-stage cultivation with glucose and soybean oil (G+S)

It was shown that the activity of extracellular lipases in the medium with 8% w/w soybean oil decreased from 2.81 μmol FFA/min×mg protein after 24 h to 0.30 μmol FFA/min×mg protein after 144 h of cultivation (Fig. 3). This relationship is interesting because it indicates that an increase in yield of glycolipids synthesis is associated with a decrease in extracellular lipases activity. This suggests negligible roles of lipases in the bioconversion of medium lipids and

234

synthesis of biosurfactants. It can be also assumed that the presence of glycolipids in the medium affects the metabolism of yeast, decreasing their predisposition to synthesis of extracellular lipases, or that glycolipids inhibit the lipases activity.

From the comparison of the surface tension of the medium before and after *C. antarctica* cultivation, it can be concluded that the surface tension of the medium after cultivation was dependent on the initial medium composition and glycolipids content. As compared to the surface tension of water (72.25 mN/m at 20.15°C), the surface tension of the post culturing medium was 33.8 mN/m, about 53% of that of water, during the two-stage cultivation in the medium with glucose and soybean oil. It is difficult to interpret the changes in surface tension of the cultivation medium, which should be associated with the glycolipids synthesis and the intake of those medium components that are responsible for an increase of surface tension by yeast. Comparing the changes in the surface tension of the media (Fig. 4) and yields of glycolipids synthesis (Fig. 2) it seems, however, that the biggest decrease in surface tension was found in the media in which yeast effectively synthesized glycolipids (Fig. 2).

3.2 Properties of biosurfactants

After 96-120 h culturing of *C. antarctica,* the culturing liquid with soybean oil or poultry fat turned to homogenous emulsion, allowing to conclude the good properties of synthesised lipids. The emulsification capability (FO) of examined biosurfactants was 79% and was higher by 50% than the emulsification capability of 1% lecithin. The critical micelle concentration of biosurfactants was achieved at its concentration of 0.6% and the value of the surface tension obtained was 34.96 mN/m. The examined glycolipids inhibited the growth of *E. coli* EC 08 and *Staphylococcus aureus* 14. The solution (0.6% w/v) of biosurfactants eluted 29.4% w/w lipids adsorbed on sand. The properties of the examined biosurfactants inform about the possibility of its utilization and low critical concentration of micelle indicates the economic reason of its use.

4. CONCLUSIONS

The yield of glycolipids synthesis by *C. antarctica* was influenced by the medium composition, aeration degree and cultivation time in the fermenter. The glycolipids present in the cultivation medium decreased its surface tension, and the lowest surface tension was 53% of that of water.

It was found that glycolipids inhibited growth of *E. coli* and/or *S. aureus* and effectively recovered lipids absorbed on sand.

REFERENCES

1. J.D. Desai, I.M. Banat, Microbiol.Mol.Biol.Rev., 61 (1997) 47.
2. I.M. Banat, Acta Biotechnol., 3 (1995) 251.
3. S.-C. Lin, J.Chem.Tech.Biotechnol., 66 (1996) 109.
4. D.Kitamoto, S. Akiba, C. Hioki, T. Tabuchi, Agric.Biol.Chem., 54 (1990a) 31.
5. D. Kitamoto, K. Haneishi, T. Nakahara, T. Tabuchi, Agric.Biol.Chem., 54 (1990b) 37.
6. D. Kitamoto, T. Fuzishiro, H. Yanagishita, T. Nakane, T. Nakahara, Biotechnol.Lett., 14 (1992) 305.
7. D. Kitamoto, H. Yanagishita, T. Shinbo, T. Nakane, C. Kamisawa, T. Nakahara, J.Biotechnol., 29 (1993) 91.

Food Biotechnology
S. Bielecki, J. Tramper and J. Polak (Editors)
© 2000 Elsevier Science B.V. All rights reserved.

The biosynthesis of *Bacillus licheniformis* α-amylase in solid state fermentation

A. Jakubowski, E. Kwapisz, J. Polak, E. Galas

Institute of Technical Biochemistry, Technical University of Łódź,
Stefanowskiego 4/10, 90-924 Łódź, Poland

The optimum conditions of the SSF process for biosynthesis of α-amylase by bacterial strain *B. licheniformis* 83 have been established. Fermentations were carried out in conical flasks and in a Swing Solid State Reactor (SSSR) manufactured by TecBio (Germany). Wheat bran and bean pods as a solid matter appeared to cause a substantial increase of α-amylase activity. The highest yield of enzyme biosynthesis was at 40% of water content and the activity of α-amylase increased by 80% compared to a medium moisture content in the range of 60-65%. The aeration conditions have been optimised by cultures in different volumes of the medium. The time course of enzyme biosynthesis observed for the SSF method was compared to the results obtained in submerged cultures. The initial characteristics of enzymes obtained by both the submerged culture and the SSF process were determined. Although both enzyme preparations expressed maximum activity at the same pH (6.0) and temperature (70°C), their response to extreme pH and temperature was different.

1. INTRODUCTION

Bacterial α-amylase is one of the most important enzymes in the food industry. Starch liquefaction under the influence of thermostable enzymes is an expensive stage in saccharification of this raw material and the cost of enzyme is a crucial factor. The enzyme produced by *B. licheniformis* is particularly suited for this process because of its high thermostability and low Ca^{2+} demand [1].

Submerged cultures are generally used for industrial production of bacterial enzymes because the classic fermenters are equipped with efficient agitation and aeration systems. However, there are some investigations showing even more cost effective methods of enzyme production. Solid state fermentation is more and more likely to become a method of choice among the cost effective processes and is used already in industrial practice for cultivation of filamentous fungi [2,3].

The technology of α-amylase production by the bacterial strain *B. licheniformis* 83 in submerged periodic and continuous cultures has been elaborated in the authors' institute [4]. The same strain was tested in an SSF process, although a special screening for microorganisms aiming to be cultured cultured by this method is suggested [5].

The aim of the presented work was to find optimum conditions of the SSF process for biosynthesis of α-amylase by the same bacteria. Special attention was focused on the solid medium, its structure and moisture content and also on mixing and aeration of the medium. Fermentations were carried out in spherical flasks and in a Swing Solid State Reactor (SSSR) manufactured by TecBio (Germany).

2. MATERIALS & METHODS

2.1. Organism
The strain *B. licheniformis* 83 is a mutant with α-amylase overproduction resulting from several mutagenisations by UV and nitrosoguanidine followed by selections on media with cycloserine, ampicillin and tunicamycin.

2.2. Media
The following three compositions of SSF media were tested in this work:

K   wheat bran – 100 g, soy bran – 22 g, lactose – 2.6 g, corn steep liquor – 0.35 g, Na_2HPO_4 - 0.02 g, $NaH_2PO_4 \cdot H_2O$ – 1.1 g, KCl – 0.53 g, $MgSO_4 \cdot 7H_2O$ – 0.02 g

WB   wheat bran – 100 g, $Na_2HPO_4 \cdot 2H_2O$ - 0.88 g, $NaH_2PO_4 \cdot 2H_2O$ – 0.49 g, $(NH_4)_2HPO_4$ - 3.3 g

WBbp   wheat bran – 100 g, bean pods – 18 g, $Na_2HPO_4 \cdot 2H_2O$ - 0.88 g, $NaH_2PO_4 \cdot 2H_2O$ - 0.49 g, $(NH_4)_2HPO_4$ – 3.3 g

The content of water was optimised in the course of experiments.
The medium W for submerged fermentation consisted of (g/l): lactose – 26.4, peptone - 12.5, corn steep liquor – 3.5, $CaCl_2 \cdot 6H_2O$ – 12.5, NaCl – 2.0, $(NH_4)_2HPO_4$ – 3.2.

2.3. Inoculum
The SSF media were inoculated with 5% (wt/wt) of bacterial suspension obtained by overnight submerged culture of *B. licheniformis* 83 in medium W at 39°C.

2.4. Culture conditions
The SSF cultures were carried out in 500 ml spherical flasks and in a Swing Solid State Reactor (SSSR) manufactured by TecBio (Germany). Flasks with different load of medium were incubated at 39°C with mixing frequency every 1, 2 or 4 hours for 48 hours.

The SSSR system consisted of specifically mixed vessel (total volume 2 l), an air conditioner, a compressor and a PC for parameter control and data collection. The following conditions were applied: medium load – 400 g, mixing frequency – 3 min., aeration – 80 l/h, temperature - 39°C, process duration – 48 h.

The submerged cultures were carried out in 500 ml spherical flasks containing 50 ml of medium W on the orbital shaker at 240 rpm and amplitude 4.5 cm at 39°C for 48 h.

2.5. Assays
Solid matter samples of 4 g were collected from SSF cultures, suspended in 10 ml of tap water and incubated on a rotary shaker at 39°C for 90 min. The α-amylase activity was assayed in the supernatant by the method of Fisher & Stein [6].

3. RESULTS & DISCUSSION

The important parameters of SSF experiments have been established in earlier experiments. In the course of the current work the composition of medium **K** was modified and to some extent also simplified. In **WB** medium soy bran, corn steep liquor and lactose were eliminated and $(NH_4)_2HPO_4$ was introduced as nitrogen supplement. To improve the medium structure dry bean pods were added in **WBbp** medium and the water content was reduced from 65% to 50%. These modifications appeared to cause a substantial increase of α-amylase activity (Fig. 1).

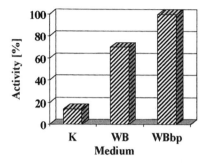

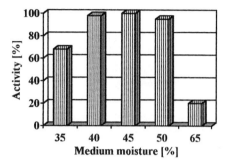

Figure 1. The effect of medium composition on α-amylase activity in solid state fermentation

Figure 2. The effect of medium moisture on α-amylase activity in solid state fermentation

Because medium moisture appeared to be an important parameter of this process, subsequent experiments were performed to further optimise it. The water content in the SSF medium was in the range from 35% to 65%. The highest yield of α-amylase was obtained when medium moisture was varied in the range 40-50% (Fig. 2). The enzyme biosynthesis was increasing with water content down to 40% and the activity of α-amylase was higher by 80% compared to earlier established medium moisture content in the range 60-65% [7] and similar effect of medium moisture was observed in spherical flasks and in the SSSR, but the production efficiency was better by 11% in the reactor. A suitable structure of the medium appeared to be even more important, to certain extent, than the water content. Apparently a dispersed structure of the medium increased the surface available to aeration.

The importance of efficient aeration was confirmed by cultures in different volumes of the medium loaded into spherical flasks and in the SSSR. The efficiency of biosynthesis increased by over 50% when a quarter of the original medium load was applied (Fig. 3).

The frequency of mixing was also a crucial factor responsible for a sufficient aeration. This parameter was optimised for cultures performed in spherical flasks. They were vigorously shaken with three frequencies: every 1 h, 2 h or 4 h. Unexpectedly, higher frequencies resulted in lower yield of the enzyme production (Fig. 4). Apparently, mixing frequency effected the structure of the medium and higher frequencies caused conglutination of solid particles and decreased the biosynthesis efficiency.

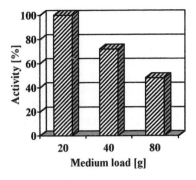

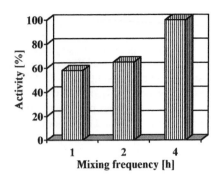

Figure 3. The effect of medium load in 500 ml spherical flasks on α-amylase activity in solid state fermentation

Figure 4. The effect of mixing frequency of the medium in spherical flasks on α-amylase activity in solid state fermentation

The time course of enzyme biosynthesis in solid state fermentation was compared to the results obtained in submerged cultures of *B. licheniformis* 83 (Fig. 5).

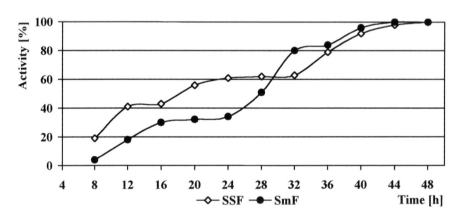

Figure 5. The time course of α-amylase activity during solid state fermentation (SSF) and submerged fermentation (SmF) performed in spherical flasks

In the case of the submerged process two apparent stages of intensive enzyme biosynthesis can be seen; first from 8th to 16th hour and second between 24 and 32 h of culture. A more even profile of α-amylase activity was observed during solid state fermentation. In this process a first plateau in the activity curve is after 20 h and lasts for the next twelve hours. Subsequently, during the last sixteen hours, the enzyme activity is elevated by 40% compared to only 20% in the submerged method. Therefore the biosynthesis of α-amylase in the SmF process seems to be faster than in the SSF method.

Ramesh & Losane [8, 9] reported about different features of enzymes produced on solid medium. Therefore the initial characteristics of enzymes obtained by submerged culture and SSF process were also determined.

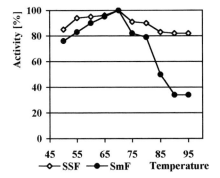

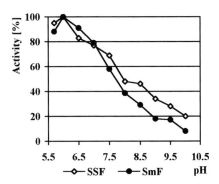

Figure 6. The effect of temperature on the activity of α-amylase obtained by solid state fermentation (SSF) and by submerged fermentation (SmF).

Figure 7. The effect of pH on the activity of α-amylase obtained by solid state fermentation (SSF) and by submerged fermentation (SmF).

Although both enzyme preparations expressed maximum activity at the same temperature (70°C) and pH (6.0), their response to extreme temperature and pH values was different (Fig. 6 and 7). The SSF enzyme expressed far better resistance to higher temperatures.

REFERENCES

1. N. Saito, Arch. Biophys., 155 (1973) 280.
2. B.K. Losane, M.P. Ghildyal, V.S. Murthy, Symposium on Fermented Food, Food Contaminants, Biofertilisers and Bioenergy, India, Mysore, 1982.
3. B.K. Losane, M.P. Ghildyal, S. Budiatman, S.V. Ramakrishna, Enzyme Microbiol. Technol., 78 (1985) 258.
4. E. Galas, A. Jakubowski, E. Kwapisz, The method of production of thermostable bacterial α-amylase, Polish Patent No. RP 114551 (1992).
5. V.S. Shankaranand, V.S. Ramesh, B.K. Losane, Process. Biochem., 27 (1992) 33.
6. E.H. Fischer, E.A. Stein, Biochem. Prep., 8 (1961) 27.
7. M.V. Ramesh, B.K. Losane, Appl. Microbiol. Biotechnol., 33 (1990) 501.
8. M.V. Ramesh, B.K. Losane, Biotechnol. Lett., 11 (1989) 49.
9. M.V. Ramesh, B.K. Losane, Starch 42 (1990) 49.

Food Biotechnology
S. Bielecki, J. Tramper and J. Polak (Editors)
© 2000 Elsevier Science B.V. All rights reserved.

The effect of nitrogen concentration in the fermentation broth on citric acid production by *Aspergillus niger*

J. Pietkiewicz and M. Janczar

Food Biotechnology Department, University of Economics,
Komandorska 118/120, 53-345 Wroclaw, Poland

The effect of initial ammonium nitrate concentration in the fermentation broth and addition of ammonium nitrate to the fermentation broth on the citric acid synthesis carried out by means of *Aspergillus niger* C-12-I/94 strain were examined.

1. INTRODUCTION

Citric acid is the most common organic acid. It is increasingly being used both in food and other industries. At present it is obtained almost exclusively by a microbiological process involving the fermentation of carbohydrates using various microorganisms, most often the *Aspergillus niger* strain [1,2].

In citric acid production by the *Aspergillus niger* strain, the nitrogen concentration is most important in the fermentation broth. The initial high nitrogen concentration causes an intensive growth of fungal biomass. An overgrowth of biomass and large consumption of carbon sources by mycelium inhibits citric acid biosynthesis and reduces fermentation yield and productivity. Intensive citric acid biosynthesis begins after almost the complete consumption of nitrogen from the fermentation broth [3,4,5]. The early studies have shown that ammonium nitrate is the best source of nitrogen in the citric acid biosynthesis. The highest productivity and yield were obtained by means of the *Aspergillus niger* B-64-5 strain in fermentations on white sugar medium with the addition of about 1.5–2.5 g/l (optimum 2.0 g/l) of NH_4NO_3 at the beginning of the fermentation [6,7].

The aim of presented work was to show the effect of initial ammonium nitrate concentration in the fermentation broth and the addition of ammonium nitrate to the fermentation broth on the citric acid synthesis by the *Aspergillus niger* C-12-I/94 strain.

2. EXPERIMENTAL

The fermentation broth contained the following substances (g/l): white sugar – 150.0; NH_4NO_3 – 2.0; $MgSO_4 \cdot 7H_2O$ – 0.2; KH_2PO_4 – 0.2; $FeSO_4 \cdot 7H_2O$ – 0.05 [6]. Fermentations were carried out on a reciprocal shaker and in the lab bioreactor BIOMER. The bioreactor was a standard CSTR with working volume of 5 l [8].

242

In fermentations carried out on a shaker the flat-bottomed flasks were used. The capacity of each flask was 750 ml and each contained 150 ml of the medium. Fermentations were carried out at 30°C and the fermentation was terminated when the acidity did not increase further. Every day, beginning from the fifth day of the fermentation, samples were taken for mycelium morphology and total acidity determination.

In fermentations carried out in lab bioreactors at 32°C, samples were taken each day and contents of biomass, sugars, ammonium ions and citric acid were determined. The fermentation was terminated when no increase of the acidity was observed. After the fermentation, the final volume of the broth was measured and the degree of evaporation calculated. This was taken into account in the final estimation of fermentation productivity and yield [4].

3. RESULTS

In the first stage of the study carried out on a shaker, the influence of initial ammonium nitrate concentrations in the fermentation broth in the range from 1.0 to 4.0 g/l was examined. Moreover, the influence of additional, single or multiple ammonium nitrate medium feed on the citric acid fermentation and its yield was investigated. The obtained results are presented in Tables 1 and 2.

Table 1
Effect of NH_4NO_3 concentration on morphology of *Aspergillus niger* C-12-I/94 strain mycelium during citric acid fermentation

Initial NH_4NO_3 concentration in medium	Additional ammonium nitrate feed		Total NH_4NO_3 in ferment. broth	Colour of ferment. broth at the end of the fermentation	Mycelium morphology at the end of the fermentation	Final biomass concentration
	day	amount				
(g/l)		(g/l)	(g/l)			(g/l)
1.0	-	-	1.0	milky-green	short filamentous	12.28
1.5	-	-	1.5	milky-green	short filamentous	18.20
2.0	-	-	2.0	milky-green	short filamentous	30.36
2.5	-	-	2.5	milky-green	short filamentous	22.24
3.0	-	-	3.0	milky-green	short filamentous	29.00
4.0	-	-	4.0	milky-green	short filamentous	22.60
2.0	4	0.5	2.5	milky-green	short filamentous	28.16
2.0	5	0.5	2.5	milky-green	short filamentous	23.92
2.0	6	0.5	2.5	milky-green	short filamentous	19.72
2.0	7	0.5	2.5	milky-green	short filamentous	18.80
2.0	8	0.5	2.5	milky-green	short filamentous	21.92
2.0	9	0.5	2.5	milky-green	short filamentous	21.80
2.0	4-13	0.5	7.0	milky-green	short filamentous	25.56
2.0	4-13	1.0	12.0	milky-green	short filamentous	40.00

Table 2
Effect of NH_4NO_3 concentrations in the fermentation broth on citric acid fermentation by the *Aspergillus niger* C-12-I/94 strain

Initial NH_4NO_3 concentration in medium	Additional ammonium nitrate feed		Total NH_4NO_3 in fermentation broth	Final sugar concent. in ferm. broth	Total acidity	Final monohydrate citric acid concentration	Citric acid yield in relation to introduced sugar
	day	amount					
(g/l)	(g/l)		(g/l)	(g/l)		(g/l)	(%)
1.0	-	-	1.0	6.5	31.85	111.5	74.6
1.5	-	-	1.5	7.9	25.00	87.5	58.5
2.0	-	-	2.0	15.7	20.80	72.8	47.8
2.5	-	-	2.5	8.8	22.00	77.0	51.5
3.0	-	-	3.0	3.2	21.25	74.4	49.8
4.0	-	-	4.0	1.7	18.75	65.6	43.9
2.0	4	0.5	2.5	14.8	19.90	69.7	46.6
2.0	5	0.5	2.5	15.8	18.35	64.2	42.9
2.0	6	0.5	2.5	12.6	21.95	76.8	51.4
2.0	7	0.5	2.5	14.4	21.85	76.5	51.2
2.0	8	0.5	2.5	15.1	21.35	74.7	50.0
2.0	9	0.5	2.5	12.7	20.75	72.6	48.6
2.0	4-13	0.5	7.0	10.4	14.60	51.1	34.2
2.0	4-13	1.0	12.0	11.2	10.90	38.2	25.6

The initial increase of ammonium nitrate content in the fermentation broth from 1.0 to 4.0 g/l, and later additional ammonium nitrate feed have no influence on the change of mycelium morphology (short filamentous) and its colour (milky-green). However, the amount of nitrogen significantly influences the level of biomass in the fermentation broth (Table 1). The higher amount of introduced nitrogen into fermentation broth resulted in the higher final concentration of biomass. The increase of initial ammonium nitrate content in the fermentation broth caused the lower final concentration of citric acid and therefore the lower fermentation yield (Table 2). The highest final citric acid concentration (111.5 g/l) and the highest yield (74.6%) were observed during the citric acid fermentation with the addition of 1.0 g/l of NH_4NO_3.

Additionally, single or multiple ammonium nitrate additions during the citric acid fermentation have a negative influence both on the fermentation yield and the final citric acid concentration.

To verify the results obtained from the fermentations carried out on a shaker, in the second stage of experiments a series of cultures in the lab bioreactor were performed.

The influence of the initial ammonium nitrate concentration in the fermentation broth on the citric acid fermentation yield and productivity are presented in Table 3.

Table 3
Effect of initial NH_4NO_3 content in the fermentation broth on yield and productivity of citric acid fermentation in lab bioreactors

Initial NH_4NO_3 concent. in medium	Time of ferm.	Final sugar concent. in fermentation broth	Final biomass concent.	Total acidity	Final monohydrate citric acid concent.	Citric acid yield in relation to introduced sugar	Citric acid productivity
(g/l)	(h)	(g/l)	(g/l)		(g/l)	(%)	(g/l/h)
1.0	192	6.0	8.42	31.0	108.5	72.5	0.565
1.5	166	6.9	11.24	35.1	122.8	82.2	0.740
2.0	178	5.7	13.10	32.3	113.1	76.3	0.654
2.5	192	6.8	15.34	30.7	107.4	72.3	0.559
3.0	204	5.2	18.85	26.5	92.8	62.2	0.455
4.0	204	4.7	22.85	22.7	79.5	53.6	0.390

With the initial increase of ammonium nitrate concentration in the fermentation broth the biomass concentration increases, what is in accordance with results obtained on a shaker.

The highest yield (82.2%) and the rate of citric acid production (0.74 g/l/h) was obtained in the fermentations carried out in the medium with 1.5 g/l of NH_4NO_3. The fermentations carried out in the medium with 1.0 g/l of NH_4NO_3 were characterised by the small amount of created biomass, by the extended fermentation time up to 192 hours, by the lower rate of production, and lower fermentation yield. On the basis of these experiments (Table 3) it transpires that 1.5 g/l was the optimal initial concentration of ammonium nitrate in the fermentation broth during the citric acid biosynthesis carried out by the *Aspergillus niger* C-12-I/94 strain.

Choe and Yoo [3] found that additional and continual fermentation broth feed by ammonium nitrate applied after 72 hours (9 mg/l/h) can increase by about twofold the citric acid biosynthesis rate. Therefore, our further experiments were carried out in the lab bioreactors with the fermentation broth containing optimal initial content of ammonium nitrate i.e. 1.5 g/l for *Aspergillus niger*. Additionally, from 72nd hour up to the end of the fermentation, ammonium nitrate was added continuously at the rate of 9 mg/l/h to the fermentation broth. The obtained results are presented in Table 4.

In spite of introducing continuously additional broth feed by ammonium nitrate there was a lack of free NO_3^- ions after 72 hours of fermentation, and free NH_4^+ ions after 120 hours of fermentation. The citric acid productivity was 0.744 g/l/h and the yield was 83.9%.

In these fermentations, compared with fermentations carried out on the medium with the same content, however, without additional feed by ammonium nitrate (Table 3) where was found a higher biomass concentration and higher biomass growth. On the other hand, the yield and citric acid productivity rate were slightly higher. These results are not in accordance with those obtained by Choe and Yoo [3].

Table 4
Basic kinetic parameters of citric acid fermentation with continuous NH4NO3 addition (9 mg/l/h) to the fermentation broth

Parameters	Unit	Results							
Time of ferm.	hour	0	24	48	72	96	120	144	168
Total acidity	ml/2 ml	0.2	0.3	2.1	8.7	18.8	28.5	33.4	35.7
Citric acid	g/l	0.7	1.1	7.4	30.4	65.8	99.8	116.9	125.0
Biomass	g/l	-	1.2	5.4	9.8	12.2	13.6	14.8	15.1
Sugars	g/l	148.9	137.6	117.8	94.6	65.2	45.6	25.1	6.3
pH	-	2.9	2.1	2.2	1.9	1.8	1.7	1.6	1.6
NH_4^+	mg/l	340	60	30	10	10	0	0	0
NO_3^-	mg/l	1200	>500	500	0	0	0	0	0

Future research should be directed to find the influence of additional fermentation broth feed by ammonium nitrate for the effectiveness of citric acid biosynthesis with *Aspergillus niger* C-12-I/94 strain by measuring intracellular concentration of NH_4^+ and NO_3 ions.

REFERENCES

1. H.Eikmeier, H.J.Rehm, Appl. Microbiol. Biotechnol., 20 (1984) 365.
2. B. Kristiansen and C.G. Sinclair, Biotechnol. Bioeng., 20 (1978) 1711.
3. Jaehoo Choe and Young Je Yoo, J. Ferment. Bioeng., 72 (2) (1991) 106.
4. J.Pintado, M.A.Murado, M.P.Gonzales, J.Miron, L.Pastrana, Biotechnol. Lett., 15 (11) (1993) 1157.
5. M. Yigitoglu and B. McNeil, Biotechnol. Lett., 14 (9) (1992) 831.
6. W. Leśniak, Prace Naukowe Akademii Ekonomicznej we Wrocławiu, Technologia, 69 (91) (1975) 94.
7. W. Leśniak and M. Kutermankiewicz, Podstawy produkcji kwasu cytrynowego, [Rudiments of citric acid production], Stowarzyszenie Techników Cukrowników NOT, Warszawa 1990.
8. J. Pietkiewicz, W. Podgórski, W. Leśniak, [in] III Ogólnokrajowa Sesja Naukowa, Postępy Inżynierii Bioreaktorowej „Łódź 97", Materiały Sesyjne, Red. H. Michalski, Wyd. Zak. Poligraf. PŁ. Łódź 1987.

Food Biotechnology
S. Bielecki, J. Tramper and J. Polak (Editors)
© 2000 Elsevier Science B.V. All rights reserved.

Induction of citric acid overproduction in *Aspergillus niger* on beet molasses

W. Podgorski and W. Lesniak

Food Biotechnology Department, Wroclaw University of Economics,
Komandorska 118/120, 53-345 Wroclaw, Poland

Two methods of induction of citric acid biosynthesis in *Aspergillus niger* on beet molasses medium based on the control of aeration rate are presented. In the first method, aeration was increased with a rate higher than demanded by the mycelium for the growth process. In the second method, aeration was used as restricting factor of mycelium development. As a result of the experiments, it was found that excess as well as deep oxygen deficiency can lead to induction of citric acid accumulation.

1. INTRODUCTION

Citric acid accumulation in *Aspergillus niger* occurs as a result of fungal overflow metabolism. It is generally accepted that for obtaining high productivity of citric acid, growth of the strain must be restricted. Among the conditions favouring citric acid production, high sugar concentration, low concentration of phosphate, low pH, absence of trace metals and high oxygen concentration are the most important [1]. Some authors working on synthetic media also claim that the highest citric acid production rate occurs under limitation of nitrogen [2].

Citric acid is commercially produced by submerged fermentation of cane and beet molasses. Those substrates cause difficulties in operation because of their non-steady chemical composition. Excess of nutrients and microelements in the molasses promotes biomass development instead of citric acid accumulation. The suitability of beet molasses is usually increased by addition of potassium ferrocyanide (PF). PF reduces the content of available heavy metals, which cause unfavourable conditions for citric acid biosynthesis. Especially the concentrations of manganese, iron and zinc have to be minimised. Those microelements, maintained at very low levels, seem to be essential for induction of citric acid overproduction. Apart from the control function of medium constituents, PF has also a toxic impact on the mycelium, modifying its growth rate. The toxic effect of free ferrocyanide ions on the growth of mycelium is also considered as an inductive factor of citric acid overproduction.

After preparation of the molasses medium, induction of the citric acid production depends mainly on aeration condition.

In the present investigation, the mechanism of respiratory quotient (RQ) changes in the growth phase as a function of different aeration profiles was studied. The aim of the conducted experiments was to compare two methods of citric acid induction by the regulation of oxygen supply.

2. MATERIALS AND METHODS

Control of citric acid biosynthesis can be continuously monitored and regulated with systems based on fermentation gas analysers. In our experiments, a Servomex 1100A Oxygen Analyser (Servomex), a Guardian II Carbon Dioxide Monitor (Edinburgh Sensors) and a Mass Flow Controller ERG 2000 (Beta Erg) were used. From exhaust gases, oxygen uptake rate (IO_2), carbon dioxide evolution rate (ICO_2) and their specific rates (qO_2 and qCO_2) were calculated [3]. The oxygen concentration (pO_2) was measured by a membrane probe and expressed as percentage of the saturated state, calibrated on molasses medium before inoculation. Redox potential (Eh) was determined by means of platinum and silver/silver chloride electrodes.

Batch experiments were carried out on beet molasses derived from Polish sugar factories. The composition of the base medium in all cases was the same, containing about 13% of sugar. Medium was supplemented with $K_4Fe(CN)_6$, KH_2PO_4 and $ZnSO_4$. The experiments were performed using dry spores of the fungus *Aspergillus niger* W78B as an inoculum.

For determination of the citric acid concentration (P), the reaction with pyridine and acetic anhydride was applied [4]. Accompanying acids were identified by thin layer chromatography. The biomass concentration (X) was determined by vacuum washing and drying to constant weight.

The fermentations were carried out in 5 and 10 l tank-type laboratory bioreactors and controlled by the computer system SysLab Bio.

3. RESULTS

After inoculation of the molasses medium with spores of the *Aspergillus niger* W78B strain, medium aeration in the growth phase was carried out in two ways. In both cases, the traditional methods of molasses treatment with potassium ferrocyanide were applied. The basis of the aeration changes during the process was determined by actual values and mutual correlation between oxygen uptake and carbon dioxide evolution rates. The commencing of citric acid overproduction resulted in changing values of IO_2 and ICO_2 according to the correlation expressed by RQ. Citric acid excretion was identified by dropping RQ values to below 1.0. Lower RQ's indicated higher rates of citric acid production [5].

In the first series of experiments aeration was increased with a rate higher than the demand of mycelium connected with the growth processes. In the exponential growth phase, an excess in supplied oxygen could be relatively easy achieved by keeping the dissolved oxygen concentration at about 40% of the saturation state and in the following phase at a level higher than 10% (Fig. 1).

It was disadvantageous to hold the oxygen uptake rate higher than the carbon dioxide evolution rate, especially at the beginning of the growth phase causing at relatively high pH an increased synthesis of accompanying acids [6,7].

In the second series of the experiments aeration was treated as a restricting factor used for mycelium respiration and growth control comparable to situations with microelements and nutrients limitation. Medium aeration was controlled by low critical aeration rate during the whole trophophase period. Restriction of the oxygen supply resulted in elimination of oxygen excess in relation to the demand coming from biomass growth and reduced excretion of citric acid as well as other organic acids to the medium (Fig. 2).

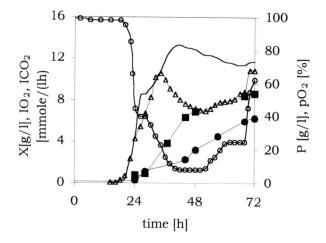

Figure 1. Time course of fermentation in the first 72 h. High oxygen uptake rate.

X (—■—), IO$_2$ (——),
ICO$_2$ (—△—), P (—●—),
pO$_2$ (—○—).

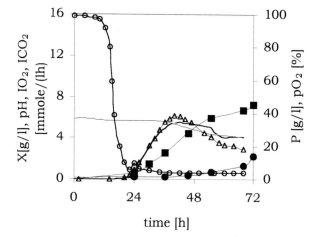

Figure 2. Time course of fermentation in the first 72 h. Low oxygen uptake rate.

X (—■—), pH (——),
IO$_2$(——), ICO$_2$ (—△—),
P (—●—), pO$_2$ (—○—).

Keeping the aeration rate constant in the transient phase increased the oxygen deficit in comparison with the biomass demand. The oxygen concentration in this case was kept at the constant level of 0%. The degree of deficiency in oxygen supply was better reflected by the redox potential (Eh), which value changed from approximately 200 mV down to 100 mV. Under the conditions of increasing inhibition of respiration, the microorganisms changed their own metabolic route in the direction of citric acid overproduction. Maintaining a very low aeration rate of about 0.017 vvm (1.0 vvh) caused a decrease in RQ as a result of a gradual falling-off carbon dioxide evolution rate. The oxygen uptake rate values did not change in time and remained at the same level. The level of respiratory inhibition was determined by specific oxygen uptake rate (qO$_2$) and specific carbon dioxide evolution rate (qCO$_2$) as well. It was found that RQ dropped below 1.0 when qO$_2$ decreased to the critical values (Fig. 3).

250

Citric acid started to accumulate at pH about 4.4 and after 72 hours reached the level of 3.6 with the total acids concentration about 12 g/l.

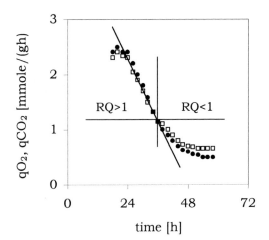

Figure 3. Changes of qO_2 (□) and qCO_2 (●) in the growth phase.

The experiments carried out with low and high aeration, showed that the low aeration reduces the biomass by 15% and total carbon dioxide evolution by 50%. Average specific carbon dioxide evolution rate (qCO_2) decreased from 0.7 to 0.4 mmole CO_2/(gh).

4. CONCLUSIONS

The obtained results indicate the possibility of the respiratory activity control and growth of the mycelium in the presence of the nutrients excess in a molasses medium. The limited aeration in the first stage of the fermentation causes induction of citric acid overproduction, decreases the biomass content and reduces amount of the accompanying acids.

REFERENCES

1. M., Rohr, C.P. Kubicek, J. Kominek, Citric acid, Biotechnology, 3, ed. H.J. Rehm, G. Reed, Weinheim, Verlag Chemie, 1983.
2. B. Kristiansen and C.G. Sinclair, Biotechnol. Bioeng., 20(11) (1978) 1711.
3. M. Reuss, S. Frohlich, B. Kramer, K. Messerschmidt, G. Pommerening, Bioprocess Eng., 1 (1986) 79.
4. J.R. Marier and M. Boulets, J. Dairy Sci., 41 (1958) 1683.
5. C.P. Kubicek, O. Zehentgruber, M. Rohr, Biotechnol. Lett., 1(1) (1979) 47.
6. M. Sobotka, V. Machon, L. Seichert, E. Ujcova, Z. Marschalkova, Folia Microbiol., 30 (1985) 381.
7. L. Seichert, E. Ujcova, M. Musilkova, Z. Fencl, Folia Microbiol., 27(5) (1982) 333.

Food Biotechnology
S. Bielecki, J. Tramper and J. Polak (Editors)

Effect of amino acids and vitamins on citric acid biosynthesis

W. Lesniak and W. Podgorski

Food Biotechnology Department, Wroclaw University of Economics,
Komandorska 118/120, 53-345 Wroclaw, Poland

It was observed that a some amino acids and vitamins have a stimulating effect on citric acid biosynthesis, particularly when it concerns *Aspergillus niger* mutants, indicating sensibility to deficiency of these substances in the fermentation medium. Therefore, the aim of the present experiment was to evaluate the effect of some amino acids and vitamins on submerged citric acid biosynthesis by the UV mutant *Aspergillus niger* B-64-5 in sucrose – mineral medium. It was found that the citric acid formation was stimulated by the amino acids valine and glutamic and aspartic acid, and by the vitamins thiamine-HCl, nicotinic acid amide and ascorbic acid.

INTRODUCTION

Some organic and non-organic compounds have a stimulating effect on *Aspergillus niger* mycelium growth and citric acid synthesis. It was observed that some amino acids and vitamins have the same effect. It applies particularly to the process of citric acid biosynthesis in synthetic medium with *Aspergillus niger* mutants, indicating sensibility to deficiency of these substances in the fermentation medium.

It was found that *Aspergillus niger* strains exposed to UV rays can loose the ability of biosynthesis of some amino acids. Among them are glutamic acid, proline and ornithine which are also needed for the production of other amino acids. The basic role is for glutamic acid, which is the first organic product of nitrogen assimilation in most microorganisms. Then, as the result of glutamic acid carbon chain reconstruction with contribution of other basic amino acids, the residual amino acids are produced. According to this, it is supposed, that decrease in the rate of essential amino acids biosynthesis leads to reduction of mycelium biomass production.

Experiments of other authors showed that from the amino acids in the medium, glutamic acid was used most, followed by aspartic acid and alanine, and then, used the least, leucine and isoleucine [1].

In *Aspergillus niger* mycelium some of the amino acids are assimilated completely, while others are only deaminated.

Deamination of amino acids leads to synthesis of pyruvic acid, α–ketoglutaric acid, succinic acid, fumaric acid and oxalic acid, which are intermediate products of metabolism.

It was shown [1] that addition of aspartic or glutamic acid to sucrose-mineral medium increases by 8 – 38% the biomass formation and by 9 – 16% the amount of citric acid formed by the UV mutant *Aspergillus niger* T.1. The minimal concentration of glutamic acid exerting a favourable effect on growth and acid formation is 0.05%. It is also well known that many vitamins play an important role in the enzymatic activities of the microbial metabolism, where they act as coenzymes for different reactions. In relation to vitamins, there is an opinion that their presence in media for *Aspergillus niger* is not necessary because of the possibility of that strain to synthesise vitamins itself. Nevertheless, it was shown that addition of small amounts of biotin to medium stimulates *Aspergillus niger* growth, as well as addition of pantothenic acid [2]. The citric acid biosynthesis is stimulated by addition of thiamine, pyridoxine and nicotinic acid [3].

Therefore, the present experiment was carried out to evaluate the effect of amino acids and vitamins on submerged citric acid biosynthesis by the active mutant *Aspergillus niger* B-64-5.

MATERIALS AND METHODS

Strain: selected UV mutant *Aspergillus niger* B-64-5.
Medium:

sucrose	- 150 g/l
NH_4NO_3	- 2 g/l
KH_2PO_4	- 0.2 g/l
$MgSO_4 \cdot 7H_2O$	- 0.2 g/l
pH	2.8 – 3.0

Fermentation was carried out in 750 ml flasks containing 125 ml of the medium on a shaker (200 rpm).

Total acidity was determined by titration of 2 ml of fermentation broth with 0.1 M NaOH against phenolphthalein. Citric acid was estimated colourometrically according to Marrier and Boulets [5].

Mycelial dry weight was determined by drying at 105°C until constant weight was attained. Reducing sugars were determined according to the Lane Eynon method as modified by Soczynski [6].

RESULTS

In the first stage of experiments, the effect of different concentrations of amino acids i.e. alanine, leucine, isoleucine, glycine, proline, serine, arginine, threonine, lysine, glycine, valine and glutamic and aspartic acid, on the growth and activity of *Aspergillus niger* was studied, (Table 1).

It was found that valine, glutamic acid and aspartic acid increased biosynthesis of citric acid by 6–8%, while glycine, serine and arginine induced an increase of 4–5%, as compared to the basic medium. In contrast, alanine, leucine and iso-leucine at a higher concentration (0.01%) decreased biosynthesis of citric acid by 24–40% (Fig. 1).

The biomass of the UV B-64-5 mutant of *Aspergillus niger* generally decreased by about 5–10% with a concentration increase of amino acids added as compared to the basic medium (Fig.1). Only for the addition of alanine at concentration 0.002–0.004% an increase of biomass by 15% was observed.

In the second stage of experiments, the basal medium was supplemented by each of the following vitamins in the proper concentration: ascorbic acid, thiamine-HCl, riboflavin, pyridoxine, nicotinic acid amide, mesoinositol, calcium pantothenate, folic acid and PABA (para-amino benzoic acid) (Table 2 and Fig. 2).

Table 1
Effect of amino acids on mycelial growth and citric acid biosynthesis

Amino acid	Biosynthesis parameters	Addition of amino acids calculated as amino acid nitrogen [%]					
	[g/l], [g d.s./l]	0.0	0.001	0.002	0.004	0.006	0.010
β–alanine	citric acid	126	126	117	110	78	72
	mycelium	12.0	11.7	13.0	13.9	11.1	10.6
L–leucine	citric acid	126	125	127	122	119	76
	mycelium	12.0	12.8	11.2	10.6	11.2	11.1
DL–isoleucine	citric acid	126	118	118	114	119	96
	mycelium	12.0	11.8	11.3	10.6	12.2	11.4
glycine	citric acid	126	125	126	132	129	115
	mycelium	12.7	12.2	11.7	11.4	11.8	10.9
DL–proline	citric acid	126	123	126	124	123	123
	mycelium	12.7	10.8	11.3	11.6	11.2	11.3
DL–serine	citric acid	126	122	131	123	120	123
	mycelium	12.7	11.4	10.6	10.9	10.8	10.8
DL–arginine	citric acid	120	120	121	121	123	126
	mycelium	13.1	12.2	11.8	11.6	12.1	12.2
DL–threonine	citric acid	120	118	122	118	119	115
	mycelium	13.1	11.3	11.7	11.3	11.6	12.2
DL–lysine	citric acid	120	116	116	120	117	114
	mycelium	13.1	12.0	11.8	11.9	12.5	13.0
DL–valine	citric acid	113	115	120	119	120	115
	mycelium	14.2	13.3	13.0	13.1	13.0	13.1
glutamic acid	citric acid	113	118	115	118	120	122
	mycelium	14.2	13.5	14.4	14.2	14.1	13.4
aspartic acid	citric acid	113	120	120	113	115	116
	mycelium	14.2	13.6	13.1	13.1	13.2	13.1

Table 2
Effect of vitamins on mycelial growth and citric acid biosynthesis

Vitamin	Biosynthesis parameters [g/l], [g d.s./l]	Addition of vitamins [mg/l]					
		0.0	0.1	0.5	1.0	5.0	10.0
thiamine–HCl	citric acid	120	120	122	126	125	124
	mycelium	13.5	12.6	11.8	12.3	11.8	11.4
riboflavin	citric acid	120	118	117	110	67	0.0
	mycelium	13.5	12.3	11.6	12.0	9.2	0.0
pyridoxine–HCl	citric acid	120	118	118	120	120	118
	mycelium	13.5	12.4	11.8	11.6	11.8	12.2
cobalamin	citric acid	113	114	116	114	110	104
	mycelium	14.6	12.8	13.6	13.0	13.2	13.7
ascorbic acid	citric acid	106	106	110	112	110	112
	mycelium	14.6	13.5	12.6	13.3	12.8	12.4
nicotinic acid amide	citric acid	106	114	118	109	110	105
	mycelium	14.6	12.0	12.1	12.2	12.0	12.5
mesoinositol	citric acid	113	115	115	113	112	112
	mycelium	14.0	13.9	12.6	12.5	12.6	12.4
calcium pantothenate	citric acid	113	112	113	113	106	105
	mycelium	14.0	12.9	13.8	13.2	13.2	13.9
folic acid	citric acid	113	110	109	105	110	108
	mycelium	14.8	13.2	12.3	12.6	12.6	12.1
para-amino-benzoic acid	citric acid	113	108	108	110	109	108
	mycelium	14.0	12.4	13.6	13.1	13.1	12.8

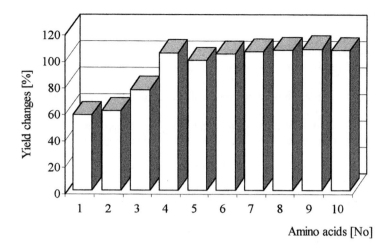

Figure 1. Effect of amino acids on citric acid production.
1 - β-alanine, 2 – l-leucine, 3 - DL-iso-leucine, 4 - glycine, 5 - DL-proline, 6- DL-serine, 7 - DL-arginine, 8 - DL-valine, 9 - glutamic acid, 10 - aspartic acid. Control = 100%.

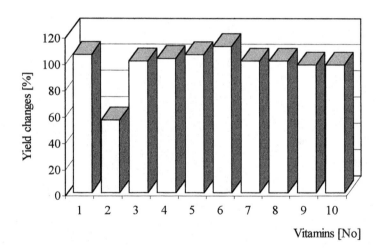

Figure 2. Effect of vitamins on citric acid production.
1 - thiamine-HCl, 2 – riboflavin, 3 - pyridoxine-HCl, 4 - cobolamin, 5 - ascorbic acid, 6 - nicotinic acid amide, 7 - mesoinositol, 8 - calcium pantothenate, 9 - folic acid, 10 – para-amino - benzoic acid. Control = 100%.

Only thiamine-HCl, nicotinic acid amide and ascorbic acid in higher concentrations (1 – 5 mg/l) increased biosynthesis of citric acid by 5 – 10%. In contrast to these vitamins, riboflavin added in a concentration 5 mg/l decreased citric acid yield to about 50% (Fig. 2) and at a concentration of 10 mg/l completely stopped growth of mycelium and therefore biosynthesis of citric acid.

Other studied vitamins showed negligible effects on citric acid biosynthesis and growth of mycelium.

We conclude that mutant *Aspergillus niger* B-64-5 does not show a strong sensibility to deficiency of amino acids or vitamins in the sucrose – mineral medium.

REFERENCES

1. J.D. Kasatkina, Mikrobiologija, 30 (1961) 3.
2. D.N. Lal and A.S. Srivastava, Zentralbl. Mikrobiol., 137(5) (1982) 381.
3. W. Mashhoor, R.F. Gamal, A.A. Refaat, S.A. Nasr, Annals Agric. Sci., 33(1) (1988) 153.
4. W. Lesniak, Prace Naukowe WSE Wroclaw, Technologia, 52(74) (1974) 49.
5. J.R. Marier and M. Boulets, J. Dairy Sci., 41 (1958) 1683.
6. S. Soczynski, Przem. Spozywczy, 9 (1955) 416.

Food Biotechnology
S. Bielecki, J. Tramper and J. Polak (Editors)
© 2000 Elsevier Science B.V. All rights reserved.

257

Suitability of *Lactobacillus* strains as components of probiotics

J. Moneta and Z. Libudzisz

Institute of Fermentation Technology and Microbiology, Technical University of Łódź, Poland,
Stefanowskiego 4/10, 90-924 Łódź, Poland.

The production of organic acids, acetaldehyde, β-galactosidase, proteolytic activity,
carbohydrate metabolism and resistance to low pH of 2.5 as well as bile of 4% salt
concentration were used as selection criteria for the assessment of 19 strains of *Lactobacillus
acidophilus*, with respect to their probiotic properties. Their activity was compared with the
activity of reference strains. Most of the strains revealed the tested features at the same or
higher level as the reference strains.

1. INTRODUCTION

In recent years there has been an increasing interest in consumption of natural food,
especially fermented dairy products containing specific strains of lactic acid bacteria known for
their health-promoting effects.

There are, however, only a few strains of clearly defined probiotic features. In addition
these specific properties usually do not exist together in the same strain. Thus, the level of
probiotic features must be checked carefully. It is necessary, when composing the starters, to
include the cultures that are complementary with regards to their probiotic features.

2. MATERIALS AND METHODS

2.1. Microorganisms

The aim of the study was to determine the range of differences in physiological properties of
19 strains of *Lactobacillus acidophilus* and of a strain *Lactobacillus delbrueckii* ssp. *lactis*
L30. We used cultures from various collections, i.e. from the collection of pure cultures of
Biolacta-Texel in Olsztyn, Poland: -336 (**2**), In3 (**3**), 1nd1 (**4**), Ros (**5**), 172 (**6**), Cz1 (**7**), 20T1
(**8**), 343 (**9**), H1 (**10**), V/74 (**11**) & CH-2 (**12**) strains; from the collection of the Institute of
Chemical Technology in Prague, Czech Republic: - CH-5 (**13**), Nestle (**14**), Bauer (**15**), Diat
(**16**), A92 (**17**) & L30 (**1**) strains; from Institute of Fermentation Technology & Microbiology,
Technical University of Łódź: - B (**18**) strain; from the National Collection of Agricultural and
Industrial Microorganisms in Hungary - NCAIMB 1075 (**19**) & NCAIMB 1152 (**20**) strains.

Moreover, 8 reference strains from the collection of Collegium Medicum, Jagiellonian
University in Cracow, i.e. *Lb. acidophilus* ATCC 4356 (**A**), *Lb. thermophilus* 094 11.78-

NCDO 489 (**B**), *Lb. casei*-Shirota (**C**), *Lb. acidophilus* LC1 (**D**), *Lb. rhamnosus* GG ATCC 53105 (**E**), *Lb. crispatus* NCFB 2752 (**F**), *Lb. casei*-NCDO 206 (**G**) & *Lb. gasseri* (**H**) were tested. The final goal was to select a strain with probiotic activity higher or comparable to the activity of the reference strains.

2.2. Analytical Methods

As a selection criteria the following features were examined: growth, lactic acid production and the level of the L (+) lactic acid in milk after 24 h incubation, production of acetic acid in milk after 24 h incubation, β-galactosidase, tolerance to low pH (2.5), survival in the presence of 4% bile salt concentration, proteolytic activity, biochemical characteristic (carbohydrate fermentation) and acetaldehyde production (a compound responsible for desirable flavour of acidophilus milks).

2.2.1. Number of bacteria was estimated using MRS agar pour plates incubated at 37°C for 48 h, under anaerobic conditions.

2.2.2. Production of acid was measured by titration with 0.1 N NaOH. Results were expressed as grams of lactic acid per 100 ml of milk. L-isomer and acetic acid were determined enzymatically (Boehringer Mannheim, Germany)

2.2.3. β-galactosidase activity of bacteria was estimated using o-nitrophenyl-D-galactopyranoside (ONPG) as substrate. Bacteria were incubated in MRS medium at 37°C for 20 h and then cells of each strain were treated with SDS (sodium dodecyl sulfate)-chloroform [1,2]. The activity of β-galactosidase was expressed as μmoles of o-nitrophenol (ONP) liberated from ONPG per miligram of cell dry weight per minute.

2.2.4. Proteolytic activity of bacteria in skim milk incubated at 37°C for 48 h was assayed using Anson's method with Folin-Ciocalteu's reagent. Results were expressed as an amount of tyrosine (mg) released by bacteria per ml of trichloroacetic acid filtrate

2.2.5. Carbohydrate metabolism was checked on the basis of API 50 CHL test system (BioMerieux).

2.2.6. Acetaldehyde was assayed by method of Lindsay and Day [3,4] and enzymatically using the test of Boehringer Mannheim.

2.2.7. Tolerance to low pH 2.5 and survival in the presence of 4% bile salt concentration. The cells, cultured in MRS medium were collected by centrifugation and suspended in sodium chloride solution (0.85 g/100 ml) to obtain an initial count of 10^8 CFU/ml. The solution was either adjusted to pH 2.5 by addition of sterile HCl or suspended in a bile salt concentration of 4% w/v. All samples were incubated at 37°C for 8 h or 74 h, respectively. The viable cell counts of each strain were enumerated at different intervals using MRS agar.

3. RESULTS

We found that the tested *Lactobacillus* strains showed different acidifying activity, accumulating from 0.6% (CH-5, Nestle strains) to 2.3% lactic acid (A92 strain) in milk, after 24 h of cultivation at 37°C (Fig.1). In the same conditions the reference strains produced from 0.68% lactic acid (NCFB 2752 strain) to 1.1% (ATCC 4356 strain) (Fig. 2). The percent of L(+) lactic acid in total lactic acid ranged for tested strains from 19% (A92 strain) to 89% (NCAIMB 1152 strain) and for most of them (85%) was from 34% to 66%. In the case of reference strains the percent of L(+) isomer ranged from 26% (NCFB 2752 strain) to 63-65% (LC1 and 094 11.78-NCDO 489 strains) (Fig.3).

a)

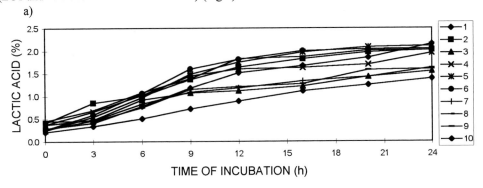

b)

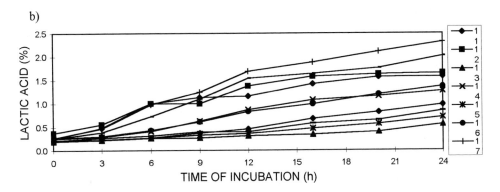

Figure 1. Acidifying activity of *Lactobacillus* strains; a)-strains 1-10, b)-strains 11-20

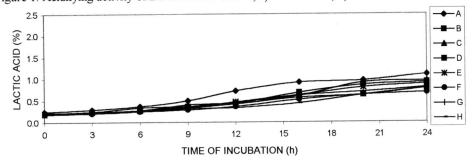

Figure 2. Acidifying activity of *Lactobacillus* reference strains.

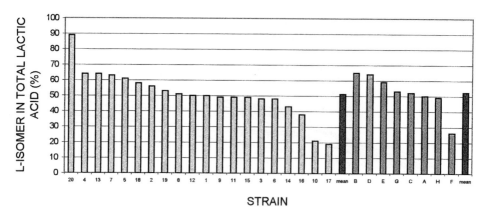

Figure 3. Production of L (+) lactic acid by *Lactobacillus* strains after 24 h incubation.

After 24 hours of incubation in milk the tested strains produced from 0.4 g/l (H-1, NCAIMB 1152 strains) to 1.4 g/l (B, L30 strains) of acetic acid. 50% of strains produced this metabolite in the amount above 1.2 g/l. The reference strains produced acetic acid at the same level, accumulating from 0.44 g/l (Shirota strain) to 1.38 g/l (ATCC4356 strain) (Fig. 4).

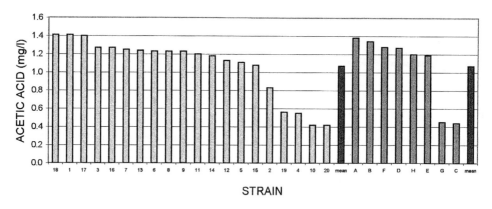

Figure 4. Production of acetic acid by *Lactobacillus* strains after 24 h incubation.

The activity of β-galactosidase of the tested bacteria ranged from 0.031 U/mg d.w. (NCAIMB 1075 strain) to 0.605 U/mg d.w. (CH-2 strain). The strains of similar activities in the range from 0.300 to 0.456 U/mg d.w. constituted the biggest group (50%). In contrast, the reference strains revealed weaker β-galactosidase activity, i.e. from 0.136 U/mg d.w. (NCFB 2752 strain) to 0.0217 U/mg d.w. (094 11.78-NCDO 489 strain) (Fig. 5).

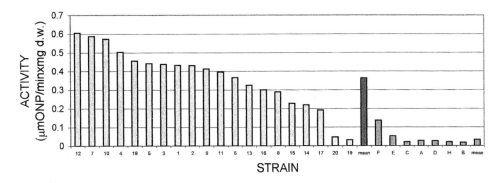

Figure 5. β-galactosidase activity of *Lactobacillus* strains.

There were significant differences in bacterial viability at pH 2.5. After 8 hours of incubation the observed decrease in the number of living cells ranged from 1.2×10^8-1.1×10^9 CFU/ml to $3.5 - 2.6\times10^6$ CFU/ml, depending on the strain. Twenty percent of strains revealed viability from 74% to 77%, while half of them viability from 36% to 66%. At the same conditions the reference strains survived at the level comparable with most tested bacteria (34-71%) (Fig. 6).

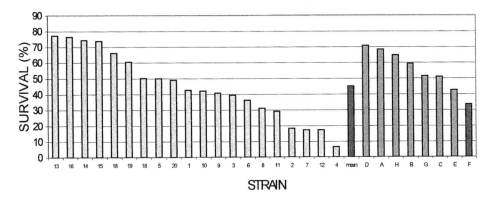

Figure 6. Survival of *Lactobacillus* strains at pH 2.5 after 8 h.

Additionally, the tested bacteria showed various sensitivity to 4% of bile salt present in the medium. After 74 hours of incubation 5% of strains (B strain) survived at the level 86%, 75% of strains expressed viability from 35% to 66% and 20% of strains from 27% to 32%. Reference strains also showed various sensitivity to 4% bile salt and their viability ranged from 28% (NCFB2752 strain) to 67% (094 11.78-NCDO 489 strain) (Fig.7).

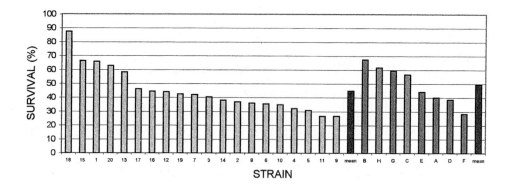

Figure 7. Survival of *Lactobacillus* strains in the presence of 4% bile salts after 74 h.

Significant differences in the proteolytic activity of tested strains and dynamics of protein degradation specific for a particular strain were observed. An increase in tyrosine level among the tested strains after 24 h of incubation in milk ranged from about 17 (NCAIMB 1152 strain) to 135 (172 strain) mg tyr./dm^3 Reference strains showed weaker proteolytic activity. Prolongation of the incubation time to 48 h caused a further increase in protein digestion for tested strains from 38 (NCAIMB 1152 strain) to 190 mg tyr./dm^3 (1nd1 strain). In the same conditions the reference strains showed an increase in tyrosine level from 26 (GG ATCC 53105 strain) to 81 mg tyr./dm^3 (ATCC 4356 strain) (Fig. 8).

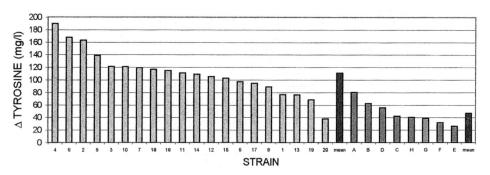

Figure 8. Proteolytic activity of *Lactobacillus* strains (increase in tyrosine level).

On the basis of API 50 CHL tests it was found that all examined strains were able to ferment glucose, galactose, fructose, mannose, maltose, lactose, saccharose and trehalose. The ability to metabolise other sugars was a feature of individual strains. Additionally, different diagnostic matching with the species pattern in the API system and according to Bergey (1986) was reported.

The amount of acetaldehyde formed in milk after 24 hours of incubation ranged from 0.8 mg/l (Bauer strain) to 16 mg/l (In3 strain) (Fig.9).

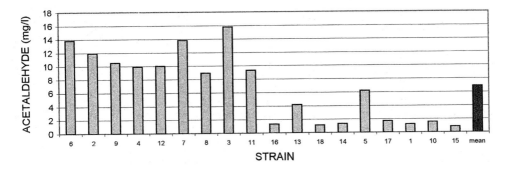

Figure 9. Production of acetaldehyde by *Lactobacillus* strains.

The tested bacteria expressed a various ability to adhere to a specified line of intestinal epithelium cells. The strongest adherence was showed by the strains of *Lb. acidophilus* 172, *Lb. acidophilus* 336, *Lb. acidophilus* Nestle, *Lb. delbrueckii* ssp. *lactis* L30, *Lb. acidophilus* A92. (studies carried out by P. Heczko, E. Marewicz, G. Kukla, Institute of Microbiology, Collegium Medicum Jagiellonian University, Cracow) (not published data).

4. CONCLUSIONS

Significant heterogeneity of the tested strains, with respect to their probiotic properties was found as a result of these investigations. Most strains were found to be able to form the tested metabolites on the same or higher level as compared to the reference strains. The most suitable appeared to be the cultures of *Lb. acidophilus* B, *Lb. acidophilus* 1nd1, *Lb. acidophilus* 172, *Lb. acidophilus* Ros, *Lb. acidophilus* 20T1, *Lb. acidophilus* Cz1, *Lb. acidophilus* CH-2, *Lb. acidophilus* CH-5, *Lb. acidophilus* A92, *Lb. acidophilus* Diat, *Lb. delbruecki* ssp. *lactis* L30, *Lb. acidophilus* NCAIMB 1152. These strains can be used in the production of probiotics or dairy products.

REFERENCES

1. J.F. Citti, W.E. Sandine, P.R. Elliker, J. Bacteriol., 4 (1965) 937.
2. N. Shah, P. Jelen, J. Food Sci., 55 (1990) 506.
3. R.C. Lindsay, E.A. Day, J. Dairy Sci., 48 (1965) 665.
4. T.M. Cogan, J. Dairy Sci., 55 (1972) 382.

Food Biotechnology
S. Bielecki, J. Tramper and J. Polak (Editors)
© 2000 Elsevier Science B.V. All rights reserved.

Viability of *Bifidobacteria* strains in fermented and non-fermented milk

I. Motyl and Z. Libudzisz

Institute of Fermentation Technology and Microbiology, Technical University of Łódź, Stefanowskiego 4/10, 90-924 Łódź, Poland.

The subject of this study was a comparison of viability of bifidobacteria, during 21 day period under storage conditions (4-5°C), in fermented and non-fermented milk. The viability was additionally tested in fermented milk at temperatures of 18°C and 30°C. Tested material included 10 strains of *Bifidobacterium* species.

High viability of bifidobacteria cells in non-fermented and fermented milk at different storage conditions resulted in least 21-day biological stability of milk product containing the bacteria of *Bifidobacterium* species. However, over-acidification of a product, especially stored at temperature 30°C, significantly deteriorated its organoleptic properties.

1. INTRODUCTION

High viability of bifidobacteria both in fermented and non-fermented milk, is a very important feature, which should be taken into account when choosing cultures for use in the production of milk products. To perform their therapeutic role, milk products should contain at least 10^6 living cells of *Bifidobacterium*/ml in the moment of consumption.

2. MATERIALS AND METHODS

The subject of this study was a comparison of viability of bifidobacteria under storage conditions (4-5°C) in fermented and non-fermented milk during 21 days period. Tested material included 10 strains of *Bifidobacterium* species from the collection of the Institute of Animal Reproduction and Food Science of the Polish Academy of Sciences in Olsztyn, Poland (B42, KNA1, B36, Bif f2), Institute of Food Biotechnology, the Academy of Agriculture in Olsztyn (558, J/III, 100J), Department of Milk and Fat Technology, Prague Technical University, Prague, Czech Republic (BN, BD) and from the National Collection of Agricultural and Industrial Microorganisms, Hungary (B.bif).

These strains were chosen from among 26 cultures as expressing the highest resistance to low pH (1.5) and bile salts (4%). After 2 hours of incubation at pH 1.5, their viability was about 10^2 CFU/ml. Tested strains were also resistant to 4% bile in the medium, and after 12 hours of incubation in such medium demonstrated viability at the level of about 10^6 CFU/ml. The initial level of bifidobacteria in those studies ranged from 10^8 to 10^9 CFU/ml.

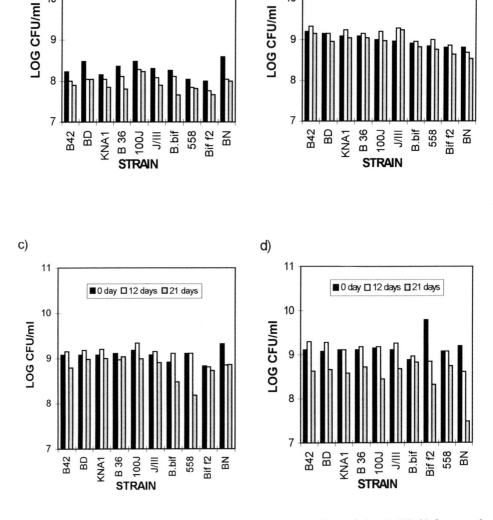

Figure 1. Number of living bacteria during storage: a) sweet milk stored at 4-5°C; b) fermented bifidus milk stored at 4-5°C; c) fermented bifidus milk stored at 18°C; d) fermented bifidus milk stored at 30°C.

267

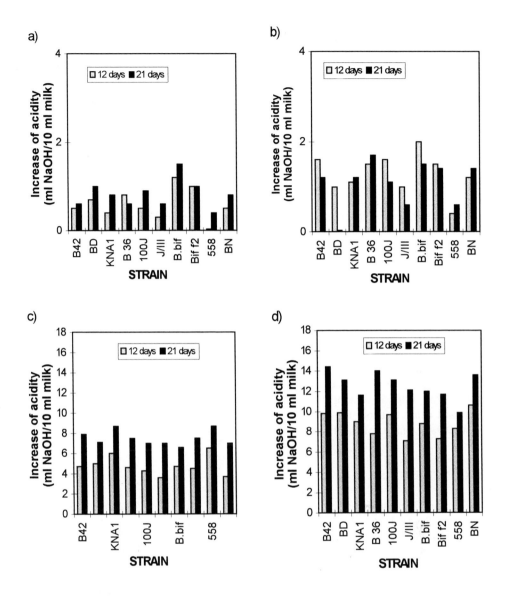

Figure 2. Change in acidity of bifidus milk during 12 and 21 days storage: a) sweet milk stored at 4-5°C; b) fermented bifidus milk stored at 4-5°C; c) fermented bifidus milk stored at 18°C; d) fermented bifidus milk stored at 30°C.

UHT milk with 2% fat was inoculated with a bacterial suspension in NaCl (10% inoculum (v/v)), and then either incubated at 37°C until acidified (fermented milk) or it was left under refrigerating conditions (non-fermented milk). In order to determine to what extent improper storage conditions might reduce the number of living bacteria, the viability of *Bifidobacterium* was also checked in fermented milk stored at 18°C and 30°C. The viability of bacteria was controlled by the plate method (Garcha'e medium after 7, 12, 17, 19 and 21 days of storage. Total lactic acid was measured by titration with 0.1M NaOH. Results were expressed in ml of 0.1 M NaOH/10 ml of milk.

3. RESULTS

As a result of these investigations it was found that all tested bacterial strains in fermented milk are characterised by high viability in refrigerating conditions. The number of living cells after 21 days of storage was almost at the initial level or slightly decreased 10^8-10^9 CFU/ml. Titration acidity of fermented milk stored at temperature 4-5°C after 21 days, irrespective of the strain, ranged from 8.1 to 11.2 ml 0.1M NaOH/10 ml of milk. The highest viability in fermented milk was expressed by strains: KNA1 and Bi36.

In non-fermented milk, the level of living *Bifidobacterium* cells after 21 days of storage at 4-5°C decreased from 10^8 CFU/ml to 10^7 CFU/ml, and titration acidity increased from about 0.4 to 1.5 ml 0.1 M NaOH. The strains 100J and 558 showed the highest viability (Fig. 1).

During storage at 18°C and 30°C the number of living bacteria was very high, and depending on the strain was recorded at the level of about 10^8-10^9 CFU/ml. Only strain BN expressed decrease in survival and the number of detected living cells at 30°C was about 10^7 CFU/ml. After 21 days of storage, depending on the strain, the acidity of fermented milk increased either to about 6.6–8.7 ml 0.1 M NaOH/10 ml milk, when stored at 18°C, or to about 9.9–14.4 ml 0.1M NaOH/10 ml milk, when stored at 30°C (Fig. 2).

4. CONCLUSIONS

Both non-fermented and fermented milk, containing bacteria of *Bifidobacterium* species, showed at 21-day long biological stability.

Additionally, it was found that the choice of strains which could survive in model conditions at pH 1.5 guaranteed very good shelf life of both fermented and non-fermented milk.

Moreover, it was observed that even improper storage of milk fermented with the tested *Bifidobacterium* resulted in survival of at least 10^6 CFU/ml, thus complying with the standard. However, over-acidification of a product, especially at temperature of 30°C, significantly deteriorated its organoleptic properties.

Food Biotechnology
S. Bielecki, J. Tramper and J. Polak (Editors)

269

Regulation of glycolysis of *Lactococcus lactis* ssp. *cremoris* MG 1363 at acidic culture conditions

M. Mercade, M. Cocaign-Bousquet, N. D. Lindley and P. Loubière

Centre de Bioingénierie Gilbert Durand, UMR CNRS 5504, UMR INRA, Institut National des Sciences Appliquées, Complexe Scientifique de Rangueil,
F-31077 TOULOUSE Cedex 4, France

Lactic acid-producing bacteria are very important organisms in the food industry, involved not only in dairy or wine products, but also in many vegetables or meat transformations. The most important carbohydrate catabolic pathway in lactococci involves glycolysis followed by the conversion of pyruvate mostly into lactate, but also in certain conditions, in other metabolic end-products (formate, acetate, ethanol). It has previously been shown that during batch growth of *Lactococcus lactis* on various sugars, the shift from homolactic to mixed-acid metabolism was directly dependent on the sugar consumption rate [1]. Under conditions of high glycolytic flux, *i.e.* when the sugar uptake rate is non-limiting, glycolysis is controlled at the level of glyceraldehyde-3-phosphate dehydrogenase (GAPDH) by a high $NADH/NAD^+$ ratio. Under such conditions, the flux limitation at the level of GAPDH leads to an increase in the pool concentrations of both glyceraldehyde-3-phosphate and dihydroxyacetone-phosphate provoking inhibition of pyruvate formate lyase activity, which together with the stimulation of lactate dehydrogenase activity by the high $NADH/NAD^+$ ratio, explains the homolactic metabolism. This ratio is an essential regulating factor shown to be proportional to the glycolytic flux.

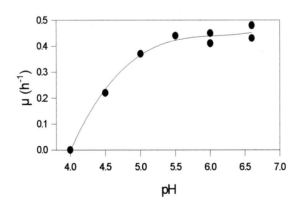

Figure 1. Maximal growth rate (μ) of *L. lactis* MG 1363 as a function of the pH of the culture
medium

This model of regulation was established for growth of the bacterium at the optimal pH for growth (6.6), while industrial food processes are characterised by a progressive acidification due to lactic acid production. In order to study the effect of pH on the growth behaviour of *L. lactis* ssp. *cremoris* MG 1363, fermentations were performed in anaerobic tube cultures, in synthetic medium with glucose as carbon substrate, and at various pH conditions. The maximal growth rate was maximal between pH 6.6 and 5.5, and decreased in more acidic media; no growth occurred at pH 4.0 (Fig 1). Independent of the pH of the culture medium, glucose metabolism remained homolactic.

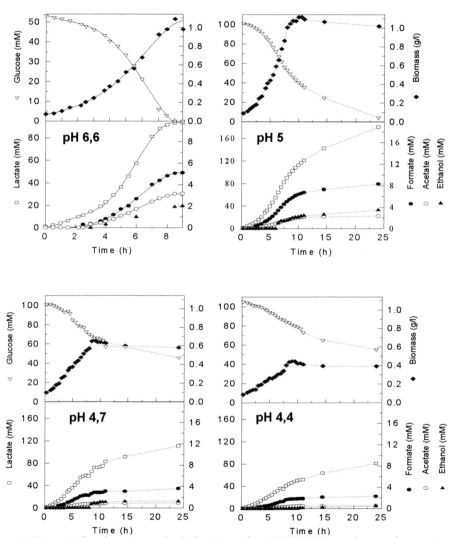

Figure 2. Fermentation time course for *L. lactis* ssp. *lactis* MG 1363 growing on glucose at different pH conditions.

Batch fermentations of *L. lactis* MG 1363 were then carried out at different pH-regulated conditions (6.6; 5.0; 4.7; 4.4), in order to study the control structure of glycolysis. The maximal growth rate observed in the different cultures decreased with the pH of the medium, as expected from the results presented above. However, the maximal specific rates of glucose consumption and of lactate production were maintained at similar levels in the range of pH from 6.6 to 4.7, and decreased only at pH 4.4. In the reference culture at pH 6.6, growth stopped when glucose had been fully exhausted, while in media of lower pH the decreased final biomass levels were attained while glucose still remained in the medium (Fig 2). Growth inhibition was most probably related to lactic acid inhibitory effect, the toxicity of the organic acid accumulation increasing in acidic conditions [2,3]. The growth phase at each pH condition was characterised by constant product yields, lactic acid being by far the major product. Hence the specific rate of ATP production calculated at different pH values was directly related to the specific rate of glucose consumption, while the specific growth rate was not, in contrast to the results obtained for different substrates or culture media at optimal pH [4]. As a consequence, the efficiency of biomass synthesis relative to the energy supply decreased by lowering the pH of the medium, as illustrated by the Y_{ATP} values which decreased from 11.5 at pH 6.6 to values between 5 and 6 for a medium pH below 5.0. It is interesting to note that during the cultures at pH 5 or lower, once the growth stopped despite incomplete use of the glucose, catabolism continued at lower rates, but homolactic fermentations were maintained.

In order to study the effect of the pH of the medium on the regulation of the central metabolism of *L. lactis* MG 1363, the concentration of glycolytic intermediates and coenzymes and the specific activities of some enzymes were measured in cell-free extracts prepared from cells growing exponentially in the different pH-regulated cultures, as previously described [1].

The internal concentration of glycolytic pools was not significantly different at the various pH conditions tested (Table 1). Only fructose-1,6-diphosphate (FDP) increased slightly when the pH was decreased. Glucose-6-phosphate (G6P) and FDP were the metabolites present at highest concentrations. On the other hand, the trioses-phosphates were present in low concentrations, though it should be noted that the GAP concentration was above the threshold sensitivity value (about 0.6 mM) and sufficiently high to saturate the GAPDH. As previously discussed [1], this level of FDP and GAP, together with the very low level of other trioses-phosphates donwstream of GAP, could be representative of a flux bottleneck at the level of the GAPDH. The intracellular NAD^+ concentration was similar for each pH condition, while the NADH concentration decreased as a function of culture acidity, resulting in a decreasing $NADH/NAD^+$ ratio (Table 1).

Among the crucial enzymes shown to be important in the control of glycolysis in *L. lactis* by Garrigues *et al.* [1], the glyceraldehyde-3-phosphate dehydrogenase (GAPDH) concentration increased 5-fold when the pH was lowered from 6.6 to 4.4 (from 900 to 4300 nmol.min^{-1}.mg protein^{-1}), while the specific activities of pyruvate kinase (PK) and lactate dehydrogenase (LDH) remained virtually unchanged (600 and 6000 nmol.min^{-1}.mg protein^{-1}, respectively).

Table 1. Glycolytic metabolite and coenzyme concentrations, expressed in mM, in cells harvested throughout the exponential growth phase of *L. lactis* MG1363 cultivated at different pH-regulated conditions.

pH	6,6	5	4,7	4,4
Glucose-1-P	1,5	1	<0.6	<0.6
Glucose-6-P	30	26	32	40
Fructose-6-P	1	<0.6	0,5	1
Fructose-1,6-diP	21	23	27	39
Dihydroxyacetone-P	<0.6	<0.6	<0.6	<0.6
Glyceraldehyde-3-P	0,8	0,8	1,8	0.6
1,3-diP-Glycerate	0,2	<0.6	<0.6	<0.6
3-P-Glycerate	<0.6	<0.6	0,5	<0.6
Phosphoenolpyruvate	2	<0.6	0,6	1,1
NAD$^+$	12,3	13,8	14,3	14,1
NADH	1,9	0,8	0,8	0,36
NADH / NAD$^+$	0,15	0,06	0,055	0,025

The activities of GAPDH and LDH of *L. lactis* were previously shown to be affected, with opposite effects, by the NADH/NAD$^+$ ratio [1]. These results were confirmed here for the enzymes of *L. lactis* MG 1363, since GAPDH was inhibited while LDH was activated (data not shown) by increasing the NADH/NAD$^+$ ratio. During the reference culture, performed at pH 6.6, the ratio value of 0.15 provoked a 60% inhibition of GAPDH, i.e., a residual activity of 40% of the maximal value, while decreasing the culture pH and consequently the NADH/NAD$^+$ ratio, led to residual activities of 68, 82 and 97% at pH 5, 4.7 and 4.4, respectively.

In order to take into account the effect of the intracellular environment on the enzyme activities, the internal pH of *L. lactis* cells growing exponentially at different pH conditions was measured by determining the internal to external gradient of ^{14}C-benzoic acid after centrifugation of the cells through silicone oil, as previously described [5]. The ΔpH value between the cytoplasm and the culture medium increased with the medium acidity, from 0.7 at pH 6.6 to 1.2 at pH 4.4. As a consequence, the internal pH values for the cultures performed at pH 6.6, 5, 4.7 and 4.4 were 7.3, 5.8, 5.6 and 5.6, respectively. This observation was consistent with previous results reviewed by Padan et al. [6] and later confirmed by other authors [7,8].

The GAPDH was strongly inhibited by decreased pH values. The residual activity taking into account the internal pH value of the bacterium decreased from 86 to 7% of the maximal value when the pH of the medium fell from 6.6 to 4.4. On the other hand, LDH activity was not affected by the pH values in the range tested (from 5.6 to 7.8).

The specific activity of GAPDH was modelled by the combined effect of the NADH/NAD$^+$ ratio and the intracellular pH to take into account the true intracellular parameters observed in each pH condition. For example, at a pH value of 6.6, the specific GAPDH activity of 900 nmol.min^{-1}.mg^{-1} was inhibited at 60% by the NADH/NAD$^+$ ratio and at 14% by the internal pH value. The resulting activity, effectively active inside the cell, was then 310 nmol.min^{-1}.mg^{-1}. For the other extreme pH condition (pH 4.4), the GAPDH activity of 4300 nmol.min^{-1}.mg^{-1} was inhibited at 3% by the coenzymes ratio and at 93% by the internal pH value. The resulting activity was 292 nmol.min^{-1}.mg^{-1}. For each condition tested the corrected GAPDH activity was quite similar to the flux passing through the pathway. While the GAPDH was previously demonstrated as a glycolytic bottleneck of *L. lactis* at neutral pH conditions [1], it was shown here that acidic conditions increased the enzyme synthesis but decreased the enzyme activity, resulting in a similar behaviour as regards the metabolic bottleneck of the glycolytic pathway. The combined effect on the GAPDH activity of the overall metabolic perturbations, level of enzyme concentration with the pH and effect of NADH/NAD$^+$ ratio and pH on the enzyme activity, was to maintain a high level of control over glycolysis at the level of GAPDH.

REFERENCES

1. C. Garrigues, P. Loubière, N. D. Lindley and M. Cocaign-Bousquet, J. Bacteriol. 179 (1997) 5282.
2. J. J. Baronofsky, W. J. A. Schreurs and E. R. Kashket, Appl. Environ. Microbiol. 48 (1984) 1134.
3. D. B. Kell, M. W. Peck, G. Rodger and J. G. Morris. B. B. R. C. 99 (1981) 81.
4. L. Novák, M. Cocaign-Bousquet, N. D. Lindley and P. Loubière, Appl. Environ. Microbiol. 63 (1997) 2665.
5. P. Loubière, P. Salou, M. J. Leroy, N. D. Lindley and A. Pareilleux, J. Bacteriol. 174 (1992) 5302.
6. E. Padan, D. Zilberstein and S. Schuldiner, Biochem. Biophys. Acta. 650 (1981) 151.
7. L. C. McDonald, H. P. Fleming and H. M. Hassan, Appl. Environ. Microbiol. 56 (1990) 2120.
8. N. L. Nannen and R. W. Hutkins, J. Dairy Sci. 74 (1991) 741.

Food Biotechnology
S. Bielecki, J. Tramper and J. Polak (Editors)
© 2000 Elsevier Science B.V. All rights reserved.

The influence of pH and oxygen on the growth and probiotic activity of lactic acid bacteria.

K.M. Stecka, R.A. Grzybowski

Institute of Agricultural and Food Biotechnology, Warsaw, Poland.

The influence of pH and presence of oxygen in the medium on the growth, lactic acid biosynthesis and probiotic activity of two lactic acid bacteria (LAB) strains, i.e. *Lactobacillus plantarum* K and *Bifidobacterium bifidum* 558 was examined.

The best growth of *B. bifidum* 558 on the level 10^9 CFU/ml was obtained at pH 6.0. The growth was considerably worse at pH 6.8 and pH 5.0. A similar dependence was obtained for biosynthesis of L (+) lactic acid. This influence of pH on the biosynthesis of D(-) lactic acid was not observed. The probiotic activity of *B. bifidum* 558 was the highest at pH 6.8.

The strain *L. plantarum* K reached the highest growth, on the level of 10^8 CFU/ml, in a culture grown at pH 5.0. The efficiency of biosynthesis of D(-) lactic acid was highest between pH 6.0 and 5.0. The probiotic activity of *L. plantarum* K, like in the case of *B. bifidum* 558, was the highest at pH 6.8, when the effectiveness of biosynthesis organic acids was the lowest.

The presence of oxygen in the culture medium (microaerobic conditions) did not influence on the growth of the strain *B. bifidum* 558 and its probiotic activity. The strain *L. plantarum* K was growing better under aerobic conditions than under anaerobic ones. The presence of oxygen lowered, however, the yield of biosynthesis of lactic acid and probiotic activity.

1. INTRODUCTION

Recently, attention has been paid to formerly known phenomenon of probiosis, which term denotes interactions between microorganisms and, at the same time, utilization of interactions between microorganisms was originated as a method of human and animal health protection. The term "probiotic" was used for the first time by Lilli and O'Sulivan [1] to describe substances stimulating growth of other organisms and produced by protozoa.

Probiotics are new generation growth stimulators and their action consists in utilization and stimulation of naturally occurring system of metabolism supporting and self-protection against infection. Investigations carried out by physiologists indicated that, actually, autochthonous microflora plays an important role in digestion process and supports the immunological system. Approximately 90% of bacteria present in alimentary system are lactic bacteria (LAB) and they are used most often to obtain probiotics [2].

Probiotics as a properly chosen qualitative and quantitative composition of microorganisms able to synthesize specific metabolites exert an influence on assimilation of fodder, simultaneously playing a role of a regulator of ecological balance of microflora in digestive system [3].

Lactic acid bacteria, which are most often used in order to obtain the probiotics are characterised by many antagonistic features in comparison with other microorganisms. The idea of their activity in digestive system mainly consists in:
- a decrease of pH production: lactic, acetic, formic and propionic acid,
- competitive elimination of pathogenic microflora,
- production of substances of antimicrobiological character,
- neutralization of pathogenic enterotoxins, especially enterotoxins of *Escherichia coli*.

Microorganisms used for production of probiotics must be recognized as safe – GRASS (Generally Recognized As Safe). For this reason, among other things, the most often utilized microorganisms are lactic acid bacteria belonging to *Lactobacillus* and some bacteria belonging to *Bifidobacterium* and *Streptococcus genus*.

In recent years, a great progress of knowledge of biology and genetics of lactic acid bacteria was accomplished. The subject matter of investigations is strictly connected with properties of bacteria, which are of great importance for lactic fermentation course, and concerns biosynthesis of bacteriocins and also widely understood probiotic activity. It has been determined that many of these properties are coded by genes located in plasmids, so by extrachromosomal genes. These features concern, for example, utilization of lactose as carbon and energy source. Beside properties conditioning proper growth of lactic acid bacteria, plasmids contain genes determining biosynthesis of bacteriocins, the albuminous substances influencing substantially the total probiotic activity.

Environmental conditions determine to what extent the genetic features of the certain microorganism are utilized. Some investigations suggest that activity of bacteriocins produced by lactic acid bacteria is influenced by composition of a culture medium and growth conditions.

Traditionally, LAB have been classified as microaerophils, so as the microorganisms well growing under anaerobic conditions and able to grow in a presence of oxygen, however, their growth is then slower. Currently, it is stated that, concerning possibility of growth under aerobic conditions, LAB constitute a very differentiated group of organisms. On examination of group of 22 LAB strains coming from Japanese collection, Sakamoto and Komagata [4] have found strains unable as well as strains able to grow in the presence of oxygen. Among strains able to grow under aerobic conditions, one has observed a very differentiated course of growth curves – some of them have grown better under aerobic conditions.

The aim of the present work was to investigate the influence of pH and presence of oxygen in a culture medium on growth, biosynthesis of lactic acid and probiotic activity of selected strains of lactic acid bacteria.

2. MATERIALS AND METHODS

The following LAB strains have been used for investigations: *L. plantarum* K and *B. bifidum* 558.

Culture medium: Bacto Lactobacilli MRS (Difco).

The cultures were carried out in New Brunswick laboratory fermentors Bioflo III of total capacity 6.5 l and working capacity 5.0–5.5 l. Culture conditions: temperature 37°C, pH 5.0-6.8, breeding time 20 h. Stirring speed 50 rpm, pH was regulated by dosing 12.5%

ammonia solution. Anaerobic conditions of the culture were assured by saturation of the culture with nitrogen and creation of nitrogen "cushion" over liquid surface.

Cultures under aerobic conditions were grown with stirring rate of 250–400 rpm – aeration was through the liquid surface.

Determination of bacterial growth density expressed as CFU/ml in post breeding liquid was carried out into Blickfeldt's medium.

Contents of L and D-lactic acid was determined by enzymatic test D-Lactat/L-Lactat by Beckman, type DU-640, at wave length $\lambda=340$ nm.

Determination of concentration of volatile organic acids (acetic, propionic, isobutyric, butyric, isovaleric and valeric) was carried out by gas chromatography with Hewlett-Packard 5890 apparatus.

The measure of probiotic activity is an amount of a substance of antimicrobial character, biosynthesized by the examined strains. It was determined by the measurement of a growth inhibition zone of the indicatory strains (*Bacillus subtilis, Listeria innocua*).

3. RESULTS AND DISCUSSION

The aim of the conducted experiments was investigation of the effect of pH and oxygen presence in the culture medium on growth, biosynthesis of organic acids and probiotic activity of the chosen LAB strains.

The cultures of *B. bifidum* 558 and *L. plantarum* K were grown at pH 5.0, 6.0 and 6.8. The obtained results are presented in Tables 1 and 2.

Table 1
The influence of pH on the growth of *B. bifidum* 558 and its organic acid production

pH	CFU/ml	Lactic acid, g/l		Acetic acid, g/l	Propionic acid, g/l
		L (+)	D (-)		
5.0	3.1×10^8	9.6	1.9	2.1	1.1
6.0	9.8×10^9	14.0	2.1	0.4	0.6
6.8	1.3×10^8	6.5	1.7	3.1	2.0

Table 2
The influence of pH on the growth of *L. plantarum* K and its organic acid production

pH	CFU/ml	Lactic acid, g/l		Acetic acid, g/l	Propionic acid, g/l
		L (+)	D (-)		
5.0	4.5×10^8	4.8	4.6	1.3	2.1
6.0	2.5×10^7	5.6	3.8	1.7	2.0
6.8	9.8×10^7	3.8	2.3	1.4	1.2

The best growth of *B. bifidum* expressed as CFU/ml was obtained at pH 6.0. The growth was considerably worse in more alkaline conditions (at pH 6.0) and similar to that obtained at pH 5.0.

A similar dependence was obtained for L-lactic acid production. The maximum yield of this acid biosynthesis was obtained in the culture at pH 6.0, slightly lower at pH 5.0; at pH 6.8 the yield of L-lactic acid biosynthesis was by 50% lower in comparison with the culture grown at pH 6.0. No effect of pH on the D(-) lactic acid biosynthesis was observed.

The ability of volatile organic acids synthesis was established by different way. The highest efficiency of acetic and propionic acids biosynthesis was at pH 6.8, lower at pH 5.0 and apparently the lowest at pH 6.0, when *B. bifidum* 558 strain achieved the maximum of growth and biosynthesis of L-lactic acid. However, it should be emphasised that the yield of of the volatile organic acids biosynthesis was small in comparison with lactic acid biosynthesis (L+D), so the total yield of organic acids biosynthesis was the highest at pH 6.0, lower at pH 5.0 and the lowest at pH 6.8 (Figure 1).

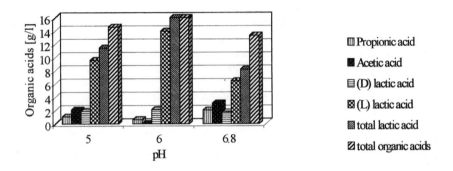

Figure 1. The influence of pH on biosynthesis of organic acids by *B. bifidum* 558

The propinic activity *B. bifidum* 558 strain increased on an increase of pH of the culture concerning both indicatory strains (Figure 2).

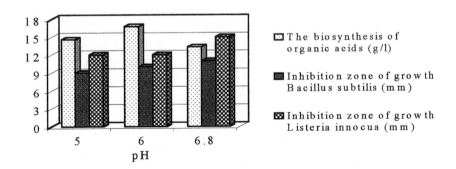

Figure 2. The influence of pH on probiotic activity *B. bifidum* 558.

The conducted studies indicated that pH 6.0 was optimal both for growth of *B. bifidum* 558 strain and for organic acids biosynthesis, however, the highest probiotic activity was determined at pH 6.8.

L. plantarum K strain achieved the best growth at pH 5.8 (Table 2). The efficiency of D-lactic acid biosynthesis decreased with the increase of pH, whereas L-lactic acid was synthesized more efficiently at pH 5.0 and 6.0, and worse at pH 6.8. The efficiency of propionic acid biosynthesis slightly decreased with the increase of pH, whereas the yield of acetic acid biosynthesis was not dependent on pH.

Summarizing, in the case of *L. plantarum* K, the total yield of lactic acid biosynthesis (L+D) decreased with the increase of pH, whereas the total biosynthesis of organic acids was the most efficient at pH 6.0 (Figure 3).

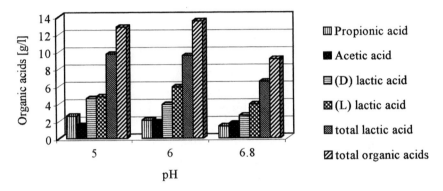

Figure 3. The influence of pH on biosynthesis of organic acids by *L. plantarum* K.

Similarly to *B. bifidum* 558 strain, *L. plantarum* K strain achieved the highest probiotic activity at pH 6.8, when efficiency of organic acids biosynthesis was the lowest (Figure 4).

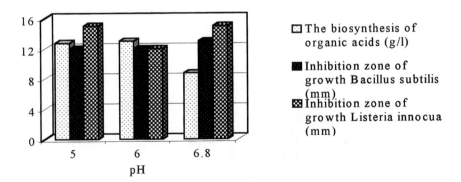

Figure 4. The influence of pH on probiotic activity *L. plantarum* K.

280

The above observations indicate that, in order to achieve higher yield of organic acids biosynthesis and better growth of bacteria, this strain culture should be grown at pH 5.0, however, *L. plantarum* K shows the highest probiotic activity in the culture at pH 6.8.

In consecutive experiments, the effect of stirring speed (the rate of aeration) on growth and probiotic activity of the examined LAB strains has been investigated.

Anaerobic conditions (corresponding to 0 rpm speed of the stirrer) were obtained by saturation of the culture medium with nitrogen and formation of "nitrogen cushion" over it. Microaerobic conditions were obtained by maintaining of the stirring speed at constant level of 50 rpm, and aerobic conditions by the increase of the stirring speed to 250–440 rpm. The results of the investigations obtained for *B. bifidum* 558 are presented in diagrams 5 and 6, and for *L. plantarum* K strain – in diagrams 7 and 8.

The *B. bifidum* strain achieved the best growth under microaerobic conditions, and the worst – under aerobic conditions, when stirring speed was increased (Figure 5).

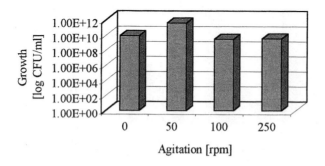

Figure 5. The influence of rotational speed of stirrer on the growth *B. bifidum* 558.

The highest probiotic activity has been determined under microaerobic conditions, the lowest – under strictly anaerobic conditions (Figure 6). It may be concluded from the conducted studies that *B. bifidum* 558 strain belongs to microaerophils.

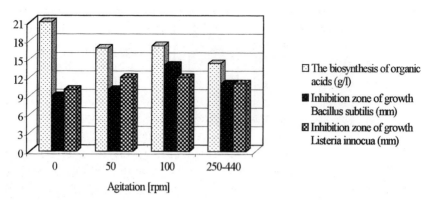

Figure 6. The influence of rotational speed of stirrer on probiotic activity *B. bifidum* 558.

The *L. plantarum* K strain belongs to these few LAB which grow with the higher yield under aerobic conditions. Similar effects for selected LAB strains was achieved by Komagata. The presence of oxygen in the culture medium unfavourably influenced the yield of organic acids biosynthesis and probiotic activity of *L. plantarum* K.

4. SUMMARY

The pH 6.0 of the culture is optimal for growth and biosynthesis of lactic acid by *B. bifidum* 558 and *L. plantarum* K strains. *B. bifidum* 558 strain belongs to microaerophils. Therefore, strict anaerobic conditions are not required for obtaining the high yield of biomass biosynthesis at a range of 10^9 CFU/ml and for the high probiotic activity.

The *L. plantarum* K strain achieves higher yield of biomass (10^8 CFU/ml) under aerobic conditions in comparison with anaerobic conditions. However, the presence of oxygen results in a decrease of yield of lactic acid biosynthesis and probiotic activities.

The presence of peroxides, hydrogen peroxide and, also, free oxygen is bactericidal for strictly anaerobic bacteria because they do not possess enzymes: catalase, peroxidase and dismutase of peroxides. LAB bacteria showing ability to aerobic growth possess, variously developed, systems of "detoxication" of environment. Although many systems of detoxication in LAB are well characterised regarding genetic aspects, molecular details of a response to oxygen stress are not explained yet and are subjected to various investigations.

REFERENCES

1. M.G. O'Sullivan, G. Thornton, G.C. O'Sullivan, J.K Collins, Trends Food Sci. Tech., 3 (1992) 309.
2. G.W. Tannock, Probiotics. A critical review, Harizon Scientific Press, 1999.
3. R. Fuller, Probiotics. The scientific basic. Chapman and Hall, 1992.
4. M. Sakamoto and K. Komagata, J. Fermentation and Bioengneering, 82 (1996).

Food Biotechnology
S. Bielecki, J. Tramper and J. Polak (Editors)

Microbiological changes in modified yoghurts during manufacture and storage

M. Bielecka, A. Majkowska, E. Biedrzycka, El. Biedrzycka

Department of Food Microbiology, Division of Food Science, Institute of Animal
Reproduction and Food Research of the Polish Academy of Sciences,
ul. Tuwima 10, 10-747 Olsztyn, Poland

The live cells of yoghurt cultures were enumerated after milk inoculation and after
fermentation at the temperatures: 30, 37 and 42°C as well as during 28 days of storage at 6°C.
Active or preserved bacteria cultures were used for milk inoculation at the level 10^5-10^7 cfu/g.
During fermentation up to pH 4.6-4.8 *Streptococcus thermophilus* and *Lactobacillus
delbrueckii* subsp. *bulgaricus* multiplied, their population numbers increased 1-2 log cycles
and reached the level of 10^8-10^9 cfu/g, whilst the numbers of probiotic bacteria *Lactobacillus
acidophilus* and *Bifidobacterium animalis* increased only 2-4 times independently of the
inoculum level, form of bacteria cultures and temperature of fermentation. Time of
fermentation was in negative correlation with temperature and ranged from 2 h 55 min – 3 h
40 min at 42°C to 4 h 35 min – 7 h 30 min at 30°C. Duration of the fermentation was also
affected by the form of bacteria cultures with one exception at the temperature of 37°C where
the time of fermentation, ranging from 4 h to 4 h 10 min, was not influenced by the bacteria
culture composition and form. The population of every species decreased slowly in limited
range during storage of the yoghurts, therefore after 28 days the live cell numbers were few
times lower than after fermentation. The results indicate the necessity of inoculation of milk
during bio-yoghurt manufacture with the number of probiotic bacteria $\geq 10^6$ live cells/g which is
recommended in the final product.

1. INTRODUCTION

Yoghurt has become the most valuable fermented milk product because of its taste as well
as nutritive, dietary and wholesome value, the latter being related to the presence of live
bacteria cultures [1-4]. Recently, emphasis has been focused on the development of yoghurts
supplemented with probiotic bacteria, most of all with *Lactobacillus acidophilus* and
bifidobacteria, which are natural inhabitants of the intestines. Due to the fact that
L. acidophilus colonises the small intestine, and *Bifidobacterium* the large intestine, the best
effects can be obtained when the two genera are used together, provided that synergistic sets
of strains are selected [16]. The benefits of consuming supplemented yoghurts (bio-yoghurts)
are well documented and have been reviewed by several authors [5-7]. In order to produce the
therapeutic effects, the suggested minimum level for probiotic bacteria in bio-yoghurts is
10^6-10^7 viable cells per g or ml [5,7,8]. Probiotic bacteria, however, multiply in milk much

slower than yoghurt cultures, so there is little probability that they might multiply during the technological process. In addition to this, their survival during storage of fermented milks is diversified and affected by a number of factors [9-15]. Although, there has been a tremendous increase in the bio-yoghurt markets of Europe, North America, Japan and many other countries over the past decade, numbers of live cells of probiotic bacteria are still too low [15,17,18].

The aim of the studies was to determine changes in the numbers of *Bifidobacterium* and *L. acidophilus*, and of yoghurt cultures in course of the production and storage of bio-yoghurts.

2. MATERIAL AND METHODS

2.1. Bacterial cultures
Experimental bio-yoghurts were prepared using the earlier developed synergistic sets of bacteria strains: *Lactobacillus delbrueckii* subsp. *bulgaricus* 151, *Streptococcus thermophilus* MK-10, *L. acidophilus* 8/4 and 43/15, *Bifidobacterium animalis* Bi30 and Bi45 [16]. The bifidobacteria strains were isolated from commercial bio-yoghurts. The following sets: C1 (151 + MK-10 + 8/4 + Bi30); C2 (151 + MK-10 + 43/15 + Bi45); C3 (151 + MK-10 + Bi30); C4 (151 + MK-10 + Bi45); C5 (151 + MK-10 + 8/4) and C6 (151 + MK-10 + 43/15) were used as active or preserved cultures. Active yoghurt cultures were multiplied in reconstituted skim milk (10% d.w.). Incubation at 42°C was terminated at pH 4.4-4.5, and at rods to the cocci proportion of 1:1 - 1:2 [16]. *L. acidophilus* and bifidobacteria were multiplied separately in reconstituted skim milk (10% d.w.) under anaerobic conditions, at the temperature of 37°C, to reach the stationary growth phase (pH 4.4-4.5). Preserved yoghurt cultures were prepared by spray drying method and probiotic cultures by lyophilisation.

2.2. Preparation of bio-yoghurt
Homogenised and pasteurised milk (2% fat) was condensed with 4% (w/v) of non-fat milk powder, pasteurised at 90°C for 10 min, cooled to required temperature, and inoculated with active cultures 10^5-10^7 cfu/g or preserved approx. 10^7cfu/g. Fermentation was carried out in plastic cups (150 ml capacity) at the temperatures: 30, 37 and 42°C to reach pH 4.6-4.8, then cooled and stored at 6°C for 4 weeks.

2.3. Analyses
Microbiological analyses were done after milk inoculation and fermentation, and during storage of bio-yoghurts in one-week intervals. Live cells of bacteria were counted using spiral plating technique performed with a Whitley Automatic Spiral Plater WASP (Don Whitley Scientific Ltd., U.K.), and the results were expressed as the colony forming units (cfu)/g. Double layer of MRS agar medium adjusted to pH 5.4, and incubation at 44°C for 72 h were used for differential enumeration of *L. delbrueckii* subsp. *bulgaricus*, while M17 agar and aerobic incubation at 37°C for 48 h were used for the enumeration of *S. thermophilus* [19]. Modified Garche's agar medium [20] (composition: Peptobak (meat peptone) 20 g, yeast extract 2.0 g, L-cysteine hydrochloride 0.4 g, lactose 10.0 g, CH_3COONa 6.0 g, KH_2PO_4 2.0 g, Na_2HPO_4 x 12 H_2O 2.5 g, $MgSO_4$ x 7 H_2O 0.12 g, agar 15.0 g, distilled water to 1000 ml, pH 6.4) supplemented with neomycin sulphate 20 mg (Sigma), penicillin G potassium

salt 50 U, (Sigma), lithium chloride 3.0 g, and anaerobic incubation (Gas Pak Anaerobic System, Oxoid, UK) at 37°C for 72 h were applied for selective enumeration of *Bifidobacterium*. BCP agar medium without glucose [21] (composition: yeast extract 2.5 g, peptone 5.0 g, Tween 80 1.0 g, L-cysteine hydrochloride 0.3 g, bromocresol purple 0.04 g, pyruvic acid 1.0 ml, K_2HPO_4 3.0 g, $MgSO_4·7H_2O$ 575 mg, $FeSO_4·7H_2O$ 34 mg, $MnSO_4·4H_2O$ 120 mg, agar 15 g, distilled water to 1000 ml, pH 6.8-7.2) supplemented with salicin 0.5 g, and anaerobic (as above) incubation was used at 37°C for 48 h for selective enumeration of *L. acidophilus*. The pH values were measured using PHM85 precision pH-meter (Radiometer, Denmark) and the proportions of rods to cocci were defined microscopically. The results of the assessments of the viability of yoghurt and probiotic bacteria were given as average values of three replications.

3. RESULTS AND DISCUSSION

3.1. Changes in the counts of bio-yoghurt bacteria

S. thermophilus population numbers increased in approximately 1-2 log cycles as the effect of their multiplication during technological process, and attained maximum numbers of 2.8×10^9 – 3.2×10^9 cfu/g, respectively to preserved or active cultures (Table 1). During 28-day storage of the bio-yoghurt at 6°C, counts of these bacteria viable cells decreased about 2-3-fold, to 9.3×10^8 – 1.2×10^9 cfu/g. *L. delbrueckii* subsp. *bulgaricus* multiplied during yoghurt manufacture at a slower rate than *S. thermophilus*, so their populations reached the lower counts of 5.5×10^8 – 1.3×10^9 cfu/g when active cultures were used for inoculation, and of 1.9×10^8 – 2.8×10^8 in the case of inoculation with preserved cultures. These values decreased during storage 2-3 times to the level of 2.0×10^8 – $6,2 \times 10^8$ cfu/g in yoghurts manufactured with active cultures and to 7.1×10^7 – 1.6×10^8 cfu/g when preserved cultures were used.

Counts of *L. delbrueckii* subsp. *bulgaricus* were about half to one log cycle lower than those of *S. thermophilus*, both in fresh and stored yoghurt. Neither form nor level of inoculum affected cell counts of *S. thermophilus*, but multiplication of *L. delbrueckii* subsp. *bulgaricus* was lower when preserved cultures were used (Table 1, Figure 1). Much lower numbers of viable cells of *L. delbrueckii* subsp. *bulgaricus* compared to *S. thermophilus* were observed also by other authors in fresh as well as stored yoghurts [22-24].

Population numbers of probiotic bacteria *L. acidophilus* and *B. animalis* increased 2-4-fold during bio-yoghurt manufacturing, and decreased at an almost the same rate during yoghurt storage. Consequently, viable cell counts of these bacteria after 28-day storage were almost the same or very close to the levels just after inoculation, amounting to 1.8×10^5 – $6,5 \times 10^7$ cfu/g for *L. acidophilus*, and to 4.4×10^6 - 8.7×10^7 for *B. animalis* (Table 1). Good survival of probiotic bacteria during storage resulted from the strain selection considering the technological properties. These results confirm the Lanhaputhra's et al. [11] observation of high diversity among bifidobacteria strains as regards to their survival during storage of acidified milk.

Table 1
Changes in number of bacteria during bio-yoghurt manufacture at 37°C and storage at 6°C (log cfu/g)

Culture	Period	Active cultures			
		ST[1]	LB[2]	LA[3]	BA[4]
C 1	0 h[5]	7.32 ± 0.17[8]	6.98 ± 0.12	5.08 ± 0 27	6.61 ± 0.42
	AF[6]	9.50 ± 0.34	9.11 ± 0.16	5.41 ± 0.42	7.26 ± 0.25
	28 d[7]	9.08 ± 0.09	8.52 ± 0.25	5.25 ± 0.27	7.11 ± 0.21
C 2	0 h	7.49 ± 0.14	7.30 ± 0.17	6.71 ± 0.19	6.60 ± 0.19
	AF	9.26 ± 0.17	8.74 ± 0.19	7.36 ± 0.26	7.14 ± 0.27
	28 d	8.96 ± 0.09	8.30 ± 0.15	6.71 ± 0.31	6.64 ± 0.18
C 3	0 h	6.57 ± 0.23	6.32 ± 0.27		7.79 ± 0.09
	AF	9.38 ± 0.15	8.99 ± 0.18		8.34 ± 0.23
	28 d	9.04 ± 0.07	8.79 ± 0.09		7.93 ± 0.16
C 4	0 h	6.62 ± 0.06	6.66 ± 0.25		7.70 ± 0.40
	AF	9.28 ± 0.12	8.85 ± 023		8.30 ± 0.27
	28 d	9.08 ± 0.24	8.72 ± 0.18		7.91 ± 0.19
C 5	0 h	6.90 ± 0.19	6.81 ± 0.16	7.54 ± 0.46	
	AF	9.49 ± 0.09	8.92 ± 0.12	8.08 ± 0.32	
	28 d	9.08 ± 0.13	8.70 ± 0.17	7.81 ± 0.08	
C 6	0 h	6.92 ± 0.16	6.86 ± 0.24	7.51 ± 0.28	
	AF	9.38 ± 0.39	8.97 ± 0.13	8.00 ± 0.28	
	28 d	8.96 ± 0.27	8.63 ± 0.15	7.56 ± 0.31	
		Preserved cultures			
C 1	0 h	7.70 ± 0.21	7.69 ± 0.66	6.94 ± 0.29	6.80 ±0.36
	AF	9.44 ± 0.49	8.44 ± 0.42	6.99 ± 0.41	7.32 ± 0.27
	28 d	9.20 ± 0.52	8.20 ±0.32	6.91 ±0.39	7.08 ± 0.42
C 2	0 h	7.64 ± 0.34	7.61 ± 0.52	6.08 ± 0.28	6.73 ± 0.25
	AF	9.34 ± 0.41	8.28 ± 0.29	6.36 ± 0.31	7.28 ± 0.41
	28 d	9.11 ± 0.37	7.85 ± 0.21	5.98 ± 0.17	6.95 ± 0.37

[1] ST – *Streptococcus thermophilus*; [2] LB – *Lactobacillus delbrueckii* subsp. *bulgaricus*;
[3] LA – *Lactobacillus acidophilus*; [4] BA – *Bifidobacterium animalis*; [5] 0 h –after inoculation;
[6] AF - after fermentation; [7] 28 d – after 28-day storage;
[8] mean value calculated from triplicates and standard deviation.

3.2. Effect of technological parameters on bio-yoghurt bacteria counts

Fermentation temperature of 30, 37 and 42°C had no significant effect on *S. thermophilus* and *L. delbrueckii* subsp. *bulgaricus* population numbers in the bio-yoghurts if the fermentation was terminated at the pH 4.6-4.8 as optimal for this product (Figure 1). Temperature of fermentation did not also influence the bacteria cultures survival during storage of bio-yoghurts.

Physiological state of the cultures (active or preserved) used for inoculation of milk had no effect on the growth and survival of particular species of bacteria with the exception of

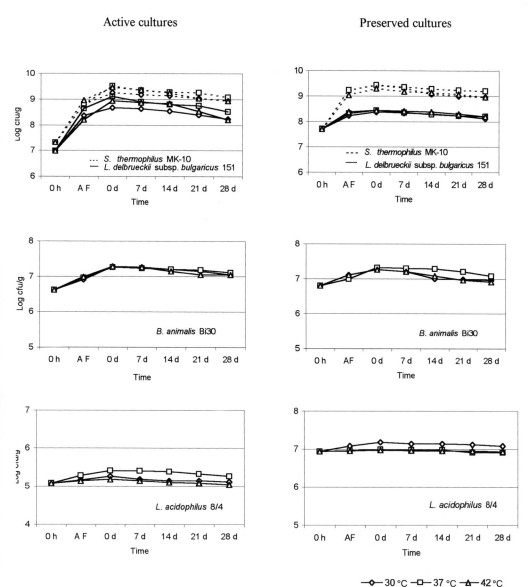

Figure 1. Effect of fermentation temperature, physiological state and inoculum of bacteria cultures on their growth during manufacture and survival during storage of bio-yoghurts.

Active or preserved cultures were used for milk inoculation; 30°C, 37°C, 42°C – fermentation temperatures; 0 h, AF, 0 d, 7-28 d = observation taken after milk inoculation, after fermentation and overnight storage, and during 7 to 28 days of bio-yoghurt storage, respectively.

L. delbrueckii subsp. *bulgaricus*. This species used in form of preserved cultures attained lower population numbers when compared with active cultures. However, reduction of viable bacteria cell counts during storage was lower than in the case of active cultures, so the cfu numbers after 28 days were similar in both cases. *L. delbrueckii* subsp. *bulgaricus* population numbers about 10 times lower than that of *S. thermophilus* in fresh yoghurt are a usual phenomenon, favourable for the taste and aroma of this product.

The inoculum level did not significantly affect the final population numbers of *L. delbrueckii* subsp. *bulgaricus* and *S. thermophilus*, because these species multiply during fermentation. Probiotic bacteria *L. acidophilus* and *Bifidobacterium* did not practically multiplied during bio-yoghurt manufacture, therefore, their live cell numbers in the end product were dependent on the inoculum. Neither physiological state of probiotic bacteria nor temperature of fermentation influenced their growth and survival.

3.3. Time of fermentation

Time of milk fermentation to pH 4.6-4.8 during bio-yoghurt manufacture was different depending on the temperature and physiological state of bacteria cultures (Table 2). The longest fermentation of 7 h 10 min - 7 h 30 min took place when preserved cultures were used and the temperature was 30^0C. At this temperature, the active cultures acidified milk during 4 h 35 min - 4 h 50 min. The shortest fermentation time of 2 h 55 min – 3 h was observed at 42^0C with active bacteria cultures, and at this temperature, the differences resulting from the physiological state of bacteria cultures were from 30 to 45 min. On the other hand, at 37°C fermentation time was almost uniform: 4 h - 4 h 10 min, and only 5 min longer with preserved cultures independently of their composition.

Table 2
Effect of bacteria physiological state and fermentation temperature on time of milk acidification to pH 4.6-4.8 during bio-yoghurt manufacture

		Time of fermentation (h.min)[1]		
Physiological state of bacteria		Temperature (°C)		
		30	37	42
C1	active	4.35	4.05	3.00
	preserved	7.30	4.10	3.30
C2	active	4.50	4.00	2.55
	preserved	7.10	4.05	3.40

[1] The mean time of 3 batches for each culture.

4. SUMMARY

During yoghurt manufacture, the *L. delbrueckii* subsp. *bulgaricus* and *S. thermophilus* cultures increased their population numbers by 1 - 2 log cycles attaining over 10^8 and 10^9 live cells/g, respectively, independently of fermentation temperature. These cultures acidified milk and created taste and flavour of the product. Cell numbers of probiotic bacteria *L. acidophilus* and *B. animalis* increased only 2-4-fold during fermentation and their numbers ranged 10^5-10^7 cfu/g due to the inoculum level. These probiotic bacteria did not affect the taste of bio-yoghurts. Negative correlation was observed between temperature of fermentation and time of milk acidification to pH 4.6-4.8, although, there was no effect on the numbers of bacteria live cells in the product. Slow, gradual decrease in live cell numbers took place during 28-day storage at 6°C, resulting in a 2-4-fold lowering of the populations. The results indicate that the probiotic bacteria in recommended numbers $\geq 10^6$ of live cells in bio-yoghurts should be added into milk during its inoculation.

REFERENCES

1. H.C. Death and A.Y. Tamime, J. Food Protect., 44 (1981) 78.
2. R.K. Robinson, Dairy Ind. Int., 54 No.7 (1989) 23-25.
3. B. Bianchi-Salvadori, P. Camaschella and E. Bazzigaluppi, Milchwissenschaft, 39 (1984) 387.
4. B. Bianchi-Salvadori, Int. J. Immunotherapy, Suppl. II (1986) 9.
5. J.A. Kurmann and J.L. Rasic, *In* Therapeutic Properties of Fermented Milks (ed. R.K. Robinson), Elsevier Applied Food Sciences Series, London (1991) 117.
6. B.K. Mital and S.K. Garg, Food Rev. Int., 8 (1992) 347.
7. N. Ishibashi and S. Shimamura, Food Technol., 47 (1993) 126.
8. R.K. Robinson, South African J. Dairy Sci., 19 No.1 (1987) 25.
9. W.E.V. Lankaputhra and N.P. Shah, 24th Inter. Dairy Congress, Melbourne, Australia, Sept. 18-22 (1994), Ha3P, 292.
10. W.E.V. Lankaputhra and N.P. Shah, Cult. Dairy Prod. J., 30 (1995) 2.
11. W.E.V. Lankaputhra, N.P. Shah and M.L. Britz, Milchwissenschaft, 51/2 (1996) 65.
12. J.H. Martin and K.M. Chou, Cult. Dairy Prod. J., 27 No.4 (1992) 21 23.
13. F.A.M. Klaver, F. Kingma and A.H. Weerkamp, Neth. Milk Dairy J., 47 (1993) 151.
14. W. Kneifel, D. Jaros and F. Erhard., Int. J. Food Microbiol., 18 (1993) 179.
15. L.M. Medina and R. Jordono, J. Food Protect., 56 (1994) 731.
16. M. Bielecka, A. Majkowska and El. Biedrzycka, Pol. J. Food Nutr. Sci., 3/44 (1994) 63.
17. M.G. O'Sullivan, G. Thornton, G.C. O'Sullivan and J.K. Collins, Trends in Food Science and Technology, 3 (1992) 309.
18. N. Shah, W.E.V. Lankaputhra, M. Britz and W.S.A. Kyle, Inter. Dairy J., 5 (1995) 515.
19. IDF Standard 146:1991.
20. Bulletin IDF, 252 (1990) 28.
21. Bulletin IDF, 252 (1990) 33.
22. W.T. Hamann and E.H. Marth, J. Food Protect., 47 10 (1984) 781.
23. N. Micanel, I.N. Haynes and M.J. Playne, Austr. J. Dairy Technol., 52 (1997) 24.
24. R.I. Dave and N.P. Shah, Int. Dairy J., 7 (1997) 31.

Food Biotechnology
S. Bielecki, J. Tramper and J. Polak (Editors)

Growth of lactic acid bacteria in alginate/starch capsules

R. Dembczynski, T. Jankowski

Department of Biotechnology and Food Microbiology, Agricultural University of Poznan, Mazowiecka 48, 60-623 Poznan, Poland

Lactobacillus acidophilus cells were encapsulated in alginate/starch liquid-core capsules and cultured in a whey-based growth medium to enable concentrated biomass production in immobilized form. The population of encapsulated viable cells reached the concentration of 6.5×10^{10} CFU/ml of capsule after 30 hours of fermentation, while the population of free cells cultured under the same conditions during control fermentation was 7.2×10^9 CFU/ml of growth medium after 45 hours. The capsules remained stable and no cell release was observed during the course of cell growth.

1. INTRODUCTION

There is a growing interest in developing inexpensive and efficient production methods for lactic acid bacteria starter cultures. In traditional suspension fermentation there are many factors which inhibit cell growth such as high pH value and concentration of oxygen, and the presence of antibiotics and bacteriophages in the culture medium. As a result, cell density rarely exceeds 10^9 CFU/ml of the culture medium and expensive separation equipment is required to concentrate the cells. Recently viable lactic acid bacteria immobilized in gels have been reported in the literature [1,2,3]. These methods avoid the disadvantages of traditional suspension cultures and make it possible to reach cell densities exceeding 10^{11} CFU/ml of a gel matrix. Biomass can easily be separated from the medium without centrifugation or filtration because of the large size of gel beads thus reducing the overall cost. The main problem is the release of cells into the culture medium as a result of substrate diffusion restrictions inside the gel [4]. Immobilizing cells inside capsules with a liquid core surrounded by a thin, crosslinked polymer membrane can eliminate these adverse effects. The semipermeable polymer membrane is much less diffusion-resistant than a compact gel. Thus biomass can grow in the whole volume of the liquid core reaching very high densities without the cell release.

In this work liquid-core capsules with alginate membranes previously developed [5] were evaluated for the production of the high-density starter culture of *Lactobacillus acidophilus*.

2. MATERIALS AND METHODS

2.1. Microorganism and growth medium

The commercially available lyophilized culture of *Lactobacillus acidophilus* (BIOMED, Lublin, Poland) was used throughout all the experiments. The inocula were obtained from the cultures in the late logarithmic growth phase, cultivated in a MRS broth at 37°C in a bath shaker.

The composition of the fermentation medium was the following: 65 g of whey powder, 10 g of yeast extract, 10 g of tryptone, 22.2 g of $CaCl_2$, and 1000 ml of distilled water. The medium was adjusted to pH 6 with NaOH and clarified with a Prostak microfiltration unit (Millipore, Milford, USA) with a pore diameter of 0.45 μm and sterilized at 121°C for 15 min.

2.2. Cell encapsulation and fermentation

Cells for inoculum were recovered from the MRS broth by centrifugation at 2600 g for 15 minutes prior to immobilization and resuspended in 200 ml of a 4% solution of hydroxy-propyl-ammonium starch (Central Potato Research Laboratory, Poznan, Poland) containing 100 mM $CaCl_2$. The mixture was extruded drop by drop into a sterile 1% Manugel DMB sodium alginate solution (Kelco, Waterfield, UK) containing 0.1% of Tween 80 at room temperature with continuous stirring. After 10 minutes of membrane forming in alginate solution, capsules were collected with a sieve and washed with a sterile physiological saline. The capsules were transferred into a 200 mM $CaCl_2$ solution and hardened for 10 minutes. The formed capsules were washed once again with a sterile physiological saline and transferred to a bioreactor.

Fermentation was carried out batchwise in a Bioflo III 5 l stirred bioreactor (New Brunswick Scientific, USA) with a working volume of 4 l. The temperature during the fermentations was maintained at 37°C, pH 6 by automatic titration with NaOH and agitation at 100 rpm. For comparison, control fermentation was carried out with free-suspended cells under the same conditions and using a similar inoculum size as for encapsulated cells when related to the volume of the fermentation medium.

2.3. Cell enumeration

For evaluation of cell concentration, immobilized cells were first released from the capsules by dissolving 30 capsules (which corresponds to 1 ml of capsule) in a known volume of sterile 3% sodium citrate. Cells from the free-suspended culture were counted immediately after the samples were taken from the bioreactor. The cell concentration was determined by plate counting on the MRS medium to which 2% of agar was added. Petri plates were incubated at 37°C for 72 h. Dilutions were carried out in a sterile physiological saline. In addition, the presence of free cells in the growth medium with encapsulated cells was examined after the experiment terminated.

3. RESULTS AND DISCUSSION

The encapsulation technique applied in this work allowed to produce the liquid-core capsules containing cells surrounded by a thin layer of alginate membrane. The mean diameter of obtained capsules was equal to 3.9±0.12 mm.

The encapsulated *Lactobacillus acidophilus* cells increased their population from 8.3×10^7 to 6.5×10^{10} CFU/ml of capsule reaching a stationary phase after 30 hours of fermentation with a maximum specific growth rate of 0.26 h^{-1} (Fig. 1). The population of cells during traditional free-cell fermentation increased from 2.8×10^7 to 7.2×10^9 CFU/ml of growth medium after 45 hours with a maximum specific growth rate of 0.25 h^{-1}. Thus the encapsulated form of bacteria enabled the production of biomass almost ten times more concentrated than under free-suspended cell fermentation. A similarity of the maximum specific growth rate figures of both fermentations indicated that the transport of substrate into the capsules was not a rate-limiting factor.

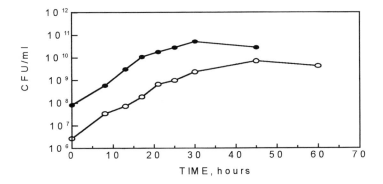

Figure 1. Growth curves of *Lactobacillus acidophilus* in alginate/starch liquid-core capsules (•) and as free-suspended cells (○). Biomass concentration is expressed as CFU/ml of capsule for encapsulated cells and as CFU/ml of growth medium for free-suspended cells.

These data are in agreement with previous observations on similar kinetics between free and gel entrapped lactic acid bacteria applied to lactic acid production [6]. However, as seen in Figure 1, the growth of encapsulated cells was halted earlier than that of free cells. This was probably caused by a different microenvironment of the encapsulated cells, especially the pH gradient across the capsule, which is generally considered as a critical parameter determining the development of lactic acid bacteria biomass in immobilization systems [7].

The concentrations of lactic acid bacteria in various polysaccharide gels reported in the literature ranged from 1.0×10^{10} to 5.0×10^{11} CFU/ml of gel beads, depending on the strain, type of fermentation, and gel support [8]. In studies especially designed for the production of highly concentrated biomass of *Lactococcus lactis* ssp. *cremoris* in gel beads a population of 2.5×10^{10} CFU/ml has been obtained in κ-carrageenan/locust bean gum [1], and 7.0×10^{10} CFU/ml in alginate beads [2]. Recently, a concentration of 4.0×10^{10} CFU/g of κ-carrageenan/locust bean gum gel beads has been reported for *Bifidobacterium longum* after 6 successive batch fermentations.

The stability of gel beads during cell growth is also an important factor, particularly when the release of free cells should be minimal [8]. In the present study, neither abrasion of capsule surfaces nor free bacteria were observed at the end of the growth cycle. However, to avoid the deterioration of capsule membranes at high lactate level, the excess $CaCl_2$ was added to the growth medium. In addition, the cells encapsulated in alginate/starch capsules grew freely in a liquid core without penetration of the alginate membrane. In contrast, in the whole gel beads due to diffusional limitations of nutrients, cell growth takes place with microcolonies close to the gel surface [9]. Consequently, the peripheral surface of the beads erodes by the forces resulting from cell growth and the cells leak from the open pores [10, 11]. Since the released cells subsequently grow in the bulk medium, their population can reach up to 60% of the total flora in the fermenter [8]. The application of the liquid-core capsules in this work for the production of concentrated cultures of lactic acid bacteria seems to offer a significant advantage over traditional gel entrapment techniques in terms of lowered diffusional resistance of the immobilization matrix and the absence of cell release.

4. CONCLUSIONS

The presented data shows that it is possible to apply encapsulation technology in the preparation of concentrated lactic acid bacterial biomass without centrifugation or filtration. The results obtained for *Lactobacillus acidophilus* cells grown in alginate/starch liquid-core capsules showed that the immobilization system allows a sufficient transport of nutrients through the capsule membrane during the growth cycle to obtain an encapsulated biomass of a high concentration without the release of cell into the growth medium.

ACKNOWLEDGMENT
This work was supported by the State Committee for Scientific Research (KBN) under grant No. P06G 03 012.

REFERENCES

1. P. Audet, D. St-Gelais and D. Roy, Milchwissenschaft, 50 (1995) 18.
2. C.P. Champagne, N. Morin, R. Couture, C. Gagnon, P. Jelen, and C. Lacroix, Food Res. Internat., 25 (1992) 419.
3. H. Maitrot, C. Paquin, C. Lacroix and C.P. Champagne, Biotechnol. Techniques, 11 (1997) 527.
4. J.P. Arnaud, C. Lacroix and L. Choplin, Biotechnol. Techniques, 6 (1992) 265.
5. T. Jankowski, M. Zielinska and A. Wysakowska, Biotechnol. Techniques, 11 (1997) 31
6. P. Audet, C. Paquin and C. Lacroix, Appl. Environ. Microbiol., 55 (1989) 185.
7. R. Cachon, C. Lacroix and C. Divies, Biotechnol. Techniques, 11 (1997) 251.
8. C.P. Champagne, C. Lacroix and I. Sodini-Galot, Crit. Rev. Biotechnol., 14 (1994) 109.
9. P.K. Walsh and D.M. Malone, Biotechnol. Adv., 13 (1995) 13.
10. J.P. Arnaud and C. Lacroix, Biotechnol. Bioeng., 38 (1991) 1041.
11. C.P. Champagne, C. Gaudy, D. Poncelet and R. Neufield, Appl. Environ. Micobiol., 58 (1992) 1429.

Food Biotechnology
S. Bielecki, J. Tramper and J. Polak (Editors)

Bacteria/yeast and plant biomass enriched in Se via bioconversion process as a source of Se supplementation in food

A. Diowksz, B. Pęczkowska, M. Włodarczyk, W. Ambroziak

Institute of Fermentation Technology and Microbiology, Technical University of Lodz
ul.Wólczańska 171/173, 90-924 Łódź, Poland

Mono and mixed populations of lactic acid bacteria of *Lactobacillus* species (*Lb. plantarum, Lb. sanfrancisco, Lb. brevis*) and the yeast *Saccharomyces cerevisiae* were tested for their ability to accumulate Se from the culture medium containing different inorganic Se sources (SeO_2, Na_2SeO_3, Na_2SeO_4). A high correlation between Se concentration in culture medium and Se content in biomass was observed. The highest amounts of Se in *Lactobacillus* species was observed with *Lb. plantarum* and the lowest one with *Lb. sanfrancisco*. Under the same conditions yeast accumulated considerable lower amounts of selenium. In all tested strains at concentrations higher than 10 µg Se/ml in the medium, there was an abrupt increase in Se content in the biomass with associated brick red color of free amorphous selenium deposit. At all Se levels a negative effect on growth parameters was observed. However, in a mixed population a stimulating effect on selenium accumulation and the growth parameters has been seen. For all strains selenite was the best and selenate the worst source of selenium.

Plant seeds and grains were germinated during 3-5 days on culture plates rinsed with water containing different concentrations (1-30 µg Se/ml) of inorganic selenium. Screening tests have shown different Se accumulation for individual plants. Similar accumulations of Se with tap, mineral and distilled water and with Se inorganic source of +4 or +6 oxidation state were seen. A one-stage fermentation process of bread production was used to test an effect of Se-enriched plant and bacteria-yeast biomass on the fermentation process. Technological parameters and bread quality of regular and Se-enriched bread (with 70-80 µg Se/250g) were comparable.

1. INTRODUCTION

Selenium is an essential micronutrient with an important and complex biological role in mammalian physiology. In the form of selenocysteine it is an integral and active part of glutathione peroxidase, iodothyronine, selenoprotein P and many others proteins which are involved in defense mechanisms against oxidative damage and immune responses on the cell level. Recently, the current interest is focused on the inverse relation between selenium status in human and many diseases (cancer, cardiomyopathy, osteoarthropathy), and on its role as an antioxidant nutrient [1,2]. In human the diet with selenium content depending on environmental factors is the main source of selenium in either organic (selenocysteine, selenomethionine) or inorganic (selenite, selenate) form [3,4]. In many countries, where dietary daily Se intake is too low, Se-status has to be improved to a proper level by oral supplementation with selenomethionine, sodium selenite, or seleno-yeast preparations [1,5]. It is known that the chemical forms of selenium in food affects its bioavailability, but precise mechanisms of selenium metabolism in human and biological activity of metabolites are not well characterized [2,3].

2. THE AIM OF THE STUDIES

Non-pathogenic lactic acid bacteria play an important role in the human nutrition. The ones that are commonly used in food preparations and germinating plant seeds and grains were investigated for their ability to accumulate selenium from the culture medium. Plant and bacteria/yeast biomass enriched in selenium via bioconversion processes were considered as alternative and natural sources of selenium organic forms with potential use in fermented food processing. Production of fermented bread with elevated levels of selenium was tested and this possibility of a widespread method of selenium supplementation in human diet was evaluated.

3. METHODS

Mono and mixed populations of lactic acid bacteria of *Lactobacillus* species (*Lb sanfrancisco, Lb. brevis, Lb. plantarum*) and yeast *Saccharomyces cerevisiae* isolated from natural sour dough were tested for their ability to accumulate Se from the culture medium containing different inorganic Se sources (SeO_2, Na_2SeO_3, Na_2SeO_4) in the concentration range of 1-20 μg Se/ml. After 24 h culture on MRS medium [6] at 28°C Se enriched biomass was centrifuged (20 min, 4000 x g) and washed with PBS buffer (phosphate buffer, pH 7.4, containing (g/l): KCl, 0.2; KH_2PO_4, 0.2; NaCl, 8; Na_2HPO_4, 1.15). Growth dynamics, biomass yield, selenium accumulation, and physiological activity of the microorganisms were analyzed. Plant seeds and grains were germinated during 3-5 days at room temperature during natural day-night cycle on culture plates rinsed with water (tap, mineral, distilled) with different concentrations (1-30 μg Se/ml) of inorganic selenium. Dynamics of selenium accumulation in the sprouting seeds and grains, effect of selenium concentrations, inorganic selenium sources and type of water used, were tested. Technological parameters and quality of bread produced with the starter culture and the selenium enriched biomass in a one-stage fermentation process, were evaluated. Selenium content in the samples was determined by the fluorimetric method of Watkinson [7] after acid mineralization.

4. RESULTS AND DISCUSSION

In all experiments with microorganisms a high correlation ($r > 0.9$) between inorganic selenium concentration in culture medium and Se accumulation in biomass was observed (Tab. 1). A similar correlation was observed in *Lactobacillus* species by Calomme *et al.* [8]. In our study among the examined strains the highest accumulation of Se in *Lactobacillus* species was observed with *Lb. plantarum* and the lowest one with *Lb. sanfrancisco*. Under the same conditions yeast *Saccharomyces cerevisiae* accumulated considerable lower amounts of selenium.

Table 1
Selenium accumulation in biomass on MRS medium containing SeO_2 (μg Se/g s.m.)
Time of cultivation 24h at 30°C

Strain	Selenium concentration in medium (μg Se/ml)				
	0	1	5	10	20
Lb. plantarum	2.1	76.7	244.0	400.9	22200.0
Lb. brevis	4.5	29.1	214.9	236.7	1881.0
Lb. sanfrancisco	3.6	20.7	100.5	204.4	1300.0
S. cerevisiae	6.4	12.5	112.9	153.9	363.8
Mixed population	4.4	80.4	280.7	679.2	4849.0

In all tested microorganisms, at concentrations higher than 10 μg Se/ml in the medium, there was an abrupt increase in Se content in the biomass with an associated brick red color of free amorphous selenium deposit. This phenomenon, observed also by other authors [8,9], proves that the examined microorganisms exhibit a detoxicating mechanism.

At all Se levels a negative effect on growth parameters and the survival rate was observed for both bacteria and yeast. The same observations were reported by Chmielowski *et al.* [10]. However, in the mixed population of *Lactobacillus* and yeast a stimulating effect on the growth parameters was seen; Se accumulation was much higher than for the single strains cultivated under the same conditions.

For all strains selenite was the best and selenate was the worst source of Se accumulation indicating that uptake of Se in the +4 oxidation state is preferred (Fig. 1).

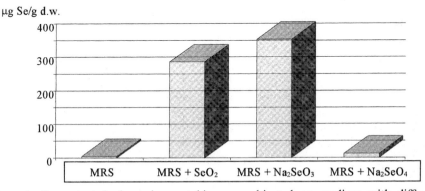

Figure 1. Se content in bacteria-yeast biomass cultivated on medium with different Se sources.

298

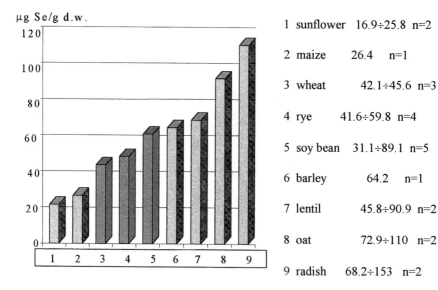

μg Se/g d.w.

1 sunflower	16.9÷25.8 n=2
2 maize	26.4 n=1
3 wheat	42.1÷45.6 n=3
4 rye	41.6÷59.8 n=4
5 soy bean	31.1÷89.1 n=5
6 barley	64.2 n=1
7 lentil	45.8÷90.9 n=2
8 oat	72.9÷110 n=2
9 radish	68.2÷153 n=2

Figure 2. Accumulation of Se by seeds and grains germinated in the presence of water containing 10 μg Se/ml (SeO$_2$).

Screening tests for different plant seeds and grains have shown Se accumulation in the range15-30 μg/g d.w. for sunflower and maize, 40-60 μg/g d.w. for rye, wheat, soy bean and 70-150 μg/g d.w. for oat and radish during 3-5 days germination with rinsing water containing 10 μg Se/ml (Fig. 2).

Wheat, rye and soy bean were selected for further investigation as the most appropriate natural additives for fermented bread production. For these sprouts similar accumulations of Se with tap, mineral and distilled water (Tab. 2) and with Se inorganic source in +4 or +6 oxidation state were seen.

Table 2
Selenium accumulation in sprouts germinated on different types of water containing 10 μg Se/ml (SeO$_2$). Results are mean ± SD (n=3).

Type of water	Sprouts		
	Wheat	Rye	Soy bean
Tape water	33.1 ± 0.4	40.1 ± 2.4	41.8 ± 0.9
Mineral water	28.7 ± 2.2	42.3 ±2.2	36.1 ± 0.8
Distilled water	27.6 ±0.4	42.2 ±1.2	34.5 ±1.3

In all experiments with germinated seeds and grains a high correlation between selenium content in rinsing water and Se accumulation in biomass was noted. At high concentrations of

Se in the rinsing water (20 μg/ml and higher) the sprouts biomass growth was considerable lower suggesting, as it was in the case of microorganisms, a toxic effect of selenium. However, no signs of free selenium deposits in plant biomass were observed.

Preliminary experiments have shown that in the protein extract of both kinds of biomass (plant and bacteria/yeast) about 25-35% of total accumulated selenium is tightly bound to the protein.

A one-stage fermentation process of bread production with a mixed population of starter culture was used to test an effect of Se-enriched plant and bacteria/yeast biomass on the fermentation process and bread quality. The control bread was produced without Se-additives. The addition of Se-enriched biomasses had practically no effect on the course of the sourdough fermentation process. Technological parameters (baking and cooling losses, bread yield) and bread quality (bread volume, porosity, acidity, water content, taste and aroma) of regular and Se-enriched bread were similar or almost identical (Tab. 3).

The selenium content in bread was calculated in reference to 250 g of bread as an average daily portion. Se-enriched bread contained 70-80 μg Se/250 g in comparison to the control 44 μg Se/250 g and even below 20 μg Se/250 g in the case of Se poor flour.

Table 3
Evaluation of bread quality

Parameter	Bread		
	Control	With rye sprouts	With bacteria/ yeast biomass
Se content (μg Se/250 g)	44.2	72.4	79.8
Baking loss (%)	7.5	7.7	7.7
Total loss (%)	9.2	10.2	10.5
Bread yield (%)	149.0	147.6	148.6
Bread volume (cm^3/100g)	256.0	254.0	250.0
Bread porosity (%)	62.6	61.5	60.2
Acidity (ml 0.1 M NaOH/10 g)	6.2	6.3	6.3
Water content (%)	41.1	41.1	41.2
Taste and aroma	Very good	Very good	Very good

5. CONCLUSIONS

On the basis of our results and preliminary technological trials we can conclude that Se enriched plant sprouts and bacteria/yeast biomass may be used as additives in many fermented food processes, preferably in Se-enriched bread, as a widespread selenium supplement. Assuming that 250 g of bread in Polish condition is consumed per one person per day, we can expect about 70% of daily recommended Se intake. Plant and bacteria/yeast biomass can be an alternative and direct source of selenium supplementation in human and animal diet and also a potential source of selenium supplementation in fermented food processing.

REFERENCES

1. J. Neve, Experimentia, 47 (1991) 187.
2. S.J. Fairweather-Tait, Eur. J. Clin. Nutr., 51 (1997) 20.
3. R.E. Litov and G. F. Combs, Pediatrics, 87 (1991) 339.
4. B. Szteke and W. Ręczajska, Zeszyty Naukowe Komisji Chemii Analitycznaj PAN, 8 (1994) 82.
5. B.A. Zachara and W. Wąsowicz, Ręczajska, Zeszyty Naukowe Komisji Chemii Analitycznaj PAN, 8 (1994) 157.
6. J.C. de Man, M. Rogosa, M. E. Sharpe, J. Appl. Bact., 23 (1960) 130.
7. J.H. Watkinson, Anal. Chem., 38 (1966) 92.
8. M.R. Calomme, K. Van den Branden, D. A. Vanden Berghe, J. Appl. Bact., 79 (1995) 331.
9. G. Falcone and W. J. Nickerson, J. Bacteriol., 85 (1963) 754.
10. J. Chmielowski, B. Kłapcińska, A. Tyflewska, Zeszyty Naukowe Komisji Chemii Analitycznaj PAN, 8 (1994) 58.

Food Biotechnology
S. Bielecki, J. Tramper and J. Polak (Editors)
© 2000 Elsevier Science B.V. All rights reserved.

The new nutritional food supplements from whey

L. V. Kirillova[a], I. P. Chernikevich[a] and V. K. Pestis[b]

[a]Laboratory of Applied Enzymology and Biotechnology, Institute of Biochemistry, National Academy of Sciences of Belarus, BLK 50, Grodno 230017, Belarus

[b]Argricultural Institute, 28 Tereshkova Str., Grodno 230600, Belarus

A technology for production of two food supplements, i.e. a protein and mineral concentrate (PMC) and 10% lactic acid, either from cheese or cottage cheese whey is proposed. The microbiological, biochemical and chemical properties of PMC and 10% lactic acid were investigated. The procedure for production of lactic acid with a high degree of purification was studied using a modified electrodialyser. The PMC can be applied for human and animal nutrition, whereas 10% lactic acid can be employed as a preservative for vegetables and as flavouring supplement. Protein deficiency can be completely annihilated by means of the concentrate. PMC contains live bacterial cells of *Lactobacillus bulgaricus* known to have a favourable effect on the functions of the intestine.

Key words: protein and mineral concentrate (PMC), *Lactobacillus bulgaricus*, whey, lactic acid, electrodialyser, cathode, anode, membrane, fermented filtrate.

1. INTRODUCTION

A wide range of essential foodstuffs is required to provide wholesale nutrition for humans. In production of cheese, casein and cottage cheese, tens of million tons of whey, which is utilized for production of dry substances, whey protein included, are accumulated in countries of the world community annually. The present data suggest that whey does not rank below ready to use dairy products in biological value (protein, sugar, vitamins, mineral salts) [1].

The present-day goal is to use whey judiciously and to preserve its essential components. In view of the absence of effective and low energy consuming methods for whey processing this problem has not yet been solved in any country [2].

Lactic acid is widely applied in food (meat and milk processing) and agricultural industries. It is generally produced either from fermented sugar-containing raw material or chemically [3,4]. All the methods of lactic acid production require subsequent purification.

2. MATERIALS AND METHODS

For the first time in Belarus at the Milk plant of Grodno there was built a pilot plant for waste-free processing of whey to obtain a dry protein and mineral concentrate (PMC) in addition to 10% lactic acid, either from cottage cheese or salt-free cheese wheys. Cheese whey (acidity of 14-30° T) or cottage cheese whey (48-60° T) was subjected to heat treatment (15 sec, 70-75°C), cooled to 40°C and a culture of *Lactobacillus bulgaricus* added and left to stay for 36-48 h. When the acidity of the mixture was below pH 4.0, the lactic acid obtained was neutralized with 10% NaOH. The fermentation processes was considered to be completed when the lactose content was decreased to 0.2%. Then the acidity of the mixture was adjusted with 10% NaOH to pH 6.0-6.2 and the fermented whey was stored for 2-3 h for complete protein coagulation.

The proteins were sedimented with a suction drum device, dried in a drier at 55-60°C and pulped with a mill.

The remaining fermented whey was transported to an electrodialyser through a pipeline and used for production of lactic acid.

We modified the electrodialysis method to obtain 10% lactic acid of a high degree of purity. Our modification consisted of the development of a dialyser with a design eliminating the contact of the anode surface with purified lactic aid. This eliminated oxidation of the product by atomic oxygen [5-7].

After the fermentation and deproteinization of the whey the fermented filtrate was directed to the cathode chamber of our electrodialyser. The dialyser represented a vertical anode chamber with the walls being made of a dielectric material, plastic, in the form of a cylinder. A cathode chamber with walls of a porous dielectric material, having the properties of ion exchange anionic membrane, was installed inside of this chamber.

A metallic cylindrical cathode was installed within each chamber. The third chamber, the walls of which were made of a porous dielectric material with the properties of an ion exchange cationic membrane, was located within the anode chamber between the outer wall of the cathode camber and the inner wall of the anode chamber. A metallic anode of an anticorrosive material was located inside of the third chamber, along the axis of it. The anode and cathode were connected directly to the electric source. The cathode was connected to the shaft of the electric engine for its rotation movement. A device to remove sedimenting mineral organic components of whey from the cathode was attached on the surface of a rib of the cathode chamber. To remove lactic acid (9-10%) from the anode chamber, a connection was installed in its lower part. For removal of the waste filtrate from the cathode chamber, pipelines were installed in its upper part. The lactic acid, entering under the action of an electric field from the cathode chamber, was concentrated in front of the cationic membrane and left the chamber without contact with the anode and, consequently, without a reaction with the atomic oxygen accumulating on the anode during the electrodialysis.

Therefore, this technology enables to obtain pure lactic acid without additional purification steps of the dialysis product.

3. RESULTS AND DISCUSSION

The technology described made it possible to obtain two biologically valuable preparations, dry PMC and pure, edible 10% lactic acid. Five to seven kg of dry PMC and 100 to 110 kg of lactic acid were obtained from 1 ton of cheese whey.

The concentrate was assayed organoleptically, microbiologically, chemically and biochemically. The results are listed in Tables 1 to 4.

Table 1
Organoleptic parameters of PMC

Parameter	Characterization
Appearance and consistence	Dry hygroscopic powder without dense lumps; a small amount of dense lumps is allowed as they are easily turned to powder
Color	Yellowish, without signs of mould
Taste	Astringent, sour
Smell	Pronounced sour milk

Table 2
Microbiological parameters of PMC

Microorganisms	Mean value, in 1 g of the concentrate
Lactobacilluss bulgaricus,	$10^7 - 10^9$
Mould and yeast,	$10^1 - 10^2$
E. coli group bacteria	not present
Pathogenic microflora	not present

According to the microbiological parameters the PMC complies with the standards for foodstuffs [8].

Table 3
Physico-chemical parameters of PMC

Parameter	Mean value
Mass proportion of total protein, %	≥ 83.0
Mass proportion of whey protein, %	≥ 75.0
Mass proportion of ashes, %,	≤ 10.0
Mass proportion of fat, %,	≥ 2.8
Mass proportion of lactose, %,	≤ 0.2
Active acidity (pH),	≥ 6.0
Titrated acidity, ^0T	≥ 70.0
Content of heavy metal salts, mg/kg of PMC :	
- copper as calculated per copper,	≤ 3.0
- tin as calculated per tin,	≤ 10.0
- content of lead salts	absent
Energy value, 418g/kJ	400

According to the physico-chemical parameters PMC complies with the standards for foodstuffs [8].

The PMC shows great biological value due to the high content of essential amino acids (up to 45% of whey protein, Table 4). Since the concentrate dissolves well in water, has high water-binding, emulsifying and foam-producing capacities, it can be applied as a supplement to products for various types of dietotherapy.

The PMC can be also utilized as a constituent in cosmetic preparations in case of protein deficiency.

The PMC antibacterial activity of live cells of *Lactobacillus bulgaricus*, which are present in PMC, allows the application of it for healing and prophylactic purposes.

Table 4
Amino acid composition of whey protein

Amino acid	Content in whey protein (%)	Amino acid	Content of whey protein (%)
Ala	2.98	*Leu	5.90
Arg	1.04	*Lys	7.48
Asp	8.49	*Met	1.07
*Val	5.42	Ser	0.79
*His	1.92	Tyr	3.15
Gly	2.21	*Thr	7.12
Glu	10.30	*Trp	3.52
*Ile	5.29	*Phe	5.21
Cys	2.11	Pro	1.00
Total			75.00

* essential amino acids

Lactic acid was assayed for some parameters [8], with the data being summarized in Tables 5 and 6.

Table 5
Organoleptic parameters of edible 10% lactic acid

Parameter	Characterization
Appearance and consistence	Transparent fluid without sediment
Color	Light-yellow, uniform
Taste	Weak, specific for lactic acid
Smell	Acid, without other flavor

Table 6
Chemical parameters of 10% lactic acid

Parameter	Mean value
Mass proportion of total lactate, %	10.0
Mass proportion of directly titrated lactate, %,	≥ 8.5
Mass proportion of anhydrides, %,	≤ 1.2
Mass proportion of reduced substances, %,	≤ 0.3
Color, %,	≤ 5
Salts of heavy metals	did not exceed standard values
Content of residual pesticides and microtoxins	did not exceed standard values
Dry substances, %	2-3
Protein, %	0.12
Ashes, %,	≤ 2.0
Lactose, %,	≤ 0.2-0.3

As the lactic acid was obtained from the natural product whey and as a technique did not employ toxic materials similar to those used in chemical methods it can be considered food grade. The edible lactic acid was indeed successfully tested as a preservative, instead of acetic acid, at a food canned factory and the tests produced good results.

REFERENCES

1. V.V. Molochnikov and P.G. Nesterenko, Production and use of whey protein, Moscow, (1983).
2. V.G. Popov et al., Problem of protein deficiency and ways of its solution, Smolensk, All - Union Research Institute for Applied Microbiology (1982) 75.
3. V. Hartmut, S. Dieter, *Lactobacillus bulgaricus* und seine Vermentung zur Herstellung von D-Milchsaure, FRG, 1983.
4. A. Mohamed Mehaia and C. Munir, Process Biochem., 22 (1987) 185.
5. G. Guerif, Lait, 64 (1984) 197.
6. T.P. Bachourina., A.I. Kozhenkow, R.N. Khandak, Nutritional and biological value of dairy produce for children's nutrition and in diethotherapy, Moscow, (1985) 65.
7. A.G. Khramtsov and P.G. Nesterenko, Waste-free technology in milk - processing industry, Agropromizdat, Moscow, 1989.
8. Standard No 490-79, Moscow, USSR, (1979) 5 - 10.

Food Biotechnology
S. Bielecki, J. Tramper and J. Polak (Editors)

The biodegradation of ochratoxin A in food products by lactic acid bacteria and baker's yeast

M. Piotrowska and Z. Żakowska

Institute of Fermentation Technology and Microbiology, Technical University of Lodz, Wolczanska 171/173, 90 – 924 Lodz
E- mail: m.piotrowska@mikrob.p.lodz.pl

This paper describes the changes in the level of ochratoxin A during a baker's sour dough fermentation using lactic acid bacteria and yeast starter cultures. It was concluded that all of the examined organisms caused a varied ochratoxin A degradation in both monocultures and the mixed population.

1. INTRODUCTION

Mycotoxins, toxic fungal metabolites, mainly produced by *Aspergillus*, *Penicillium* and *Fusarium* species, contaminate raw materials and food products. According to many previous studies, the ochratoxin A (OA) is the most important mycotoxin polluting raw materials in Polish climate conditions. This toxin is often found in food of both plant and animal origin [1]. There is considerable interest in research aimed at mycotoxins elimination either by the use of chemical or physical methods or by inhibiting the udesired growth of toxin producing fungi [2,3]. Unfortunately these methods are currently not suitable for food application. The great interest is therefore focused on a natural mycotoxin degradation by chosen microorganisms, both during biotechnological processes as well as in a human organism. There are only a few brief reports on microbial degradation of ochratoxin A [4-6]. Most of the information on mycotoxins biodegradation concerns aflatoxin B_1 [7-9]. There is only e few brief reports on microbial degradation of ochratoxin A.

Among the microorganisms capable of mycotoxins degradation, lactic acid bacteria and yeast are particular interest, mainly because of their wide application in the biotechnology. Moreover, the lactic acid bacteria play a very important physiological role in a human organism [7,8,9].

Degradation ability of ochratoxin A should be taken into account as an additional criterion of microorganisms selection for biotechnological processes. The using of strains with these feature should reduce the risk of human exposure to ochratoxin A, the extremely toxic and carcinogenic compounds.

The aim of the study was to examine the ability of ochratoxin A degradation by lactic acid bacteria and yeast acting as starter culture in bakery.

2. MATERIALS AND METHODS

2.1. Microorganisms

Pure cultures of lactic acid bacteria: (*Lactobacillus plantarum, Lactobacillus sanfrancisco, Lactobacillus brevis*) and yeast (*Saccharomyces cerevisiae*). The mixed population of these microorganisms and commercial bakery lyophilised starter culture were also examined.

All organisms were obtained from BioStarPlus Company, Lodz, Poland.

2.2. Methods

Fermentation was carried out in Erlenmeyer flask containing medium consist with wheat flour (45 g/l), glucose (7,5 g/l) and water (to 1 l). The medium was supplemented with ochratoxin A (Sigma). The level of ochratoxin A was 300 μg/kg.

Medium was inoculated with 1% of active cultures on MRS (bacteria) and YPG (yeasts) media and incubated at 30°C. The initial level of microorganisms ranged from 10^6 cfu/ml for yeasts to 10^7 cfu/ ml for lactic acid bacteria. Media were obtained from BTL Company (Lodz, Poland).

During the fermentation the concentration of ochratoxin A was monitored using ELISA kit – RIDASCREEN OCHRATOXIN A (R-Biopharm GmbH, Damstadt, Germany). The procedure for extraction of ochratoxin A from medium was in accordance with test instruction.

The percent of degradation was calculated as:

(initial OA – final OA concentration / initial OA concentration)×100

3. RESULTS

3.1. Lactic acid bacteria and yeasts

The biodegradation ability of pure cultures and mixed population is presented in Tab. 1. The results of the experiments showed that both the strains of lactic acid bacteria and yeast monocultures caused a varied ochratoxin A biodegradation. After 10 hours of fermentation the ochratoxin A degradation level was: 34 % for *L. brevis* strain, 29 % for *L. sanfrancisco* and 40 % for *L. plantarum*. After 24 hours incubation the degree of degradation increased to 37 % for *L. brevis*, 50 % for *L. sanfrancisco* and 54 % for *L. plantarum*. For the yeast strain the biodegradation was at the level of 23 % (after 10 hours incubation) and 41 % (after 24 hours incubation). The fermentation carried out by the mixed culture population proved to be the most efficient in the OA biodegradation (33 % after 10 hours incubation and 66 % after 24 hours incubation).

Table 1
Biodegradation of ochratoxin A in flour medium by lactic acid bacteria and yeasts

Organism	Incubation time [h]	Degradation %
	0	0
Lactobacillus plantarum	10	40
	24	54
	40	56
	0	0
Lactobacillus brevis	10	34
	24	37
	40	38
	0	0
Lactobacillus sanfrancisco	10	29
	24	50
	40	51
	0	0
Saccharomyces cerevisiae	10	23
	24	41
	40	41
	0	0
Mixed population	10	33
	24	66
	40	68

3.2. Commercial starter culture
The biodegradation ability of bakery commercial starter culture is presented in fig. 2.

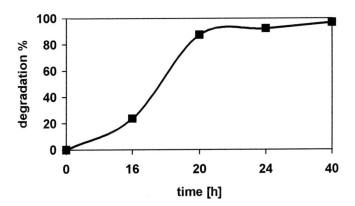

Figure 2. Degradation of ochratoxin A in sourdough by lyophilised starter culture.

It was found that during the sourdough fermentation the ochratoxin A biodegradation degree was elevated from 25 % to 97 %. The strongest biodegradation activity was observed between 16 and 20 hours of incubation.

3. CONCLUSIONS

All investigated mocroorganisms proved capable of ochratoxin A degradation. Mixed culture population was found to be most efficient in the toxin degradation, which could be explained by the synergistic relationship between these microorganisms in the baker's sourdough.

The fate of OA from food products may be a results of metabolic transformation or adsorption to the cell wall. The futures study should provide information concerning the mechanism of microbial removed of ochratoxin A.

REFERENCES

1. J. Chełkowski (eds.), Cereal grain. Mycotoxins, fungi and quality in drying and storage. Elsevier, Amsterdam, 1991.
2. M.P. Doyle, R.S. Applebaum, R.E. Brackett, E.H.J. Marth, Food Prot., 45 (1982) 964.
3. E.H. Marth, M.P. Doyle, Food Technol., January (1979) 81.
4. M. Piotrowska, Z. Żakowska, IV Symp. Mycotoxins in food and feed, Bydgoszcz, Poland, 1998 Abstracts book 126.
5. M. Škinjar, J.L. Rašič, V. Stojic, Folia Microbiol., 41 (1996) 26.
6. I. Štyriaka, E. Conkovab, E. Razzizac, J. Bohmc, IV Symp. Mycotoxins in food and feed, Bydgoszcz, Poland, 1998 Abstracts book 101.
7. A. Ciegler, Lillehoj E.B., Petrson R.E., Hall H.H. Appl. Microbiol., 14 (1966) 934.
8. S.E. Megalla, A.H. Hafez, Mycopathologia, 77 (1982) 89.
9. S.E. Megalla, M.A. Mohran, Mycopathologia, 88 (1984) 27.

Food Biotechnology
S. Bielecki, J. Tramper and J. Polak (Editors)
© 2000 Elsevier Science B.V. All rights reserved.

311

The use of *Geotrichum candidum* starter cultures in malting of brewery barley

E. Dziuba[a], M. Wojtatowicz[b], R. Stempniewicz[b], B. Foszczyńska[a]

[a]Department of Food Storage and Technology, Agricultural University of Wrocław, Norwida 25, 50-375 Wrocław, Poland

[b]Department of Biotechnology and Food Microbiology, Agricultural University of Wrocław, Norwida 25, 50-375 Wrocław, Poland

The aim of the study described in this paper was to determine the effect of two strains of *Geotrichum candidum*, used as starter cultures in the malting process, on the quality of final malt. Results showed that the use of *G. candidum* resulted in a greater extract yield of the malt, a lower viscosity of the wort, but also affected adversely proteolytic modification of the malt. Having tested the grain of four barley varieties grown in Poland of low usability for malt production, we obtained best results for Mobek variety.

1. INTRODUCTION

In the process of barley malting, the microorganisms present on kernels develop in abundance, finding good conditions for growth, which also contributes to unfavourable changes [1]. Some moulds, e.g. *Fusarium* or *Aspergillus* have an ability to produce mycotoxins [2-5]. Considering that the said metabolites damage a germ, they thus affect negatively the process of barely germination and the synthesis of α-amylase. Furthermore, mycotoxins may cause lower growth and activity of yeast cells [6]. The toxic metabolites also undermine human health causing some diseases, e.g. skin or pulmonary cancer [7,9]. Some species of *Fusarium*, i.e. *F. graminearum*, *F. moniliforme*, *F. culmorum* and *F. avenaceum* and the strains of the genera *Rhizopus*, *Stemphylium*, *Penicillium*, *Aspergillus*, and *Alternaria* produce polypeptides responsible for gushing [3,8,9].

In order to improve malt quality and safety, spontaneous development of microorganisms can be controlled during malting by applying starter cultures which grow faster, and which thus limit the development of undesirable microflora, in particular fungi [10-14].

The objective of our studies was to define the effect of *G. candidum* strains, selected at random, used as starter cultures during barley malting, on malt quality.

2. MATERIALS AND METHODS

2.1. Experimental material

Two strains, i.e. *G. candidum* 1 and *G. candidum* SC12, were used as starter cultures in the process of barley malting. The strains originated from the culture collection of the Department of Biotechnology and Food Microbiology, Agricultural University of Wrocław.

The raw material, grain of four varieties of brewery barley: Maresi, Mobek, Orlik and Polo, originating from the Field Experimental Station of the Institute of Plant Protection in Trzebnica (harvested in 1997) was applied in the malt production. The malts were produced under laboratory conditions with and without a starter culture, which was added to the first steeping water at the concentration of $5.0 - 5.8 \times 10^5$ CFU/ml.

2.2. Microbiological analysis

The number of microorganisms on the surface of barely kernels, and green and kilned malt was determined by the plate method on selective media, i.e. yeast and moulds on OGY agar (Merck), aerobic mesophilic bacteria on nutrient agar (BTL) supplemented with 100 ppm of cyckloheximide, and lactic acid bacteria on MRS agar (BTL).

2.3. Analytical methods

For control and starter malts diastatic power [15], and viscosity of wort [16] were determined; extract yield of the malt [16] as well as Kolbach index [16] were also calculated.

2. RESULTS AND DISCUSSION

The barley grain of Maresi, Mobek, Orlik and Polo varieties was tiny and characterized by low weight of thousand kernels (33.8–38.4 g d.w.), low percentage of kernels of size above 2.5 mm (38.6–70.8%) and, except for Mobek variety, the protein content above 12% d.w. [17]. Therefore, it was not a good raw material for malt production. Its microbiological quality did not meet the grain standards [8] and in terms of filamentous fungi contamination, including *Fusarium, Aspergillus, Mucor* and *Rhizopus* genera, the norm for grain was exceeded from 3 to 12 times. Also, the viable bacteria count was 14 to 20 times higher then the acceptable limit (Fig.1).

The changes in the number of microorganisms on the kernels during steeping and kilning process were similar in the case of all tested varieties of barley. In Fig. 1 results obtained for Maresi and Mobek varieties are presented. During steeping, the total viable counts increased both in starter and control samples, especially in terms of lactic acid and aerobic bacteria, whose number grew by 2 up to 3 logarithmic cycles. The number of filamentous fungi altered slightly. The kilning process reduced the level of microorganisms, nevertheless the kilned malt showed higher counts than those of barley grain. Starter cultures of *G. candidum* prevailed in green malt (75–100% of the total fungi number) and stayed high on kilned malt as well (64.7–98.5% of the total fungi number). The inoculation of barley with starter cultures eliminated filamentous fungi of *Fusarium, Aspergillus* and *Rhizopus* genera.

The starters influenced significant biotechnological parameters of malt in different ways. They did not lower diastatic power of malts (Fig. 2a), which showed 220 to 330 WK-units. Only starter malts obtained from the Maresi variety had high values of diastatic power (450-460 WK-units).

MARESI

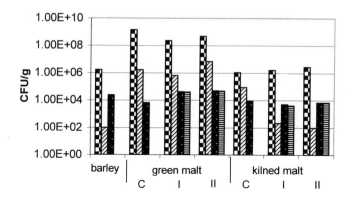

MOBEK

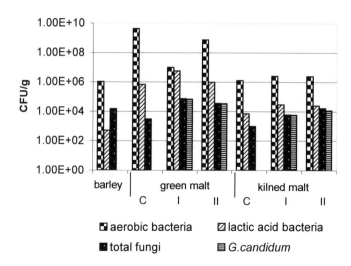

Figure 1. The bacterial and fungal viable count on malts of two barley varieties after inoculation with starter culture of *G. candidum*

C - control; I - *G. candidum* 1; II - *G. candidum* SC12

314

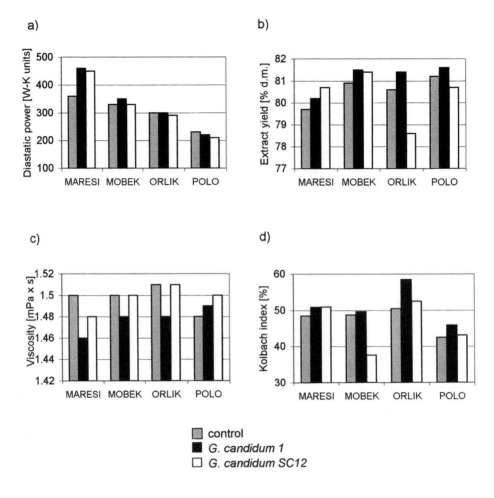

a) diastatic power
b) extract yield

c) viscosity of wort
d) Kolbach index

Figure 2. Effect of starter cultures of *G. candidum* on some features of malts derieved from various varieties of barley
a) diastatic power
b) extract yield
c) viscosity of wort
d) Kolbach index

Compared to the control, an increase in extract yield by 0.2–1.0% d.w. was recorded, except for Orlik and Polo grain malted with the *G. candidum* SC12 strain (Fig. 2b). One of the parameters affecting mash separation (lautering) is wort viscosity. Differentiated wort viscosity was generated by the strains used (Fig. 2c). *G. candidum* 1 lowered wort viscosity in the case of most barley varieties by 0.02–0.04 mPaxs whereas *G. candidum* SC12 did not affect favourably the changes in viscosity. Starter malts were characterized by stronger proteolytic modification, which resulted in higher Kolbach index (Fig. 2d).

Application of starter cultures in the process of malting of brewery barley had been of interest to Finnish and French research workers. The former group used for this purpose lactic acid bacteria [4,13,14] whereas the latter made use of *G. candidum* [10,11].

The results of our work on the improvement of malt quality reflect the findings quoted in the literature. The inhibition of mould development on barley grain is related closely to specific features of starter culture, barley variety, and initial contamination of the grain. Mobek variety grain malted with *G. candidum* 1 strain produced malts of the best biotechnological parameters.

In conclusion, replacing chemical agents, used traditionally for grain disinfection in the brewery industry, by starter cultures applied in the malting process, offers the possibility for improved and unified quality malts, eliminating filamentous fungi metabolites unfavourable for human health.

The results obtained by us so far [12,18] speak for further studies on the selection of strains producing the best effects in limiting the development of undesirable microflora, typical for Polish barley and climatic conditions.

This biological method will prove to be advantageous when physical, chemical and organoleptic features of the beer produced with starter cultures are comparable with those of the beer obtained in traditional way.

REFERENCES

1. J. Chełkowski, K. Trojanowska, A. Pawłowski, Przem. Ferm., 7 (1985) 3.
2. J. Chełkowski, W. Tobiasz, W. Karwowska, Przem. Ferm., 12 (1980) 6.
3. D. Czajkowska, Przem. Ferm. 10 (1996) 12.
4. Haikara, Proc. Eur. Brew. Conv. Congr., London 1983, 401.
5. L. Kelly, D.E. Briggs, J. Inst.Brew., 98 (1992) 395.
6. M. P. Whitehead, B. Flannigan, J. Inst. Brew., 95 (1989) 411.
7. J. Chełkowski, K. Trojanowska, Przem. Ferm., 4 (1979) 4.
8. Campbell, in: Materiały I Szkoły Technologii Fermentacji „Drożdże w technologiach fermentacyjnych" (TEMPUS S JEP-09770-95), 1996, 7.
9. F.G. Priest, I. Campbell I (eds.), Brewing Microbiology, Chapman and Hall, London, 1987.
10. P. Boivin, M. Malanda, Proc.Eur. Brew. Conv. Congr., Oslo 1993, 95.
11. P. Boivin, M. Malanda, MBAA Techn. Quarter., 34 (1997) 96.
12. E. Dziuba, B. Foszczyńska, R. Stempniewicz, M. Wojtatowicz, Materiały Konferencji Naukowej „Ograniczenie stosowania dodatków do żywności - za i przeciw", Sielinko 1998, 14.
13. Haikara, A. Laitila, Proc. Eur. Brew. Conv. Congr., Brussels 1995, 249.
14. Laitila, K.-M. Tapani, A. Haikara, Proc. Eur. Brew. Conv. Congr., Maastricht 1997, 137.

15. Golachowski , W. Leszczyński, Przem. Ferm., 2 (1980) 1.
16. PN-67/A-79083. Słód browarowy i słód diastatyczny. Pobieranie próbek i metody badań.
17. J. Błażewicz , (personal information, 1998).
18. E. Dziuba, R. Stempniewicz, B. Foszczyńska, M. Wojtatowicz, Materiały XXIX Sesji Naukowej KT i CHŻ PAN, Olsztyn 1998, 165.

Food Biotechnology
S. Bielecki, J. Tramper and J. Polak (Editors)
© 2000 Elsevier Science B.V. All rights reserved.

Enzymes as a phosphorus management tool in poultry nutrition*

K. Żyła[a], J. Koreleski[b] and D.R. Ledoux[c]

[a]University of Agriculture, Department of Food Biotechnology,
29-Listopada Ave. 46, 31-425 Kraków, Poland

[b]Institute of Zootechnics, Research Farm of Balice, Department of Animal Nutrition in Brzezie, 32-080 Zabierzów, Poland

[c]112 Animal Science Research Center, University of Missouri-Columbia,
Columbia, MO 65211, USA

An in vitro multidigestion technique that simulated conditions of the poultry intestine and an experimental design module of a statistical software package were used to determine optimal dosages of phosphorolytic (phytase, acid phosphatase) and cell-wall-degrading (crude pectinase) enzymes and of a fungal mycelium preparation for dephosphorylation of wheat-based feeds (total P:0.41%; available P: 0.17%) fed to growing broilers. Seventy six percent of phosphorus retention and 68% reduction in amounts of P excreted were observed in 21 day-old broilers that were fed the optimal combination of enzymes and mycelium. This was accompanied by body weight gains and bone mineralization that were superior to values observed in chicken fed the positive control diet supplemented with calcium phosphate (total P: 0.71%; available P: 0.41%).

1. INTRODUCTION

Salts of phytic acid (*myo*-inositol hexakis-dihydrogenphosphate, phytate) are known to be a principal storage form of phosphorus in plant seeds and pollen. Phytates comprise about two-thirds of the phosphorus in cereal grains and oil seed meals, the two major components of poultry feeds. Phytate P is poorly available to simple-stomached animals. Inorganic P supplements in feeds, necessary to fulfil birds' nutritional requirements, result in high P content in poultry manure and eventually stimulate eutrophication of waters. This creates a serious environmental concern especially in areas of intensive animal production. Phytate dephosphorylation by microbial phytase (EC 3.1.3.8.) has been studied throughly, but the extent of phytate hydrolysis in the intestine of poultry (expressed as phosphorus digestibility, availability or utilizability) has been reported to be no more than 60 % [1-4]. As a result of phytase addition inorganic phosphorus supplements to poultry feeds can be reduced from 4.5 g/kg to 3.3 g/kg without negative influence on performance or on bone mineralization. The problem is therefore that *even if phytase is added to poultry feeds a substantial portion of* extent of phytate hydrolysis in the intestine of poultry (expressed as phosphorus digestibility, availability or utilizability) has been reported to be no more than 60 % [1-4]. As a result of phytase addition inorganic phosphorus supplements to poultry feeds can be reduced from 4.5 g/kg to 3.3 g/kg without negative influence on performance or on bone mineralization. The problem is therefore that *even if phytase is added to poultry feeds a substantial portion of phytates remains indigestible*. Different strategies have been suggested to cope with the problem. Quian et al. [4] confirmed that feeding megadoses of cholecalciferol increases the efficacy of phytase. Denbow et al. [5] reported that soybeans transformed with a fungal phytase gen were more effective than microbial phytase. Another approach was based on the genetically modified high available phosphorus (HAP) corn [6]. Finally, Żyła et al. [7] suggested that *mixtures of phosphorolytic and tissue degrading enzymes* added either as isolated proteins or in the intracellular form (fungal mycelium preparation) may exert a profound stimulating effect on phosphorus utilization by poultry. In 1996 we tested that hypothesis on corn-soybean meal feeds fed to growing turkeys. The objective of the present study was to apply the strategy to wheat-based diets fed to growing broilers.

2. MATERIALS AND METHODS

Commercial preparations of phytase, acid phosphatase, pectinase and glucose oxidase were used throughout the study. A selected *Aspergillus niger* strain was cultivated on an appropriate medium for 8 days at $30^{\circ}C$. The mycelium was separated from the medium and used as a source of intracellular enzymes after appropriate processing. Dephosphorylation of commercial-type poultry diets was determined by an *in vitro* technique that simulated intestinal tract of poultry [8,9]. Enzymes and mycelium dosages were optimized in a series of *in vitro* experiments using the Experimental Design Module of Statgraphics Plus for Windows. Factors improving dephosphorylation were selected by a 2^3 factorial screening design and their levels optimized by a central composite design. Results of *in vitro* experiments were validated in a feeding experiment with growing broilers (days 1-21 posthatching). Chicken were fed on a low phosphorus negative control diet (0.41% total P; 0.17% available P; „C-") that comprised wheat (55%), soybean meal (37.1%), rapeseed oil (5.15%), limestone (1.2%), lucerne meal (1%), salt (0.3%), methionine (0.125%), vitamin and mineral premix (0.125%). Experimental treatments included also a positive control diet supplemented with calcium phosphate (0.71% total P, 0.41% available P; „C+"), and the „C-" diet enriched with different enzymes and/or mycelium preparation. Levels of dietary additions were those established in the *in vitro* part of the study. Dietary calcium level was 0.69% in each of the treatments. The following criteria of response were determined: body weight gain (BWG, g), bone mineralization expressed as a percentage of ash in the toes, phosphorus retention (%) and mg of phosphorus excreted per g of BWG.

3. RESULTS AND DISCUSSION

3.1. In vitro studies

Results of the preliminary *in vitro* experiments are depicted in Figure 1. The addition of tiered phytase concentrations (0-1250 FTU/kg in 250 FTU/kg increments) caused substantial increases in amounts of phosphorus released from feed, but the response plateaued at around

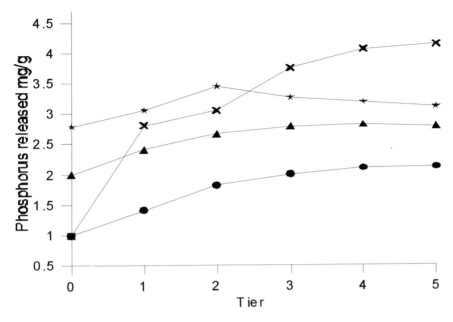

Figure 1. The influence of tiered concentrations of phytase (-●-); phytase (750 FTU/kg) and tiered activities of acid phosphatase (- ▲ -); phytase (750 FTU/kg), acid phosphatase (3156 AcPU/kg) and tiered amounts of pectinase (-✳-), as well as of graded levels of mycelium preparation (-x-) on feed dephosphorylation

750 FTU/kg. Acid phosphatase added to feed in concentrations ranging from 0 to 7890 AcPU/kg (increments: 1578 AcPU/kg) along with 750 FTU/kg of phytase produced similar effect with the level of saturation occurring at 3156 AcPU/kg. The two phosphorolytic enzymes combined (750 FTU + 3156 AcPU/kg) released 2.8 g of phosphorus per kg of feed that was equal to 68% of the total phosphorus content in feed. Pectinase, that was added on top of the two phosphorolytic enzymes in amounts varying from 0 to 7600 PGU/g enhanced feed dephosphorylation most efficiently at 950 PGU/g. It was found however, that citric acid interacted with pectinase when both phosphorolytic enzymes were added to the feed. The two phosphorolytic enzymes, pectinase (1900 PGU/g) and citric acid (3%) brought dephosphorylation rate to around 90% (data not shown). The preparation of fungal mycelium added to feed in concentrations ranging from 0-5% (1% increments) proved to be a highly efficient dephosphorylating agent. Several enzyme activities and substances were tested to find those that positively influence dephosphorylation of feed that had been supplemented with 4% of mycelium preparation. The most significant results are summarized in the Table 1. In order to find optimal dosages of the chosen factors the screening design was augmented by adding centerpoints and star points to form a central composite design. Then, the path of steepest ascend was followed. It was established in that way that 1300 FTU of phytase, 2% of ascorbic acid and 1% of glucose oxidase were the optimal dosages.

Table 1

Optimization of feed dephosphorylation with mycelium preparation (2^3 factorial screening designs)

Nb	Constants	Variables	Units	Lower level (-1)	Upper level (+1)	Significance	Lack of fit (p)	R^2	P. released (%)
1	none	Mycelium	g/kg	20	40	***	0.24	0.98	93±0.4
		ascorbic acid	g/kg	0	30	***			
		Xylanase	FXU/kg	0	4000	NS			
2	mycelium	Phytase	FTU/kg	0	750	*	0.38	0.51	101±0.7
		Pullulanase	µL/g	0	5	NS			
	ascorbic acid	glucose oxidase	g/kg	0	10	*			
3	mycelium 4%	Phytase	FTU/kg	500	1500	*	0.99	0.87	101±0.8
		Ascorbic Acid	g/kg	6	30	**			
		Glucose Oxidase	g/kg	3	15	*			

NS- not significant

3.2. Feeding experiment with growing broilers

Experimental treatments comprised the C- and C+ diets, the C- diet supplemented with phytase (Ph; 750 FTU/kg), the C- diet enriched in phytase and acid phosphatase (Ph+Phs; 750 FTU + 3156 AcPU/kg), the C- diet supplemented with phytase, acid phosphatase, pectinase (1900 PGU/g) and 3% of citric acid (Coc), the C- diet with 4% of fungal mycelium added (Myc), the C- diet supplemented with 4% of fungal mycelium and also with glucose oxidase (1%), phytase (1300 FTU/kg) and 2 % of ascorbic acid (Myc+).

There was a significant influence of dietary treatments on each of the measured parameters (Table 2). As expected, phytase addition increased BWG significantly above those observed in chicks fed on the C- diet. Acid phosphatase brought BWG above the levels observed with phytase fed as the sole supplemental enzyme. In the Coc treatment BWG were superior to those observed in birds fed the two phosphorolytic enzymes and similar to those observed in chicks receiving the Myc diet. There were no differences between the Myc and Myc+ treatments in terms of BWG. It should be emphasized that BWG of chicken fed the Coc, Myc and Myc+ diets were superior to those observed in the C+ diet.

Table 2
Summary of the results from the feeding trial with growing broilers

Treatment	C-	Ph	Ph+Ps	Coc	Myc	Myc+	C+	p <
Body gain (g)	289^a	389^b	430^c	484^d	500^d	500^d	443^c	0,001
Toe ash (%)	$8,73^a$	$9,07^a$	$11,02^b$	$11,96^c$	$11,88^c$	$12,63^d$	$11,93^c$	0,001
Phosphorus retention (%)	$48,8^a$	$57,5^b$	$58,8^b$	$71,5^c$	$73,0^{cd}$	$76,0^d$	$47,3^a$	0,001
Phosphorus excreted mg/g gain	$4,33^d$	$3,53^c$	$3,34^c$	$2,38^b$	$2,04^a$	$1,98^a$ 32%	$6,30^d$ 100%	0,001

[a-d] Means within rows with no common superscript differ significantly (p< .05)

Bone mineralization measured as a percentage of ash in the toes was the highest in the Myc+ treatment, and it was superior to that observed in the C+ diet. Acid phosphatase supplementation stimulated mineralization of bones whereas the enzymic cocktail (phytase + acid phosphatase + pectinase + citric acid) and the mycelium were more effective than the two phosphorolytic enzymes.

Chicken fed the Myc+ diet retained as much as 76% of total phosphorus consumed. To our knowledge, *seventy six percent of phosphorus retention* observed in this group of chicken *is the highest value reported ever for poultry fed diets based on wheat. This was accompanied by 68% reduction in amounts of phosphorus excreted.* Similar strategy applied to corn-soybean meal based diets resulted in 79% phosphorus retention in growing turkeys [7]. Comparable levels of P utilization (72-73%) were observed in birds fed the Coc and the Myc diets. Surprisingly, no influence of acid phosphatase on P utilization or excretion was noticed. Similar observations however, were reported by Näsi et al. [10], who studied the efficacy of phytase along with graded levels of acid phosphatase in growing swine.

3.3. Conclusions

1. In spite of high endogenous activities of phytase and acid phosphatase in wheat, microbial enzymes improve dephosphorylation of wheat-based feeds.
2. Acid phosphatase (phytase B) significantly enhances dephosphorylation of feeds supplemented with phytase. Addition of acid phosphatase to a diet containing phytase improved phosphorus utilization in respect to performance and bone mineralization but did not increase phosphorus retention.
3. A crude pectinase preparation and citric acid significantly improve dephosphorylation of feeds supplemented with phosphorolytic enzymes (phytase + acid phosphatase). This results in further improvements in BWG, bone mineralization and phosphorus retention in growing broilers.
4. A preparation of fungal mycelium showed high dephosphorylating capacity towards wheat-based feeds. At the supplementation level of 4% (w/w), its ability to free phosphorus was comparable to that observed with a cocktail of phosphorolytic and cell-wall-degrading enzymes.
5. Dephosphorylating ability of mycelium may further be enhanced by the addition of phytase (1300 FTU/kg), ascorbic acid (2%) and glucose oxidase (1%). When this composition is added to feeds for growing broilers it improves bone mineralization and phosphorus retention as compared to mycelium fed as the sole feed supplement.

322

Table 3
The comparison of different strategies employed for optimal P utilization in poultry

Literature	Strategy	P utilization	BWG	Bone mineralization
[4]	660 μg cholecalciferol + phytase 600 FTU	68%	16% over NC	8% over NC
[5]	soybeans transformed with phytase gen (1200 FTU/kg)	62%	16% over NC	29% over NC
[6]	a genetically modified, H A P corn + phytase	?	2% over PC	?
[11]	a genetically modified, H A P barley	52%	?	?
[7]	enzyme cocktails, turkeys	79%	9% over PC	4% over PC
Present work	enzyme cocktails, broilers	76%	73% over NC 13% over PC	45% over NC 6% over PC

PC- positive control diet, NC - negative control diet, HAP - high available phosphorus

In Table 3 results of the present research are compared with those known from the literature in terms of P utilization, BWG and bone mineralization. The enzyme cocktails strategy seems to be a highly competitive one in enhancing P utilization by poultry.

REFERENCES

1. T.S Nelson, T.R. Shieh, R.J. Wodzinski, J.H. Ware, Poultry Sci., 47 (1968) 1842.
2. F.J. Schöner, P.P. Hoppe, H. Schwarz, J. Wiesche, Anim. Physiol. Anim. Nutr., 69 (1993) 235.
3. E.T. Kornegay, D.M. Denbow, Z.Yi, V. Ravindran, Br. J. Nutr., 75 (1996) 839.
4. H. Quian, E.T. Kornegay, D.M. Denbow, Poultry Sci., 76 (1997) 37.
5. D.M. Denbow, E.A. Grabau, G.H. Lacy, E.T. Kornegay, D.R. Russell, P. F. Umbeck, Poultry Sci., 77 (1998) 878.
6. W.E. Huff, P.A. Moor, P.W. Waldroup, A.L. Waldroup, J.M. Balog, C.R. Huff, N.C. Rath, V. Raboy. Poultry Sci., 77 (1998) 1899.
7. K. Żyła, D.R. Ledoux, M. Kujawski, T.L. Veum, Poultry Sci., 75 (1996) 381.
8. K. Żyła, D.R. Ledoux, A. Garcia, T.L. Veum, Br. J. Nutr., 74 (1995) 3.
9. K. Żyła, D. Gogol, J. Koreleski, S. Świątkiewicz, D.R. Ledoux, J. Sci. Food Agric., 79 (1999) 1832.
10. M. Näsi, J. Piironen, K. Partanen, Anim. Feed Sci. Technol., 77 (1999) 125.
11. D.R. Ledoux, Y.C. Yi, T.L. Veum, V. Raboy, K. Żyła, A. Wikiera, Proceedings of the 12th European Symposium on Poultry Nutrition, Veldhoven, The Netherlands, August 15-19, 1999, WPSA-Dutch Branch, pp. 51-53.

Food Biotechnology
S. Bielecki, J. Tramper and J. Polak (Editors)
© 2000 Elsevier Science B.V. All rights reserved.

Application of bacterial cellulose for clarification of fruit juices

A. Krystynowicz, S. Bielecki, W. Czaja, M. Rzyska,

Institute of Technical Biochemistry, Technical University of Lodz,
ul. Stefanowskiego 4/10, Łódź 90-924, Poland

The aim of our studies was to estimate the usefulness of bacterial cellulose synthesised by *Acetobacter xylinum* E_{25} for the clarification of apple juice. The cellulose was applied either as a membrane formed under stationary culture conditions or as a suspension obtained by disintegration of such a membrane. The membrane was used as a filter bed. The results of our studies confirm that cellulose membranes may be used as filtration ones. After filtration through cellulose membranes, apple juice displayed very high clarity, higher than its sample clarified under standard conditions, even after a stability test. Juice clarification only by filtration through cellulose membrane does not assure its stable colour, which darkens under the conditions of a stability test performance. This phenomenon is also observed for other products clarified by filtration only. Bacterial cellulose application as an adsorbent for fruit juice clarification yielded the product with demanded stability.

1. INTRODUCTION

Bacterial cellulose synthesised by acetic acid bacteria, especially strains of *Acetobacter xylinum*, has attracted attention because of its unique properties, distinguishing this kind of cellulose from the plant one, and determining its practical application. The differences lie first of all in thickness of elementary microfibrils comparable to thickness of fibrils present in primary walls of higher plants and equal to 3,0 nm [1,2]. The fibrils originating from 50-80 adjacent pores localised in bacterial cell membrane combine into bundles which form ribbons with the width of 40-60 nm, thickness of 10 nm and length of 10 μm [3]. These ribbons synthesised by many bacterial cells form a pellicle on the surface of a liquid but not agitated growth medium. Both shape and area of this pellicle depend on the size of the horizontal cross section of a culture vessel. The pellicles contain almost pure (about 97%) and highly crystalline (about 65%) α-cellulose [4].

The cellulose synthesis yield depends on the activity of a strain-producer, composition of growth medium and culture conditions, and may amount up to about 28g per 1l under stationary culture conditions or up to about 9g per 1l under submerged culture conditions [5].

The mechanism of cellulose fibrils formation results in a high dynamic durability of bacterial cellulose, equal to 54000 kg/mm^2. This factor is 3-4 times higher in comparison to Kebler fibrils and comparable to glass fibrils [6]. Pressing, squeezing or rolling in a definite direction may increase mechanical durability of membranes. After drying, cellulose membranes resemble a parchment paper with a thickness of 0,01-0,5 nm, and display porous

324

structure. Pores with diameters of about 3µm may constitute up to 50-93% of the membrane surface [7,8].

Because of such structure, the cellulose membrane may be applied as a filter cloth [6,9]. Moreover, bacterial cellulose may be used in the form of a defibered suspension for the production of filter media, which due to the size of pores meet the requirements for dialysis, micro- or ultrafiltration membranes [10,11]. Bacterial cellulose structure resulting from the mechanism of its biosynthesis is characterised by a very well developed internal surface, markedly larger in comparison to the plant pulp surface, which gives it high absorption capacity.

Since bacterial cellulose has been recognised to be safe for humans, it seems reasonable to use it in food industry [12]. The above mentioned advantages of bacterial cellulose prompted our research into its application for clarification of fruit juices.

2. MATERIALS AND METHODS

2.1. Materials

Apple (*Malus domestica, Cortland* variety) juice after extrusion was depectinized with Pektopol PT (1ml/l) for 4 hours at 45°C and pasteurised for 1 hour. Clarifying agents: gelatine of food quality for clarification SIHA and silica gel for beverage clarification SHIASOL-30 were purchased from Begerow. Bacterial cellulose in the form of membranes produced under stationary culture conditions by *Acetobacter xylinum* E_{25} from the collection at the Institute of Technical Biochemistry, Technical University of Lodz. The following growth media were used:

(A) The modified Schramm medium [4] containing glucose (2%), yeast extract (0.5%), Bactopeptone (0.5%), $NaHPO_4$ (0.27%), $MgSO_4 \cdot 7H_2O$ (0.05%), ethanol (1.0%), pH=5.5.
(B) The modified Schramm medium in which glucose was replaced with 2% of fructose.

Acetobacter xylinum E_{25} was cultured under stationary conditions at 30°C for 6 days. The synthesised cellulose pellicles were washed with tap water, boiled in 1% NaOH solution and again washed with water till NaOH removal.

2.2. Analytical methods

Determination of water permeability was based on the measurements of time required for the filtration membrane under constant pressure. Determination of thickness and porosity was performed by the computer image analysis method, using the Carl Zeiss Jena microscope with a program IPS-512 (Imal system).

Mechanical testing was performed with cellulose membranes in the form of 20 mm x 160 mm strips partly dehydrated by squeezing. The degree of squeezing off was determined by the per cent content of cellulose. The sample tensile strength was measured with an INSTRON TMM-1111 machine under the following conditions for all membrane variants under investigation:
-   distance between jaws 10 cm,
-   tension rate 10 cm/min.,
-   force measurement range 50 N,
-   recording paper rate 50 cm/min.,
-   temperature 20°C

The tensile strength σ was determined, using the breaking force F in relation to the surface area of dry membrane cross-sections.

Determination of total polyphenols content was based on measurements of the absorbance at 700 nm of a complex formed in a reaction of polyphenols and a tungsten-molybdate reagent. Total extract was assayed according to the Polish standard PN-90/A-75101/02, sugars and sugars-free extract – PN-90/A-75101/07, total acidity - PN-90/A-75101/04, clarity - PN-A-75957 and colour in CIE system – PN 65/N01252.

3. RESULTS AND DISCUSSION

For mechanical properties testing of the cellulose membranes, 3 samples of the biomaterial were used. They were synthesised in media containing glucose, fructose or sucrose as a sugar component. The factors characterising mechanical properties of the examined membranes are presented in Table 1. The values of these factors confirm that physicomechanical properties of the membranes are markedly influenced by the composition of a growth medium, especially by sugar being the source of glucose.

Table 1
Mechanical properties of cellulose membranes

Sample	Cellulose content [%]	Thickness of dried membrane [μm]	F [N]	σ [MPa]	E [%]	E [MPa]	F_r [N]
Membrane from glucose containing medium	3.0	33.9	20.7	30.6	30.1	30.8	0.69
Membrane from fructose containing medium	2.9	34.9	42.4	60.8	24.7	41.8	1.6
Membrane from sucrose containing medium	4.1	25.2	6.96	13.8	16.0	48.4	0.72

The differences between the values of the examined parameters probably point to differences in structural density of the membranes, and are particularly distinct in the case of values of the longitudinal tearing force (F) and the tensile strength (σ). The values of these factors are relatively high for the membranes produced in the fructose-containing medium.

Cellulose membrane porosity and water permeability are factors limiting its application as filter media (Table 2). Because drying destroys the primary structure of cellulose pellicles which are converted into a parchment-like paper, the permeability of dried membranes is lower in comparison to that of wet pellicles, despite minor differences of pore dimensions observed for membranes produced in glucose containing medium. The lowest factors of water permeability and the smallest pores were observed in the membranes synthesised in the medium containing fructose. This fact probably confirms the conclusion that in this pellicle the cellulose microfibrils are packed more tightly. So one can modify properties of cellulose membranes by means of selection of the growth media composition.

Table 2
Determination of the porosity of cellulose membranes produced in glucose containing medium

Sample	Water permeability coefficient [l/m² min]	Pore diameter [μm]
Wet membrane	6.4	1.4
Air-dried membrane	2.1	1.4
Membrane dried at 80°C	0.8	1.5
Freeze-dried membrane	2.2	1.4
Membrane dried at 80°C *	0.2	0.4

*from the fructose containing medium

The cellulose membranes produced in the glucose-containing medium were applied in the studies on apple juice filtration. The properties of the juice obtained in this way were compared to the properties of the juice clarified under standard conditions, by using gelatin (0,5 ml of 1% gelatin solution per 100 ml of the juice) or sol of silicic acid (0,5 ml of 3% sol per 100 ml of the juice), followed by filtration on the diatomaceous earth Hyflo Super Cel. Bacterial cellulose in the form of suspension was applied for juice clarification by adsorption using 3 g of the cellulose dry mass per 100 ml of the juice.

Table 3
Characteristics of the apple juice after filtration

Assay	Before clarification	Filtration through membrane		Clarified by adsorption	Standard clarification
		wet	Dried		
Extract [%]	13	11.5	11.6	11.5	11.5
pH	3.8	3.9	3.9	3.9	3.8
Total acidity (g of malic acid per 1 l)	3.5	3.22	3.38	3.35	3.35
Total polyphenols (g/l)	0.39	0.27	0.28	0.22	0.24
Clarity (%T$_{620nm}$)	56	99	98	97	96
Colour (A$_{420nm}$)	0.68	0.21	0.21	0.22	0.23
Colour CIA (λ- the lenght of maximal absorbance)	560	578	580	574	577
Y - paleness	0.88	0.88	0.88	0.91	0.84
Pe – purity of stimulation	0.14	0.19	0.20	0.24	0.14

The results of the studies are presented in Table 3 and 4. They prove that the application of this form of bacterial cellulose for filtration or adsorption in the process of juice filtration enables yielding of a product, which displays higher clarity than a juice sample clarified in a standard manner. The parameters characterising the colour saturation and paleness point to the fact that the juice clarified using bacterial cellulose was decolourised to a higher extent. It was particularly distinct in the case of the sample clarified by means of adsorption, in which a decrease of polyphenol concentration (about 45%) was the highest.

The loss of polyphenol stability confirmed the stabilisation of clarity of the examined juice samples. Darkening of juice samples clarified by means of filtration through cellulose membranes only was observed. This phenomenon points to the incomplete stability of polyphenols. When bacterial cellulose was used as an adsorbent for filtration stable juice was obtained. Since the application of adsorption resins for food production is still questionable [14], bacterial cellulose because of its purity and physicochemical properties seems to be a suitable material for fruit juice clarification.

Table 4
Characteristics of the apple juice after stability test.

Assay	Standard clarification	Filtration through membrane		Clarified by adsorption
		wet	dried	
Clarity (%T_{620nm})	92	95	94	94
Colour (A_{420nm})	0.24	0.28	0.28	0.24
Colour CIA (λ- the lenght of maximal absorbance)	579	575	574	576
Y - paleness	0.91	0.89	0.88	0.91
Pe – purity of stimulation	0.14	0.19	0.18	0.21

REFERENCES

1. C.H. Haighler, R.M. Brown, M. Benziman, Science, 210 (1980), 903.
2. K. Kudlicka, Postępy biologii komórki, 16 (1989), 197.
3. K. Zaow, J. Cell Biol., 80 (1979), 773,
4. S. Yamanaka, K. Watanabe, Kitamura N., Iguchi M., S. Mitsuhashi, Y. Nishi, M. Uryu, J. of Materials Science, 24 (1989), 3141.
5. E.J. Vandamme, S. De Baets, A. Vanbaelen, K. Joris, Polymer Degradation and Stability, 59 (1998), 93.
6. Patent JP 1199604, (1989).
7. Patent EP 86308092.
8. Patent JP 1193335, (1989).
9. Patent JP 4326994, (1992).
10. Patent EP 416470, (1991).
11. Patent JP 3032726, (1991).
12. A. Okiyama, Motoki M., Yamanaka S., Food Hydrocoll., 6 (1992), 479.
13. M. Schramm, S. Hestrin, J. Gen. Microbiol., 11 (1954), 123.
14. R. Lyndon, Fruit Process 6,4 (1996), 130.

Food Biotechnology
S. Bielecki, J. Tramper and J. Polak (Editors)
© 2000 Elsevier Science B.V. All rights reserved.

The Effect of culture medium sterilisation methods on divercin production yield in continuous fermentation

A. Sip, W. Grajek

Department of Biotechnology and Food Microbiology, Agricultural University,
48 Mazowiecka St. 60-623 Poznan, Poland

1. INTRODUCTION

Lactic acid bacteria (LAB) are able to produce some proteinaceous compounds exhibiting antibacterial activity [1]. Among bacteriocins, nisin, lactacin and pediocin have been the most studied. Bacteriocins of LAB can be used as biological food preservatives in many branches of food industry, particularly in the dairy and meat industries.

Recently, there have been many reports on micro-organisms producing bacteriocins as well as on the chemical structure and biological activities of these substances [2]. However, it should be noted that information on the production of bacteriocins in batch and continuous fermentations is rather scarce [3-7]. These studies refer mostly to batch cultures carried out in small volumes, usually in Erlenmeyer flasks. There are only a few descriptions of the experiments performed in laboratory bioreactors [3,4,8,9]. Little research has been done on increasing bacteriocin productivity by using immobilised cells [10,11] or using a bioreactor with a microfiltration module [6].

Bacterial cultures for bacteriocin production are usually cultivated in complex media containing such components as meat and yeast extracts as well as peptone and protein-rich hydrolysates [3,4]. The authors have studied the effect of some environmental factors on cell growth and bacteriocin production. Among these factors temperature, pH and media composition have been most studied.

One of the most interesting bacteriocin produced by the bacterium *Carnobacterium divergens* is divercin. It is a thermostable, proteinaceous substance with a molecular weight of about 4.5 kDa. It is resistant to heating at 100°C for 30 min and at 121°C for 10 min and is deactivated only when autoclaved for 15 min. Divercin is also as stabile as SDS, Tween 80 and urea. Divercin shows bactericidal activity against *Listeria monocytogenes, Listeria innocua, Enterococcus faecalis*, same strains of *Carnobacterium piscicola, Carnobacterium divergens* and *Clostridium tyrobutyricum*. It demonstrates no antibacterial activity against lactic acid bacteria belonging to genera *Lactobacillus, Lactococcus, Pediococcus* and *Leuconostoc* as well as against Gramm-negative bacteria [9]. These properties make divercin a potential preservative for the fish, meat and dairy industries.

The aim of this study was to determine the effect of methods of medium sterilisation on the yield of divercin production.

2. MATERIALS AND METHODS

2.1. Bacteria

Divercin was produced with the use of the *Carnobacterium divergens* AS7 strain, isolated from the sea salmon digestive tract. The *Carnobacterium piscicola* NCDO 2765 strain (Professor X. Dousset, E.N.I.T.I.A.A. Nentes, France), was utilised as indicator bacteria for the determination of divercin activity. The strains were stored as a suspended in 5% v/v glycerol at - 70∞C. The stored strains were cultured at 30°C for 24 h and then transferred into 30 ml of fresh medium and cultured in relatively anaerobic conditions in 50 ml Erlenmeyer̀s flasks at 30∞C for followed 24 h. 10 ml of liquid was sampled from the cultures, transferred into flasks with 90 ml of fresh medium and cultivated at 30°C for 14 h. The resulting cultures were used to inoculate the culture media.

2.2. Media

The MRS medium, modified through removal of Tween 80 and by replacing triptone with casein peptone, was used for *Carnobacterium divergens* bacteria fermentation. Organic components of the medium were provided by BTL Łódź, and the minerals originated from POCh Gliwice. The medium was sterilised at 121°C for between 15 to 60 minutes as well as by the membrane filtration technique ($\varphi=0,22$ μm, Millipore, USA).

The indicator bacteria *Carnobacterium piscicola* was cultured in the medium composed of 1% glucose (BTL, Łódź), 1% NaCl (POCh, Gliwice) and 0.5% yeast extract (BTL, Łódź) [12,13]. The medium was sterilised at 121°C for 15 minutes.

2.3. Continuous fermentation

Continuous fermentations were carried out in the 5 dm^3 Bioflo III bioreactors (New Brunswick Sci., USA). The bacteria were cultured in anaerobic conditions in the modified MRS medium at 30°C at pH 6.5, with an agitation of 80 rpm and dilution rate (D) ranging from 0.03 to 0.12 h^{-1}. The dilution rate was increased after reaching the chaemostate for the previous D value. The fermentation medium was sterilised with the use of two techniques such as steam heat and microfiltration. A stable pH level was automatically adjusted with 5M NaOH. The cultures were inoculated with a 2% v/v 14-hour-old inoculum. The continuous fermentations were preceded by 12-14 hour batch fermentations. During the fermentation, the divercin activity and biomass concentrations were assayed.

2.4. Determination of divercin antibacterial activity

The divercin activity versus the indicator bacteria *C. piscicola* NCDO 2762 was determined in an active extract obtained by removing of bacteria cells from liquid media by centrifugation and pH adjusting to 6.5. The technique of critical dilutions, described by Pilet et al. [9] was used for the assays. The divercin activity present in the analysed samples was calculated in the conventional units (AU ml^{-1}) which express the inverse of the lowest dilution exhibiting no ability to inhibit the control bacteria growth.

2.5. Biomass determination

The wet biomass was dried in two steps, firstly at 80°C for 5 h and then dried to achieve a stable mass at 105°C. The dry bacteria mass (x) was expressed in grams per litre of the analysed fermentation liquid.

2.6. Technological parameters

Based on the fermentation material balance, the values of the following parameters were calculated: divercin volume yield $Y_{Pd/v}$ [AU h^{-1}] and divercin specific yield $Y_{Pd/x}$ [AU l^{-1} g^{-1} h^{-1}]. Formulas and correlations given by Sinclair and Cantero [14] and Brown [15] were used for the calculations.

3. RESULTS AND DISCUSSION

3.1. Continuous fermentation using the heat sterilised medium

The dilution rate within the range from 0.03 to 0.085/h had a slight effect on the divercin activity released to the medium. At D 0.03 and 0.05/h the divercin activity in the active extract was 819 200 AU/ml whereas at D 0.07 and 0.085/h it remained at the level of 409 600 AU/ml. During the fermentation with the increased dilution rate, the activity of the released divercin was much lower. At D=0.12/h it reached a minimum of 25 600 AU/ml (Table 1).

Table 1
Parameters describing the course and the yield of continuous fermentations in the thermal sterilised medium

D	x	P_d	$Y_{Pd/v}$	$Y_{pd/x}$
$[h^{-1}]$	[g d.m./l]	[AU/ml]	10^8 [AU/lh]	10^7 [AU/gh]
0.03	1.55	819 200	1.22	52.85
0.05	1.50	819 200	2.05	54.61
0.07	2.00	409 600	1.43	20.48
0.085	2.25	409 600	1.74	18.21
0.12	2.90	25 600	0.15	0.88

The dilution rate had an effect on the divercin biosyntheses yield. This is shown by the results presented in table 1. The divercin volume yield increased with the increase of D within the range of 0.03 to 0.05/h. At D= 0.05/h it reached the value of 2.05×10^8 AU/lh. At higher D values the divercin biosynthesis was lower. The lowest divercin volume yield was achieved for D=0.12/h and amounted 0.15×10^8 AU/lh. Comparing the divercin biosynthesis yield in continuous and batch fermentations (the first 12 h of culturing), it was observed that the maximum divercin volume yield in the continuous fermentations was 2.5×10^3 times higher than the maximum divercin productivity in the batch fermentations.

During the fermentation the divercin specific yield also changed. Within the range of D-0.07 and 0.12/h the bacteria ability to produce divercin was inversely correlated to the average dilution rate (Table 1). At D=0.12/h the divercin specific yield was 60 times lower than the yield achieved at D=0.03/h.

Activity changes of the divercin released into the fermentation environment were related to the biomass concentration changes. The data illustrated in Figure 1 shows the above.

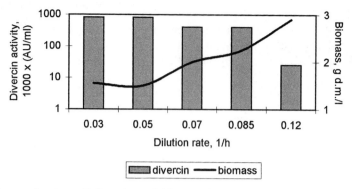

Figure 1. The production of divercin and biomass in continuous fermentation in the heat sterilised medium.

The divercin activity in the active extract was the highest at the lowest biomass concentration. It was observed that the higher the biomass concentration within the range from 1.5 to 2.0 g d.m./l, the lower the divercin activity. At biomass concentrations within the range from 2.0 to 2.25 g d.m./l, the divercin activity maintained a stable level. A dramatic divercin activity drop in the active extract occurred when the biomass concentration increased to 2.9 g d.m./l.

3.2. Continuous fermentation using the medium sterilised by microfiltration

The dilution rate had an effect on the divercin activity released to the medium. At D within the range from 0.03/h to 0.07/h the divercin activity in the active extract was constant and equalled 3 200 AU/ml. At higher dilution rates its value was many times lower. The data presented in Figure 4 show that the divercin activity in the active extract decreased along the increase of D between 0.085 and 0.12/h. At D=0.12 it reached the lowest value (400 AU/ml). In the whole D range the divercin activity in the extract was lower than the maximum divercin activity achieved during the period of batch fermentations.

Comparing the results of the continuous fermentations carried out in the thermal sterilised medium and in the medium sterilised by microfiltration, it was claimed that the fermentation medium sterilisation technique has a significant effect on divercin activity in the active extract. During continuous fermentations in the thermal-sterilised medium in the whole tested range of D, the divercin activity in the active extract was much higher than the divercin activity released to the medium sterilised by microfiltration.

The divercin volume yield depended on the average dilution rate. This fact is illustrated in Table 2. The divercin biosynthesis yield increased along the increase of the dilution rate within the range of 0.03 and 0.07/h. At D=0.07/h it reached the maximum value. It was, however, almost 180 times lower than the maximum divercin volume yield achieved in continuous fermentations in the thermal sterilised medium. It means that the thermal-sterilised medium was a better medium for divercin production.

Table 2
The parameters describing the course and the yield of the continuous fermentations carried out in the medium sterilised by microfiltration

D [h⁻¹]	x [g d.m./l]	P_d [AU/ml]	$Y_{Pd/v}$ 10^6[AU/lh]	$Y_{Pd/x}$ 10^6[AU/gh]
0.03	2.80	3 200	0.48	1.14
0.05	3.00	3 200	0.80	1.07
0.07	3.20	3 200	1.12	1.00
0.085	3.30	800	0.34	0.24
0.12	3.40	400	0.24	0.13

Bacteria ability to produce divercin was also correlated to the dilution rate. The highest value was achieved at D=0.03/h and then along the dilution rate increase it gradually decreased (Table 2). During continuous fermentations in the medium sterilised by microfiltration the divercin specific yield was significantly lower than the divercin yield for the thermal sterilised medium. The medium sterilisation technique had a similar effect on the bacteria ability to produce divercin in the batch fermentations.

The divercin activity in the active extract was constant at biomass concentrations between 2.8 and 3.2 g/l and while it increased, the divercin activity dramatically dropped. At D=0.03/h the biomass concentration was 2.8 g/l and was comparable to the biomass concentration achieved in the thermal- sterilised medium at D=0.12/h (Figure 2).

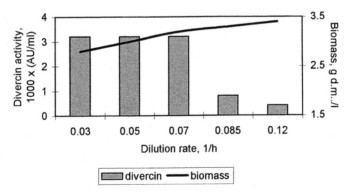

Figure 2. The production of divercin and the biomass in the continuous fermentations in the medium sterilised by microfiltration.

After analysing the results of the continuous fermentations in the thermal sterilised medium and in the medium sterilised by microfiltration, it was claimed that at the same biomass concentration the activity of the divercin released to the thermal-sterilised medium was 8 times higher than the divercin activity released to the medium sterilised by microfiltration. This fact suggests that during thermal sterilisation some compounds are formed in the medium that stimulate divercin production. It is also probable that lower divercin activity in the medium sterilised by microfiltration resulted from a larger synthesis of proteolytic enzymes which have a destructive influence on divercin.

4. CONCLUSIONS

Method of medium sterilisation has significant effect on the divercin and biomass production. In continuous culture the divercin production was manifold higher in heat sterilised medium than in medium sterilised by microfiltration. It means that heat treatment of the medium enhanced divercin production

REFERENCES

1. C.M. Holz and U. Stahl, Food Biotechnol., 9 (1995) 85.
2. . D.H. Hoover and L.R. Steenson, (eds.), Academic Press Inc, New York, 1993.
3. R.S. Biswas, B. Ray, P. Ray and M. C. Johnson, Appl. Environ. Microbiol. 57 (1991) 1265.
4. L. De Vuyst and E.J. Vandamme, J. Gen. Microbiol., 138 (1992) 571.
5. A.L. Kaiser and T.J. Montville, J. Appl. Bacteriol., 75 (1993) 536.
6. M. Taniguchi M., K. Hoshino, H. Urasaki. and M. Fujii, J. Ferment. Bioeng., 6 (1994) 704.
7. R. Yang and B. Ray, Food Microbiol., 11 (1994) 281.
8. . P.M. Muriana and T.R. Klaenhammer, Appl. Environ. Microbiol., 57 (1991) 114.
9. M.F. Pilet, X. Dusset, R. Barre, G. Novel, M. Desmazaeud and J.Ch. Piard, J. Food Protect., 58 (1995) 256.
10. N. Zezza, G. Pasini, A. Lombardi, A. Mercenier, P. Spettoli, A. Zamorani and M.P. Nuti, J. Dairy. Res., 60 (1993) 581.
11. J. Wan, M.W. Hickey and M.J. Coventry, J. Appl. Bacteriol., 79 (1994) 671.
12. W. Grajek, T. Lassocinska, A. Sip, Pol. J. Food Nutr. Sci., 5 (1996) 74.
13. A. Sip and W. Grajek, Pol. J. Food Nutr. Sci., 3 (1997) 1.
14. C.G. Sinclair and D. Cantero, IRL PRESS at Oxford University Press, New York, 1990.
15. A. Brown, In: Fermentation a practical approach. Eds.: McNeil B., Harvey L. M.; IRL PRESS at Oxford University Press, New York, 1990.

Food Biotechnology
S. Bielecki, J. Tramper and J. Polak (Editors)
© 2000 Elsevier Science B.V. All rights reserved.

Production of *Carnobacterium divergens* biomass

A. Sip, W. Grajek

Department of Biotechnology and Food Microbiology, Agricultural University,
48 Mazowiecka St., 60-623 Poznan, Poland

1. INTRODUCTION

Advances in fermentation technology have resulted in predictable quality and standardisation of fermented products. Starter cultures of lactic acid bacteria have an important influence on the product quality in food industry. They exhibit the bacteriocin activity, probiotic properties and introduce attractive sensorial features into food products [1-4]. Starter cultures are produced as mono- or heterogenic cultures. One of potential components of them are *Carnobacterium divergens* bacteria producing lactic acid and antilisterial bacteriocin divercin [5].

The use of the culture of high cell density makes the technology of starter production easy. The use of membrane bioreactor allows to obtain high concentrated cell suspension as a result of the continuous separation of bacteria by microfiltration. The high performance of this process determines economic production of cell biomass and a suitable product quality [6-13]. This paper describes the conditions, which influenced high cell density culture of *C. divergens* produced in membrane bioreactor with cell recycling.

2. MATERIALS AND METHODS

2.1. Bacteria

The *C. divergens* AS7 strain, isolated from the digestive tract of marine salmon, was used in this study. This strain was stored as cell suspension in 5 % v/v glycerol at -70°C.

The stored strain was cultured at room temperature and then transferred into 30 ml of fresh medium and cultured for 24 h in relatively anaerobic conditions in 50 ml Erlenmeyer's flasks at 30°C. 10 ml of liquid was sampled from the cultures and transferred into flasks with 90 ml of fresh medium. The bacterial culture was prepared at 30°C for 14 h. The resulting culture was used to inoculate the culture media.

2.2. Media

The MRS medium [14], modified by the removal of Tween 80 and the replacement of triptone with a casein peptone, was used for *C. divergens* culturing. Organic components of the medium were provided by BTL Łódź (Poland), and the minerals originated from POCh Gliwice (Poland). The medium was sterilised at 121°C for 60 minutes.

2.3. Batch fermentations with pH adjustment

Batch fermentations with pH adjustment were carried out in 5 dm³ Bioflo III bioreactors produced by Brunswick Sci. (Edison, N.J., USA) with 3 dm³ of the modified MRS medium in anaerobic conditions, at 30°C and pH 6.5 and agitation at the level of 80 rpm. The culture pH was controlled by automatic injection of 5 M NaOH. The desirable medium pH was reached with the use of 5 M NaOH and then blown with nitrogen (15 min.) and inoculated with 2% v/v 14 hour inoculum. The initial bacteria cell concentration was approximately 10^6 CFU/ml of the medium. The fermentation was run for 35 hours.

2.4. Continuous fermentation

Continuous fermentations were carried out in the 5 dm³ Bioflo III bioreactors (New Brunswick Sci., USA). The bacteria were fermented in anaerobic conditions in the modified MRS medium at 30°C at pH 6.5, with an agitation of 80 rpm and a dilution rate (D) ranging between 0.03 and 0.12 h⁻¹. The dilution rate was increased after reaching the chemostat for the previous D value. A stable pH level was maintained with of 5M NaOH. The cultures were seeded with a 2% v/v 14-hour-old inoculum. The continuous fermentations were preceded by 12-14 hour batch fermentations.

2.5. Continuous fermentation with cell recycling

Continuous fermentations with cell re-circulation were carried out in a membrane bioreactor including the 5 dm³ Bioflo III bioreactor (New Brunswick Sci., USA), a peristaltic circulation pump and a microfilter (Prostak, Millipore, USA) with a flat canal membrane with a pore diameter of 0.22 μm made of polyvinilidene fluoride (CVDF). The filtration module was equipped with a microprocessor to control the seed of fermentation liquid flow in the system as well as the inlet, outlet and transmembrane pressures. Continuous fermentations with cell recycling were carried out in anaerobic conditions at 30°C in a modified MRS medium at pH 6.5 which was maintained constantly by automatic injection of 5M NaOH. The following are the experimental variables: medium dilution rate (D) ranging between 0.14 and 0.36 h⁻¹, fermentation liquid recycling rate (Vcirc) ranging between 60–120 l h⁻¹ and biomass excess outlet (bleeding, CB) from 0 to 100% of the dilution rate. The fermentation parameters were modified only after reaching chemostat in the previous step. The culture medium (5.5 dm³) pH after the introduction into a bioreactor was adjusted to pH 6.5 and inoculated with 2% v/v 14 hour inoculum. Bacteria were cultured for 14 hours in a batch fermentation in optimum growth conditions (anaerobic conditions, temperature 30°C, pH 6.5, agitation 80 rpm). The anaerobic conditions were achieved through blowing nitrogen through the medium. Afterwards, the agitation was switched off and the fermentation liquid recycling was initiated in the bioreactor and microfilter system. At the same time 1.5 dm³ of fresh medium was supplemented to have 7 dm³ of total liquid volume. Batch fermentations with medium recycling were carried out for 8 h. In the 22ⁿᵈ hour, continuous fermentation with cell recycling was initiated. After reaching the turbidostat the cell excess was started.

2.6. Biomass determination

The centrifuged biomass was rinsed with a phosphorous buffer (0.01 M, pH 6.5) and then dried in two phases, first at 80°C for 5 hours and then dried up to the constant mass at 105°C. The dry bacteria mass was calculated per 1 litre of the analysed fermentation liquid. The number of live *C. divergens* AS 7 cells was determined with the use of a classical plating technique. The results were given in cell units per 1 ml of the analysed sample.

3. RESULTS AND DISCUSSION

3.1. Batch fermentations with pH adjustment

C. divergens AS7 bacteria effectively multiplied until the 20th hour of the fermentation. Their biomass concentration reached the maximum value of 0.8 g D.M./l in the 20th hour and was twice as high as the maximum biomass concentration achieved in batch fermentations without pH adjustment. The biomass concentration remained at the maximum level until the 25th hour. During further fermentation the biomass concentration decreased. In the 35th hour it was 6% (0.05 g D.M./l) lower than the highest biomass concentration achieved in the 20th hour of fermentation. The changes of bacteria dry matter were related to the change in the amount of live bacteria. The percentage of the live cells in the bacteria population in a batch fermentation with pH adjustment was larger than the percentage of live cells observed in batch fermentations without pH adjustment (Figure 1).

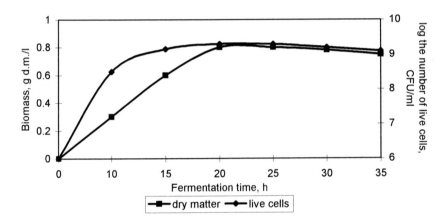

Figure 1. The dynamics of *C. divergens* AS7 growth in batch fermentation at 30°C at pH 6.5.

3.2. Continuous fermentations

Continuous fermentations on a thermal sterilised medium were initiated at a medium dilution rate of 0.03/h. In the 84th hour of the fermentation the medium dilution rate was increased up to 0.05/h. Fermentations at D=0.05/h were carried out for 76 hours. In the 160th hour of the fermentation, the medium dilution rate was increased up to 0.07/h and after 60 hours up to 0.085/h. Fermentations at D=0.085/h were carried out for 88 hours and then the fresh medium supply was initiated at the dilution rate of 0.12/h.

The medium dilution rate had a significant influence on the bacteria growth yield. The bacterial growth yield increased along the increase of D between 0.05 to 1.12/h (Figure 2). At D=0.12/h the biomass concentration reached the maximum value. It was 2.9 g D.M./l and it

338

was almost 4 times higher than the maximum biomass concentration achieved in the batch fermentations. At two lowest dilution rates the bacteria growth yield was similar (1.5-1.55 g D.M./l).

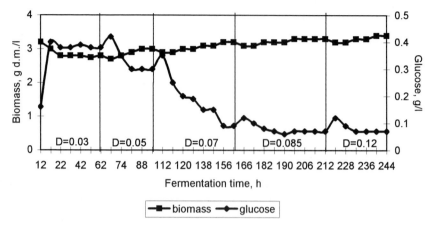

Figure 2. Biomass production and glucose utilisation by the *C. divergens* AS7 in the continuous fermentation on the thermal sterilised medium.

During the fermentation the *C. divergens* AS7 cells showed a strong aggregation. In the initial fermentation period the aggregates were small. After 100 hours of fermentation their size began to increase immediately. Multicell bacteria aggregations were visible to the naked eye in the form of fibres. Their size increased in time. The analyses showed that the fibres were stable in the medium of between pH 4.5 and 6.5, whereas they were easily split with the agitation increase from 80 to 300 rpm. After the 300th hour the cell flocculation phenomenon disappeared and the fermentation revived its suspension state.

3.3. Continuous fermentation with cell recycling
Batch fermentation carried out for 14 hours in the optimum conditions for *C. divergens* AS7 growth was the first phase of the continuous fermentation with cell recycling. In the 14th hour of the fermentation 1.5 l of fresh medium was introduced and the fermentation liquid recycling was initiated in the bioreactor-microfilter system with the rate of 60 l/h. The fermentation with fermentation liquid recycling was carried out in the batch manner for 8 hours. In the 22nd hour of the fermentation the cell recycling through the filtration module, permeate outlet and fresh medium supply at the dilution rate of 0.14/h were initiated. At the initialisation of the filtration process, the glucose concentration in the medium was 0.14 g/l and the biomass concentration equalled 0.8 g D.M./l ($4x10^9$ CFU/ml). In the 54th hour the medium dilution rate was increased up to 0.22/h. Fermentations at D=0.22/h were carried out for 24 hours. In the 78th hour of the fermentation the medium dilution rate was increased to 0.28/h and after the 16th hour up to 0.36/h.

The *C. divergens* AS7 growth yield in the continuous fermentations with cell recycling depended on the dilution rate and increased along its increase (Figure 3). At D=0.14/h the biomass concentration stabilised at the level of 4.4 g D.M./l. The medium dilution rate increase to 0.22/h resulted in 400% biomass concentration increase. After the following change of D value the biomass amount increased by 6.5 g D.M./l. At D=0.36/h the biomass concentration reached the maximum value of 48 g d.m./l (7x10^{10} CFU/ml) and was 80 times higher than the concentration of the biomass achieved in the batch fermentations. While analysing the information included in the publications of Hayakowa [6], Aeschlimann and Stockar [7], Major and Bull [8], it was found that the *C. divergens* AS7 growth yield in fermentations with cell recycling was comparable to the growth yield of other lactic bacteria cultured in a membrane bioreactor at a similar medium dilution rate. Higher biomass concentrations than 48 g D.M./l have only been achieved so far in fermentations with higher medium dilution and recycling rates than the rates applied in this study. For example, fermentation at D=0.95/h gave 80 g D.M. of *Lb. delbrueckii* [9], whereas the *Lb. plantarum* biomass was concentrated to the level of 60 g D.M./l at D=1.27/h [10].

In fermentations with cell recycling it was observed that the *C. divergens* AS7 dry matter increase is accompanied by an increase in live cell count (Figure 3). Biomass concentration increased however, faster than the count of live cells which resulted in the decrease in their percentage in the bacteria population along an increase in medium dilution rate. The living cell count in 1 g of biomass at D= 0.14/h was almost 6 times higher than the count achieved at D=0.36/h. The decrease of living cell count in the biomass could have resulted from the effect of mechanical stress or their autolysis. The completed studies, however, showed that at the medium recycling rate between 0 and 120 l/h the percentage of living cells in the population was constant. It also means that any mechanical force resulting from stress due to cell recycling through the peristaltic pump and microfilter did not have any effect on *C. divergens* survival. In regards to bacteria resistance to a destructive mechanical force during re-circulation, it could be suspected that the decrease of living bacteria cells percentage in the total bacteria population was caused by their autolysis.

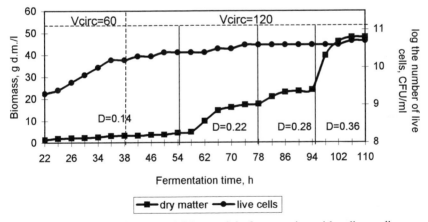

Figure 3. Dynamics of *C. divergens* AS7 growth in fermentation with cell recycling.

C. divergens AS7 bacteria during the fermentation with bacteria recycling underwent significant aggregation. Similar phenomenon was also observed in the fermentation without bacteria recycling. The microscopic analysis showed that the size of aggregates increased along the increase of biomass concentration. During the fermentation, the bacteria rods shortened and became oval.

The membrane bioreactor worked correctly when the biomass concentration was lower than 30 g D.M./l. At higher biomass concentrations the filtration yield decreased. The filtration yield decrease resulted from biomass aggregation at the filter surface. After reaching the biomass concentration level of 48 g D.M./l, the filter was completely blocked. These observations were confirmed by the increasing transmembrane pressure.

The data gathered during the fermentation also suggested the necessity of cell autolysis reduction through partial biomass bleeding.

3.4. Continuous fermentations with cell recycling and bleeding

Fermentations with biomass bleeding (CB) were carried out at D=0.14/h. The biomass bleeding in the amount of 10% of the medium volume introduced to the bioreactor was initiated in the 54th hour of fermentation. After 41 hours the CB was increased and equalled 20% D. In the 135th hour bleeding was increased to 30% D, and after 48 h up to 50% D. In the 247th hour the pump collecting permeate was switched off and the bleeding was increased to 100% D. It was a return to the classic continuous fermentation.

Bleeding during the fermentation had a strong impact on *C. divergens* AS7 growth yield. The data in Figure 4 shows that the bacteria growth yield increased along with the increase of bleeding within the range from 10 to 30% D. At CB=30% D the biomass concentration reached its maximum value. It was 8.1 g d.m./l and was twice as high as the concentration for fermentations without biomass excess collection. At CB higher than 30%D cells were washed out from the bioreactor which resulted in a bacteria population decrease. At CB=50%D the biomass concentration was similar to the biomass concentration achieved for fermentations without bleeding, whereas at CB=D it was half as much.

In fermentations with cell recycling and bleeding within the range of 10 and 30 %D the dry matter concentration (*C. divergens* AS7) increase was accompanied by an increase in the living bacteria count (Figure 4). The bacteria count at CB=30%D reached the level of 2.5×10^{11} CFU/ml and remained unchanged during fermentations at higher levels of cell bleeding. It means that the living cell percentage in the bacteria population increased along with the increase of CB. At CB=D it was over 8 times higher than in the fermentations without bleeding.

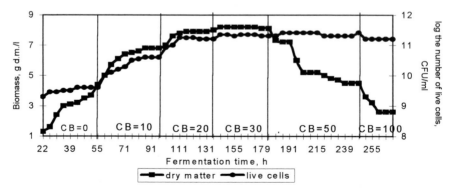

Figure 4. Dynamics of *C. divergens* AS7 growth in continuous fermentation with cell recycling and bleeding at D=0.14/h.

This means that bleeding effectively reduced cell autolysis and improved the speed of bacteria growth. Microscopic analysis indicates that the biomass bleeding prevents cell aggregation.The completed studies showed that cell bleeding reduced the process of cell autolysis.

3.5. Alternating fermentations with periodical cell recycling

Fermentations were carried out with periodical cell recycling at D=0.14/h. The continuous fermentation without cell recycling was initiated in the 54th hour of the fermentation. After 24 hours, the continuous fermentation was switched to the fermentation with cell recycling. Fermentations with cell recycling were carried out for 24 hours. The fermentation type was changed again after the 54th hour. Further fermentation with cell recycling was started in the 152nd hour. After 40 hours of the fermentation the pump receiving permeate was switched off and further continuous fermentation was initiated, which was run for 68 hours. Each change of fermentation type triggered the change in biomass concentration (Figure 5).

After each microfiltration period the biomass concentration occurred. During the alternating fermentations the living bacteria count also changed. The changes were inversely proportional to the dry matter concentration changes. During fermentations with cell recycling, the number of living cells dropped, whereas during the fermentation without cell recycling it went up.

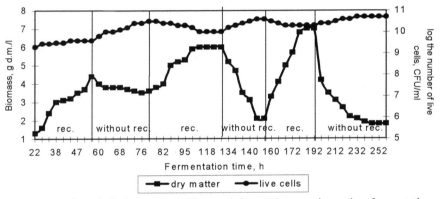

Figure 5. Dynamics of *C. divergens* AS7 growth in continuous alternating fermentation with periodical recycling at D=0.14/h.

Changes of living bacteria count were however, significantly smaller than the changes in biomass concentration which resulted in the increase of the live cell percentage in bacteria population during each following continuous fermentation (with or without cell recycling).

4. CONCLUSIONS

In batch process biomass yield was significantly lower as compared with continuous process. Biomass production in continuous process was positively correlated to dilution rate. The highest biomass yield was occurred at D values 0.12 and reached 2.9 g D.M./L. The higher biomass yield was obtained when the membrane bioreactor was used. In continuous process with cell recycling the highest biomass concentration was observed at dilution rate 0.36/h and reached 48 g D.M./L. It was 16 times higher then obtained in the process without cell recycling. Increase of dilution rate resulted in a decrease of viable cell density. The cell bleeding was performed to prevent cell autolysis. The fluid circulation rate in range from 60 to 120 L/h had not effect on the cell viability.

At two step culture, with cell recycling and then without cell recycling, biomass production was not stabile. When the separate module was switch on again the higher concentration level of cells was obtained.

REFERENCES

1. M.J. Gasson, FEMS Microbiol. Rev., 12 (1993) 3.
2. U. Schillinger and F.K. Lucke, Fleischwirtsch., 70 (1990) 1296.
3. U. Schillinger, and F.K. Lucke, Fleisch. Int., 1 (1991) 3.
4. P.A. Vandenbergh, FEMS Microbiol. Rev., 12 (1993) 221.
5. M.F. Pilet, X. Dusset, R.G. Novel, M. Desmazaeud, J.Ch. Piard, J. Food Protect., 58 (1995) 256.
6. K. Hayakawa, H. Sansawa and T. Nagamune, J. Ferment. Bioeng., 70 (1990) 404.
7. A. Aeschlimann and U. von Stockar, Microb. Technol., 13 (1991) 811.
8. N.C. Major and A.T. Bull, Biotechnol. Bioeng., 34 (1988) 592.
9. E. Ohleyer, H.W. Blanch and C.R. Wilke, Appl. Biochem. Biotechnol., 11 (1985) 317.
10. M.T.O. Barreto, E.P. Melo, J.L. Moiriera, M.J.T. Carronde, J. Ind. Mikrobiol., 7 (1991) 63.
11. P. Boyaval, Ch.M. Corre and N. Medec , J. Chem. Tech. Biotechnol., 53 (1992) 189.
12. P. Boyaval, Ch.M. Corre and N. Medec, Biotechnol. Lett., 14 (1992) 589.
13. A. Olmos-Dichara, F. Ape, J.L. Uribelarrea, A. Pareilleux and G. Goma, Biotechnol. Lett., 19 (1997) 709.
14. J.C. De Man, Appl. Bacteriol., 23 (1960) 130.

MEASUREMENT AND QUALITY CONTROL

Food Biotechnology
S. Bielecki, J. Tramper and J. Polak (Editors)
345

Towards a new type of electrochemical sensor system for process control

BGD Haggett,[a] A Bell,[b] BJ Birch,[a] JW Dilleen,[a] SJ Edwards,[c] D Law,[c]
S McIntyre[c] and S Palmer[b]

[a]Sensor and Cryobiology Research Group, Research Centre, The Spires,
2 Adelaide Street, Luton, LU1 5DU, United Kingdom

[b]Oxley Developments Company Limited, Priory Park, Ulverston, Cumbria,
LA12 9QG, United Kingdom

[c]Kodak European Research, Headstone Drive, Harrow, Middlesex,
HA1 4TY, United Kingdom

Conventional chemical measurement systems usually incorporate static sensors that are intended for multiple measurements over extended periods of time. Unfortunately, such sensors are prone to fouling and to drift. In consequence systems for process control must include facilities for calibration and for in situ cleaning. Frequently, the foregoing requirements need to be supplemented with a need for addition of one or more reagents. Thus, such chemical measurement systems may require complex engineering solutions for sample dilution, reagent delivery and mixing, sensor calibration and washing, as well as possible automated sensor removal during pipe or vessel cleaning.

For biotechnological applications, in particular, biosensors may offer some advantages - such as direct analyte measurement in complex matrices. However, it should be recognised that it is easier to invent a biosensor than it is to turn it into a robust, reliable and commercially viable sensing system! There is a wider range of chemical analysers available, but in the context of process control, these systems have a reputation for unreliability and high cost of ownership. Experience shows that many such instruments are installed but inadequately maintained. In consequence, the devices produce inconsistent and unreliable results that are rightly ignored – the machine may even be switched off and remain idle – users then manage without any information or rely upon data from samples analysed in a conventional laboratory! or take inferential information from physical measurements.

The ideal measurement system should be cost effective, simple, maintenance free and without need for calibration. Disposable, single-use sensors might offer some of these characteristics and in appropriate embodiments they may be valuable for applications in process control. Such sensors are being developed - these are calibrated in manufacture, incorporate necessary reagents and do not require maintenance since each sensor is used only once. Instrumentation that is correspondingly simple and reliable is being developed both for hand-held (single-use) and process control (repeated single-use). At present, the emphasis is on development of chemical sensors, but it is believed that the approach is applicable to other types of sensing system.

1. BIOSENSORS

The first commercial biosensor was the Yellow Springs Instruments Model 23 blood glucose analyser introduced in 1974. The measurement principle was based on immobilized glucose oxidase (GOD):

$$glucose + O_2 \xrightarrow{GOD} gluconolactone + H_2O_2$$

Oxygen was scavenged from the measurement solution and hydrogen peroxide was measured at a platinum electrode. GOD was immobilized in a membrane sandwiched between polycarbonate and cellulose acetate layers. The polycarbonate excluded large molecules from the sensor and restricted the flux of glucose to GOD so that the enzyme was not saturated at concentrations in the range of interest. The cellulose acetate excluded many interfering molecules but allowed hydrogen peroxide to diffuse to the measurement electrode.

YSI have used the same measurement principle, with different enzymes, to produce a diverse range of instruments including the 2700 Select Biochemistry Analyser capable of simultaneously measuring two selected analytes, Table 1.

The analyser has been successfully applied in the food and beverage industries, as well as in biotechnology. In addition to the usual facility for off-line monitoring, it has capabilities for on-line process control. Nevertheless, the instrument requires a minimum of maintenance – replacement of expired membranes, calibration solutions and wash bottles. The platinum measuring electrode is replaced on an infrequent basis.

2. THICK-FILM ELECTRODES

Disposable, single-use electrodes have been widely described [2-11]. These devices are most usually produced by printing conducting inks through a patterned screen, Fig. 1. Insulating dielectrics are similarly printed over those parts of the conductors that are not required to contact the test solution or to make contact with an external measurement circuit. Planar sensor structures can be fabricated in this way - Fig. 2 shows a simple device with carbon working, silver-silver chloride reference and carbon counter electrodes. Fabrication of these screen-printed sensor devices has several advantages:
- tens or even hundreds of devices can be printed simultaneously
- the printing process can be automated
- very small amounts of substrate material and printed inks are required for each sensor

Table 1
Sensors available for the YSI 2700.
Numbers in brackets indicate typical membrane working life [1].

Glucose (21)	Glutamine (5)	Lactose (10)
Sucrose (10)	Glutamate (7)	Ethanol (5)
Lactate (14)	Choline (7)	Blank – H_2O_2 (21)

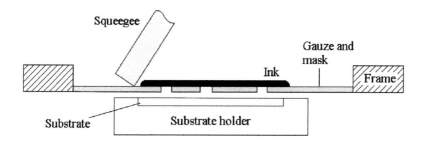

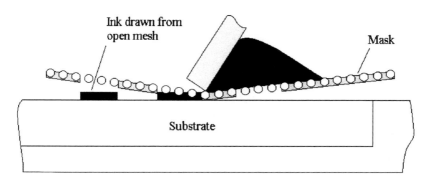

Figure 1. Schematic representation of the screen-printing process. (Upper picture) The substrate (*e.g.* plastic, ceramic, *etc.*) to be printed is held in a holder beneath a patterned mask - on top of which is spread the ink to be printed. (Lower picture) A squeegee blade is slid over the mask so as to squeeze ink through gaps in the mask and onto the substrate where it leaves the required pattern.

- the resultant sensors can be sensibly small (*e.g.* commercial sensors for blood glucose monitoring are typically <1 mm thick × a few mm wide × ten or twenty mm long) or large (*e.g.* 1 mm thick × one or two cm wide × a few cm long)
- the volume of sample required by such sensors can be sensibly small (*e.g.* a few tens of microlitres)
- correspondingly, the amounts of reagents required to carry out an analytical measurement can be small (*e.g.* a milligram).

When the above factors are considered together, it may be seen that screen-printed sensors are potentially sufficiently cheap to be considered disposable items. Millions of single-use blood glucose sensors are manufactured for diabetics to monitor and control their condition. Several manufacturers use electrochemical measurement approaches, and the necessary analytical reagents are incorporated into a cell formed by a small gap between two pieces of laminated polymer. The gap also contains the measurement electrodes. Blood glucose measurements are quick (say <1 min) and can be accomplished by diabetics in the privacy of their homes.

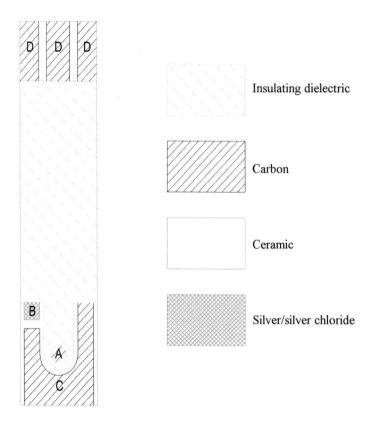

Figure 2. Schematic representation of a planar screen-printed electrochemical sensor printed on a ceramic substrate. (A) Carbon working electrode, (B) Silver-silver chloride reference electrode, (C) carbon counter electrode, and (D) electrical connections to external circuit.

This type of sensor has some of the attributes desired of ideal sensors for process control in industrial applications: cost-effective, simple and maintenance free. The sensors may not be entirely without need for calibration - but, since they can be used only once, the sensors must be manufactured reproducibly within batches so that the sensor need not be calibrated more than once each time a new batch of electrodes is used.

The authors are collaborating in order to produce these thick-film devices for application in various industries. The approach is to demonstrate much of the required technology with individual single-use sensors, using hand-held instrumentation (Figs. 3 & 4) while developing arrays of single-use sensors that may be used in sequence by appropriate automated instrumentation (Fig. 5).

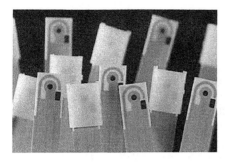

Figure 3. Commercial prototype single-use sensors.

Figure 5. Laboratory prototype system using an array of single-use sensors.

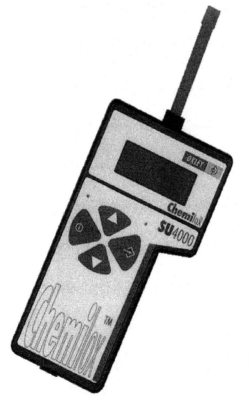

Figure 4. Commercial prototype hand-held measurement instrument for single-use sensors.

3. EXAMPLES OF SENSORS BEING DEVELOPED

At present, sensors are being developed for silver (for photoprocessing and for environmental monitoring applications, Fig. 6), for colour developers (a range of redox active organic molecules, Fig. 7) for other photographic developing agents, and for trace metals (for process control and environmental monitoring applications, Fig. 8).

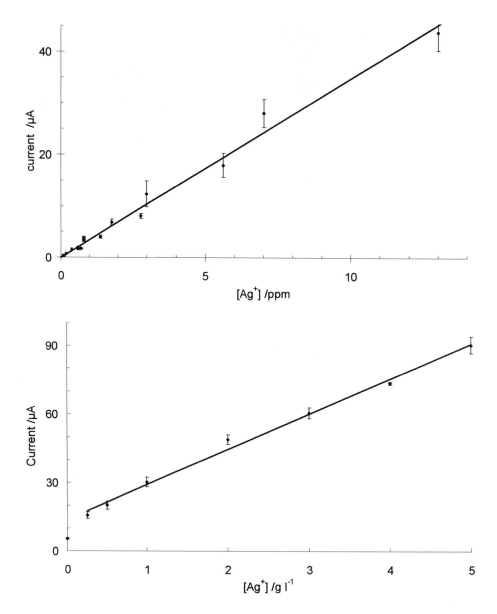

Figure 6. Calibration data from prototype single-use sensors for silver ion measurement. (Upper) Low concentrations of silver - for effluent analysis and environmental monitoring. (Lower) High concentrations of silver - for photoprocessing applications.

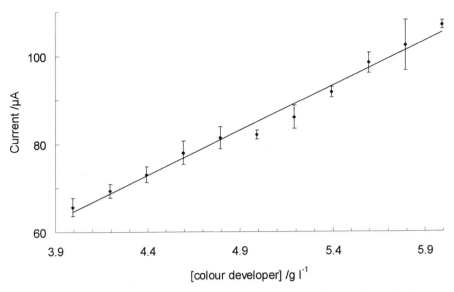

Figure 7. Calibration data from prototype single-use sensors for a colour developer - for process control applications.

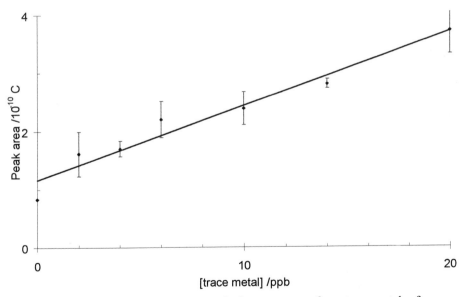

Figure 8. Calibration data from prototype single-use sensors for a trace metal - for process control and environmental monitoring.

4. CONCLUSION

Disposable single-use sensors provide an invaluable tool for thousands of individuals that suffer from some types of diabetes. Such sensors are of increasing interest to scientists and technicians in a wide range of applications, from analysis of lead in paint to environmental monitoring to veterinary practice. The objective of the work outlined above is to provide users in industry with disposable sensors that incorporate necessary reagents, buffers, *etc.* so that it is required only to introduce the sample to the sensor. The chemistry is hidden from the user - there should be no need to pre-treat the sample, by dilution or addition of reagents. In this way it is thought possible to provide sensors that are simple to use, require only a simple batch calibration (or no calibration at all), require no maintenance, and are cost effective in use. Furthermore, by deploying multiple copies of the single-use sensors in suitable formats with appropriate instrumentation it is possible to build automated sensing systems with few moving parts and many of the attributes of the ideal measurement system.

REFERENCES

1. YSI Inc. (1998) *YSI 2700 Select Biochemistry Analyzer* brochure.
2. Dilleen JW, Sprules SD, Birch BJ and Haggett BGD (1998), *Analyst* **123**, 2095-2097.
3. Hart JP, Abass AK, Cowell DC and Chappell A (1999), *Electroanalysis* **11**, 406-411.
4. Reeder GS and Heineman WR (1998), *Sensors & Actuators B* **52**, 58-64.
5. Wang J, Tang B and Rogers K (1998), *Anal. Chem.* **70**, 1682-1685.
6. Collier WA, Lovejoy P and Hart AL (1998), *Biosens. Bioelectron.* **13**, 219-225.
7. Hart AL, Matthews C and Collier WA (1999), *Anal. Chim. Acta* **386**, 7-12.
8. Sarkar P, Tothill IE, Setford SJ and Turner APF (1999), *Analyst* **124**, 865-870.
9. Martinelli G, Carotta MC, Ferroni M, Sadaoka Y and Traversa E (1999), *Sensors & Actuators B* **55**, 99-110.
10. Shih Y and Zen JM (1999), *Electroanalysis* **11**, 229-233.
11. Marrazza G, Chianella I and Mascini M (1999), *Anal. Chim. Acta* **387**, 297-307.

Food Biotechnology
S. Bielecki, J. Tramper and J. Polak (Editors)
© 2000 Elsevier Science B.V. All rights reserved.

The use of enzyme flow microcalorimetry for determination of soluble enzyme activity

V. Štefuca*, M. Polakovič

Slovak University of Technology, Faculty of Chemical Technology, Department of Chemical and Biochemical Engineering, Radlinského 9, 812 37 Bratislava, Slovakia

The possibility of using flow microcalorimetry for enzyme activity testing is demonstrated. Two model enzymes, urease and invertase, were studied with urea and sucrose as substrates, respectively. The technique is based on the measurement of the temperature change as result of the heat of an enzyme reaction. A precooled sample of enzyme solution is mixed with the substrate solution and then immediately injected through an injection valve into a carrier buffer pumped continuously into the flow microcalorimeter column packed with glass particles. A temperature change in the column is observed as a heat peak. The peak area depends on the sample volume and of the enzymatic reaction rate.

The enzyme activity in a sample was determined from the peak height using a microcalorimeter calibration for the activity of both tested enzymes. The linearity range of enzyme activity and the influence of substrate concentration on the method sensitivity were investigated. The main advantage of the proposed technique is its versatility due to the versatility of the detection principle. The technique can be used, for example, for regular enzyme activity measurement in enzyme inactivation studies or in monitoring of chromatographic enzyme purification.

1. INTRODUCTION

An assay for enzyme activity can be arranged generally in two steps. The first one is the enzyme reaction step. Then, after stopping the reaction, an analytical step is needed for the reactant concentration determination. These two steps can be effectuated simultaneously when one of the reactants can be monitored directly. Changes typically utilized for monitoring enzyme catalyzed reactions are optical properties of the solution, either absorption or emission, concentration of ions, most often H^+, detectable electrochemically.

There is a lack of universal methods that allow monitoring of a wider range of enzyme reactions. One of the possible solutions is calorimetry, and it can be used for the kinetic investigation of chemical and enzymatic reactions with a significant heat of reaction. The advantages of kinetic calorimetry are obvious. Firstly, the rate of a chemical reaction can be measured without any special requirements being imposed on the reaction medium (solid, viscous, multicomponent systems). Second, the high efficiency, meaning a large amount of kinetic information from one experiment with a non-destructive character. Third, the chemical conversion is recorded directly at the time of its occurrence [2].

There are two main experimental configurations of calorimetry used for the determination of enzyme activity, i.e. batch [3–7] and flow [8-10]. In this article, a simple procedure based on the flow injection analysis principle is described.

2. MATERIALS AND METHODS

2.1. Materials
Urease (Type III from Jack Beans, Sigma, St. Louis, USA, 31 Units/mg solid) and invertase (Grade VII, from bakers' yeast, Sigma, St. Louis, USA, 400 Units/mg solid) were used as model enzymes. All chemicals used as buffer components and enzyme substrates were of analytical grade and provided by Sigma Co.

2.2. Enzyme flow microcalorimetry
The enzyme flow microcalorimetric measurement is based on the registration of the temperature change provoked by the heat of the reaction catalyzed by an enzyme. The calorimetric measurements were performed using the 3300 Thermal Assay Probe, Advanced Biosensor Technology AB, Lund, Sweden. The experimental set-up used is depicted in Fig. 1. The sample of enzyme mixed with the substrate solution is injected through an injection valve into the buffer stream. The enzyme reaction takes place in a small column packed with glass beads (diameter of 0.15 mm). The column is placed in a thermostated block and is used as a minireactor with packed bed having standard dimensions of 2 cm in length and 0.4 cm in inner diameter. The temperature change is measured by thermistors connected to a Wheatstone bridge and the signal amplified and registered by the personal computer. A typical registration from a single assay of urease activity is shown in Fig. 2. As indicated in the figure, the peak height was used as a measure of enzyme activity for the measurement evaluation.

2.2. Assay procedure
The urease activity was measured by mixing 0.2 ml of the enzyme solution in phosphate buffer (0.1 M, pH 7) with 0.2 ml of urea solution in the same buffer and its rapid injecting into the flow microcalorimeter. Both enzyme and buffer solutions were pre-cooled at 4°C prior to mixing and injecting for the calorimetric analysis. The flow rate of the carrier buffer solution was 1 ml/min and the reaction temperature was 30°C. The injection loop volume was 0.1 ml.

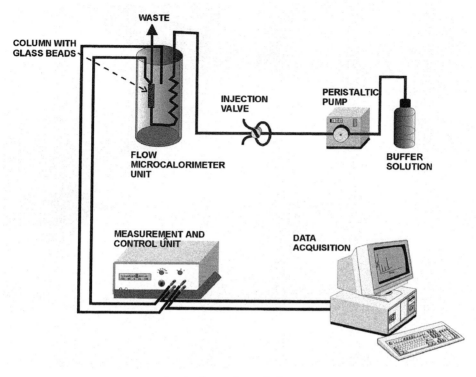

Figure 1. Experimental set-up for the flow microcalorimetry.

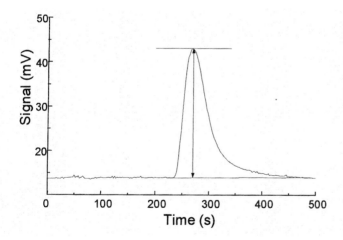

Figure 2. Example of thermometric registration from a single injection of urea–urease sample.

3. RESULTS AND DISCUSSION

In the present work, the possibility of using flow microcalorimetry for the soluble enzyme activity measurement is demonstrated. Urease and invertase were studied as the enzyme models. Optimum analytical conditions, such as enzyme and substrate concentrations were tested. Calibration curves in the form of thermometric signals (in milivolts) vs. total activity of injected sample were obtained using varying enzyme activities (Fig. 3 and 5). The experimental results in Fig. 3 show three calibrations for different urea concentrations in the samples.

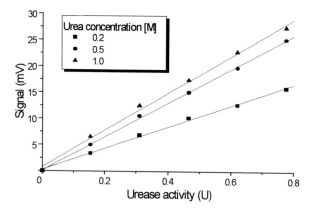

Figure 3. Calibration results for urease.

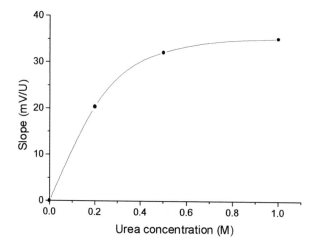

Figure 4. Slopes of calibration lines for urease vs. urea concentration.

Urea concentrations above the K_m - value for urease, about 6 mM [11], were used in order to supply a sufficient excess of substrate. In the case of invertase, 1 M of sucrose was used, which is also above the K_m - value being approximately 40 mM for this type of invertase [12]. Independent of the urea concentration, the calibration lines for urease were linear, while the slopes increased with increasing urea concentration.

It is clear from the values of slopes in Tab. 1 and their trend depicted in Fig. 4 that the increase of urea concentration from 0.5 to 1 M is not significantly beneficial for the analysis sensitivity. Using similar analysis it was found that, in the case of invertase, 1 M of sucrose was a sufficiently high concentration for the maximum analysis sensitivity. In contrast to urease, a nonlinear course of calibration line was observed (Fig. 6) above certain experimental concentrations of invertase. It was probably caused by substrate depletion due to the relatively high enzyme concentration. The K_m - value of invertase is, moreover, nearly seven times higher than that of urease. Therefore, the reaction kinetics for invertase will shift more significantly than for urease (this effect is even amplified by product inhibition).

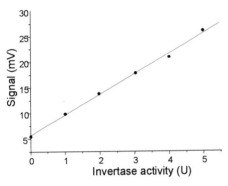

Figure 5. Calibration results for invertase.   Figure 6. Calibration results for invertase – broader activity range

Table 1
Linear regression data of microcalorimeter calibration for urease and invertase activities

Substrate concentration (M)	Slope (error) (mV U^{-1})	Intercept (error) (mV)	Correlation coefficient
Urease			
0.2	20.35 (0.53)	0.27 (0.24)	0.9987
0.5	32.13 (0.43)	0.13 (0.20)	0.9996
1.0	35.10 (1.02)	0.79 (0.47)	0.9983
Invertase			
1.0	4.00 (0.12)	5.68 (0.37)	0.9987

Concerning the intercepts, the value obtained for invertase is not negligible compared to urease. A certain signal was in fact observed at injection of blank samples (samples without the enzyme), provoked probably by the mixing heat. Therefore, this effect must be taken into account for a routine analysis.

The comparison of slopes (Tab. 1) leads to the conclusion that the urease assay is much more sensitive. This is in agreement with the molar reaction enthalpies of urease (61 kJ mol^{-1} [13]) and invertase (14.5 kJ mol^{-1} [14]) catalysed reactions.

4. CONCLUSION

Experimental conditions for the analysis of activities of urease and invertase by the flow injection microcalorimetry were determined in this preliminary work. This technique can be useful in the cases when a rapid and simple measurement is required, e.g. during enzyme purification, for effluent enzyme activity monitoring during chromatographic separation, in inactivation studies, or in screening of enzymes. The presented technique has been used for inactivation studies of invertase and ureasae. The goal of future studies is to determine a more general strategy for the optimization of equipment configurations and conditions of analysis with regards maximum analysis sensitivity. This will be achieved using mathematical modeling based on mass and heat balances.

ACKNOWLEDGEMENT

The work was partially financed by the Slovak Grant Agency for Science (grant VEGA 2/4149/97).

REFERENCES

1. I.M. Barkalov and D.P. Kiryukhin, Internat. Rev. Physical. Chem., 13 (1994) 337.
2. J. Grime and K.B. Tan, et al., Anal. Chim. Acta, 109 (1979) 393.
3. M. Kurvits and E. Siimer, Thermochim. Acta, 103 (1986) 297.
4. B.A. Williams and E.J. Toone J. Org. Chem., 58 (1993) 3507.
5. Y. Liang, C.X. Wang, et al., Thermochim. Acta, 268 (1995) 17.
6. G. Salieri, G. Vinci, et al., Anal. Chim. Acta, 300 (1995) 287.
7. B. Danielsson and K. Mossbach, FEBS Lett., 101 (1979) 47.
8. B. Danielsson and K. Mosbach, Methods Enzymol., 137 (1988) 181.
9. M. Beranand and V. Paulicek, J. Thermal Anal., 38 (1992) 1979.
10. E. Kohutovičová, Diploma Thesis, Slovak University of Technology, Bratislava, 1998.
11. V. Štefuca, P. Gemeiner and V. Báleš, Enzyme Microb. Technol., 306 (1988) 10.
12. R.K. Owusu, M.J. Trevhella and A. Finch, Thermochim. Acta, 111 (1987) 1.
13. R.N. Goldberg, Y.B. Tewari and J.C Ahluwalia, J. Biol. Chem., 264 (1989) 9901.

Food Biotechnology
S. Bielecki, J. Tramper and J. Polak (Editors)

Study of an ELISA method for the detection of *E. coli* O157 in food

P. Arbault, V. Buecher, S. Poumerol and M.-L. Sorin

Diffchamb Technical Centre, 8 Rue Saint Jean de Dieu, 69007 Lyon, France.

A new ELISA assay, so-called Transia Plate *E. coli* O157, was developed for monitoring *E. coli* O157 contamination in food samples. The performances of this new method were compared to the gold standard method (immunomagnetic separation combined with agar plate) based on an enrichment step in buffer peptone water, followed by an immunoseparation with magnetic beads, then streaking onto Cefixime Tellurite Sorbitol Mac Conkey agar (CT-SMAC).

The ELISA was composed of 3 steps, with incubation times of 20, 10 and 5 minutes respectively. This immunoassay was demonstrated specific for *E. coli* O157 and showed a limit of detection of around 5×10^4 CFU/ml with several strains. Among the different other strains evaluated (47 *non-E. coli* O157 strains), only two *Salmonella*, belonging to the serogroup N, were crossreacting with the ELISA. The best enrichment protocol was combining either mTSBn or mECn, with an incubation time between 16 to 20 hours at a temperature of 41.5°C. This temperature allowed an easier confirmation of the positive ELISA samples once isolated onto CT-SMAC plate. The limit of detection with artificially contaminated samples was assessed in the range of 1 to 3 CFU/25g of food for both methods (ELISA and gold standard). Finally, naturally contaminated samples (n=47, raw meat and dairy products) were analysed with the Transia Plate *E. coli* O157 method. Two positive samples were detected in the raw meat samples, and were confirmed as *E. coli* O157 without any verotoxin production.

The Transia Plate *E. coli* O157 method was representing a good alternative to the IMS method as it was highly specific for *E. coli* O157 and proposed an objective analysis of food samples, without any cumbersome step such as immunocapture. This method improved also the safety as the ELISA was done with heat treated samples. It allowed an effective screening of food samples within 45 minutes, and avoided streaking and reading of agar plates for all the samples.

1. INTRODUCTION

Since its first description in 1982 in USA, *E. coli* O157:H7 has been clearly recognised as a major foodborne pathogen over the world [1]. During the last 15 years, several outbreaks were reported in USA, Japan and Europe [2]. This bacteria has been implicated as a causative agent of mild non bloody diarrhoea, haemorrhagic colitis and more serious diseases such as haemolytic uremic syndrome which can cause death of young children and elderly people [1]. The food matrices implicated in *E. coli* O157 :H7 contaminations are firstly undercooked ground beef, but also dairy products and acidic food like apple cider [2]. The main reservoir of these pathogenic bacteria seems to be farm animals.

The laboratory methods [3-6] routinely used for detecting *E. coli* O157 in food combine an enrichment step during 20-24 hours either in buffer peptone water (BPW) or trypticase soya broth (TSB) or modified *E. coli* broth (mEC), followed by plating onto sorbitol Mac-Conkey agar (SMAC) with or without addition of cefixime and tellurite [7] and incubation during 24 hours at 37°C. As most of the *E. coli* O157 strains do not ferment sorbitol, they appear colourless colonies onto SMAC agar. But such cultural methods present some major drawbacks such as overgrowth of competing flora due to a limited selectivity of the media used during enrichment and plating, and a lack of specificity for *E. coli* O157 [8-10]. The use of immunomagnetic separation (IMS) for extracting the *E. coli* O157 cells from the enrichment broth has improved the method reliability as the competing flora is partly eliminated [3,11-14]. But IMS is tedious and time-consuming to carry out, especially when analysing simultaneously several samples, and still it requires SMAC agar for isolating *E. coli* O157.

Other methods, such as immunoassays [10,11,15] and PCR-based [16,17] systems, have offered a good alternative to the cultural method by improving the specificity relying on the use of antibody and DNA and allowing a first screening for isolating the presumptive positive samples before further identification.

The following study presents the performances of a new ELISA-based method, the Transia Plate *E. coli* O157, and analyses some critical parameters of the enrichment step, such as nature of the broth, incubation time and temperature. The performances of the ELISA-based method are also compared to the IMS method combining enrichment in BPW, followed by immunocapture and streaking onto SMAC agar.

2. MATERIAL AND METHOD

2.1. Material
Buffered Peptone Water (BPW), modified *E. coli* broth with novobiocine (mECn), modified Trypticase Soya Broth with novobiocine (mTSBn), Trypticase Soya Yeast Extract media (TS/YE) and Cefixime Tellurite-Sorbitol Mac Conkey (CT-SMAC) agar were purchased from Merck (Darmstadt, Germany). The Transia Plate *E. coli* O157 kit was manufactured by Diffchamb (Lyon, France). Dynabeads anti-*E. coli* O157 were purchased from Dynal (Oslo, Norway). For agglutination, the anti *E. coli* O157 latex was bought from Oxoid (Basingstoke, UK). Bacteria identification was carried out using the bioMérieux API 20E gallery (Marcy l'Etoile, France).

2.2. Enrichment protocol for the Transia Plate *E. coli* O157 method
→ For the study of both incubation times (16h and 20h):
After blending 25 g of sample into 225 ml of mECn or mTSBn broth, the samples were incubated for 16 and 20 hours at 37°C or 41.5°C. One millilitre of each enrichment broth was transferred into a tube and boiled during 20 minutes, then cooled to room temperature before the ELISA test.
→ For the study of both incubation temperatures (37°C and 41.5°C):
Seventy five grams of sample were homogenised in a Waring blendor then divided in two parts of 25 g. After blending 25 g of sample into 225 ml of mECn and mTSBn, the samples

were incubated either at 37°C or at 41.5°C for 20 hours. Then sample treatment was done as described above.

2.3. ELISA test:

Into a 96-well microplate coated with a rabbit polyclonal anti-*E. coli* O157 antibody, 100μl of sample per well were dispensed. For each run, 2 wells for the negative control and 1 well for the positive control were included.

The microplate was incubated during 30 min at room temperature (RT, 18-25°C). Then each well was washed 5 times with diluted washing buffer (450 μl/well). The second antibody (goat anti-*E. coli* O157 antibody conjugated to peroxidase) was dispensed into each well (100 μl/well) and incubated 10 min at RT. Each well was then washed 6 times. Then 100 μl of urea peroxide/tetramethylbenzidine mixture (commercially available from Intergen, Oxford, UK) was dispensed into each well. After 5 min of incubation at RT, the coloured reaction was stopped by adding 100 μl of sulphuric acid solution into each well. The optical densities were read at 450 nm after a blank against the air.

The cut-off was calculated as the mean OD of both negative controls + 0.15. Over the cut-off the sample is considered positive for *E. coli* O157.

Any presumptive sample was streaked onto CT-SMAC agar, then a maximum of 5 characteristic colonies were analysed by the serotyping agglutination latex from Oxoid in order to confirm the presence of *E. coli* O157. Positive samples were further characterised with the biochemical identification galleries.

2.4. Dynabeads method (AFNOR approval n° DYN-16/2-06/96)

Into a stomacher bag, 25 g of sample were mixed with 225 ml of BPW and incubated 6 hours at 37°C. Then 1 ml of BPW was taken up and transferred to tubes containing 20 μl of anti-*E. coli* O157 Dynabeads (Magnetic beads coated with anti-O157 antibodies). Tubes were incubated during 30 minutes with shaking, and then beads were washed twice. The magnetic beads were mixed into 100 μl of PBS-Tween solution, and the mixture was streaked onto CT-SMAC agar. Agar plates were read after 18-24 hour incubation at 37°C.

2.5. Study of the specificity for both Transia Plate and Dynal methods

Ten *E. coli* O157 and 47 other bacteria strains including 5 *Citrobacter*, 1 *Enterobacter*, 3 *Hafnia*, 3 *Klebsiella*, 5 *Proteus*, 2 *Serratia*, 10 *Salmonella*, 3 *Pseudomonas*, 5 *Staphylococcus*, 2 *Bacillus* and 5 non O157 *Escherichia coli*, were tested by the ELISA and the Dynabeads.

From a 24-hour culture in TSYE at 37°C, each strain was subcultured (0.1 ml into 10 ml) in BPW for 3 hours minimum at 37°C. Then the growth was estimated by reading the optical density at 470 nm. For the ELISA test, *E. coli* O157 cultures were adjusted at 10^8 cells/ml in BPW and the other bacteria were diluted at 10^9 cells/ml, after a heat shock of 20 minutes at 100°C.

The strains analysed by ELISA were tested by the Dynal method. After immunocapture of the alive cells, the Dynabeads were streaked simultaneously onto CT-SMAC.

2.6. Limit of detection of the ELISA for *E. coli* O157 strains

Four *E. coli* O157 strains were tested. After a 24 hour incubation into TSYE at 37°C, each strain was subcultured into mTSB during 3 hours at 37°C. Then dilutions at 10^8, 10^7, 10^6, 5×10^5, 10^5, 5×10^4 and 10^4 cells/ml were performed in mTSB to be tested by ELISA after

362

boiling. Accurate numerations of the bacteria were done in parallel. Each concentration was tested twice.

2.7. Study of the enrichment step with artificially contaminated food matrices

The food matrices were raw minced meat, raw milk and pasteurised soft cheese: about 1 kg of each food matrix was mixed and then divided into 25g samples.

One *E. coli* O157 strain (1074) was used for the contamination at 3 theoretical inoculum levels: 3, 10 and 30 CFU/25 g. The contaminating cells were added into each stomacher bag after blending the sample into 225 ml of enrichment broth. The concentration of the contaminating suspension was firstly estimated by reading the optical density at 470 nm and secondly measured more accurately from a numeration performed in TSYE agar plate.

For each level and each incubation protocol, the experiment was repeated three times simultaneously.

During each experiment and for each matrix, three non-contaminated control samples were tested simultaneously.

For each matrix, the level of competition flora was estimated before starting and after the enrichment period from a numeration performed in TSYE agar plate.

The Dynal method was only tested in the study of the incubation times.

2.8. Study of natural samples

Thirty commercially available raw meat products and seventeen dairy products (3 raw milks and 14 cheeses produced with raw milk), were tested after enrichment in mECn broth for 16 h and 20 h.

3. RESULTS AND DISCUSSION

3.1. Study of the specificity for both Transia Plate *E. coli* O157 (ELISA) and Dynabeads (IMS+CT-SMAC agar) assays.

Both assays detected the 10 *E. coli* O157 strains (Table 1). The ELISA did not cross react with any non O157 *E. coli*, but showed cross-reaction with 2 *Salmonella* strains belonging to serogroup N, certainly due to the presence of a common LPS antigen to *E. coli*. Such crossreactivity was previously reported for other immunochemical methods (latex agglutination and immunoassays) with bacteria like *E. hermanii* [18,19]. As *E. coli* O157 shares common surface antigen with other enterobacteria, this crossreactivity is nearly impossible to fully eliminate. The use of monoclonal antibody instead of polyclonal antibody may contribute to decrease the rate of cross-reactivity because of a higher specificity. Nevertheless, such cross-reactivity has never been reported as causing a lot of false positive results when working with naturally contaminated food samples.

Among the 47 non-*E.coli* O157 strains, 25 were growing onto CT-SMAC and 3 gave colourless colonies (1 *E. coli*, 1 *Salmonella* and 1 *Klebsiella*): this underlined that Dynabeads were possibly able to capture or entrap non-specific strains. This was confirmed by streaking the entire IMS fraction onto a non-selective agar like TSYE: all samples showed bacterial growth onto the agar plate (Data not shown). As the immunoseparation was performed with pure strains, it showed that it was not possible to eliminate all the non-specific bacteria with the washing step done after IMS.

Table 1
Specificity of both immunoassay and Dynabeads/CT-SMAC agar with pure bacterial strains

| Strains [a,b] | N | Transia Plate *E. coli O157* | | Colourless colonies on CT-SMAC |
		Positive	Negative	
E. coli O157	10	10	0	10
Other *E. coli*	5	0	5	1
Citrobacter freundii	5	0	5	0
Enterobacter cloacae	1	0	1	0
Hafnia alvei	3	0	3	0
Klebsiella	3	0	3	1
Proteus	5	0	5	0
Serratia	2	0	2	0
Salmonella	10	2[c]	8	1
Pseudomonas	3	0	3	0
Staphylococcus	5	0	5	0
Streptococcus	3	0	3	0
Bacillus	2	0	2	0

[a] : *E. coli* O157 strains were tested at 10^8 cells/ml.

[b] : Other strains were tested at 10^9 cells/ml.

[c] : *Salmonella urbana* and *ramatgan*, serogroup N.

3.2. Limit of detection of the ELISA

For the 4 *E. coli* O157, the ELISA showed a limit of detection in the range 5×10^4 and 6×10^5 cells/ml (Table 2). This range was quite similar to those usually announced for all ELISA kits detecting pathogenic bacteria. With such a limit of detection, *E. coli* O157 was expected to be detected after a16 hour enrichment step.

Table 2
Limit of detection of the immunoassay with pure *E. coli* O157 strains

Strain Code	Reference	Serotype	Limit of detection (cells/ml)
1074	Institut Pasteur	O157. H7	7.5E+04
1154	Food	O157. H7	5.0E+04
1177	Clinical stool	O157	6.0E+04
1257	Food	O157	6.0E+05

3.3. Comparison of 2 incubation times (16 and 20 hours) for both mECn and mTSBn broths at 37°C

For the total number of samples (n=30), the enrichment step before ELISA was running just as well in mECn or in mTSBn. Both broths, combined with Transia Plate *E. coli* O157, gave results equivalent to the reference method (Table 3). Many authors reported the effectiveness of both mEC and mTSB broths, supplemented with novobiocine, for recovering *E. coli* O157 contamination in food [4-6,20,21].

For the low levels of contamination (1-5 CFU/25 g), all the negative ELISA samples were confirmed negative after streaking onto CT-SMAC. They could be explained by the lack of *E. coli* O157 cells when contaminating the samples.

Among the positive ELISA samples, totally three results (2 with mECn and 1 with mTSBn) were not confirmed after streaking onto CT-SMAC agar and agglutination tests. Two of them (1 mTSBn and 1 mECn) gave optical densities over 1.0 for both incubation times (16 and 20 hours) which were also identical to the ODs obtained for the confirmed positive samples.

Table 3
Influence of the incubation time during the enrichment step

Conta level (CFU/25 g)	Number of samples	mECn		mTSBn		Reference method
		16h	20h	16h	20h	
0	9	0	0	0	0	0
1-5	9	6[a]	6[a]	8[b-c]	8[b-c]	7
10-19	12	12	12	12	12	12
30-57	9	9	9	9	9	9
Total of positive samples	30	27	27	29	29	28

For ELISA results, the numbers refer to the total number of positive ELISA samples. For the reference method, the numbers refer to the total number of positive *E. coli* O157 samples after streaking onto CT-SMAC and agglutination tests.
[a] Whatever the incubation time (16 h/20 h), the same sample showed positive ELISA result but was not confirmed positive after streaking, despite an OD similar to the other confirmed positive ELISA samples.
[b] One sample showed positive ELISA result whatever the incubation time (16 h/20 h) but was not confirmed positive after streaking, despite an OD similar to the other confirmed positive ELISA samples.
[c] One sample showed positive ELISA result whatever the incubation time (16 h/20 h) but *E. coli* O157 strains were isolated only from the incubation done during 20 h.

The presence of an elevated competing flora in all these samples was also noticed and contributed to the difficulties in isolating *E. coli* colonies onto CT-SMAC. According to their high OD, we could suggest to classify these non confirmed positive ELISA samples as true positives. This hypothesis was reinforced by the fact that the third sample, showing a "false positive" result after 16 hours in mTSBn, was finally confirmed as true positive from the 20 hour incubation. Both incubation times gave equivalent ODs (0.73 after 16h versus 0.76 after 20 h). Our results confirm the difficulties to isolate and identify *E. coli* O157 onto CT-SMAC agar, especially when the competing flora is high.

Such limitations were previously described for SMAC agar [8] and also for haemorrhagic coli agar (HC,10). Feldsine *et al.* [10] showed a very high bacterial population precluded the ability to isolate *E. coli* O157 by plating onto HC agar. In order to improve the confirmation procedure, he suggested to produce 10-fold serial dilutions of the enriched broth and then to streak these dilutions onto agar plates. This protocol was briefly tested during our study but none improvement for *E. coli* O157 identification was noticed. Rocelle *et al.* reported that stressed *E. coli* O157 cells were difficult to recover according to the plating agar [8]. A comparison between HC and SMAC agars did not highlight a strong difference in performance when working with stressed and unstressed cells in whole milk and ice cream. In order to improve the confirmatory step, more selective enrichment broth could be used. Weagant *et al.* [21] used mTSB broth supplemented with a cocktail of antibiotics composed of cefsulodin, cefixime and vancomycin. But if such a cocktail may be effective for reducing the competing flora during enrichment, it may also inhibit the growth of stressed *E. coli* O157 cells and so decrease the sensitivity of the method. In order to improve *E. coli* O157 recovery from the presumptive positive samples, the use of immunomagnetic bead could represent a good alternative, as it would allow the capture of *E. coli* O157 and partly the elimination of the competing flora. Moreover it would be done only for the presumptive positive samples. Nevertheless, the selectivity of the confirmatory agar used for *E. coli* O157 needs to be improved in order to facilitate the confirmatory step.

3.4. Study of two incubation temperatures (37 and 41.5°C) during the enrichment step
Table 4
Study of both temperatures 37 and 41.5°C during the enrichment step

Conta level (CFU/25 g)	Number of samples	mECn / 20h		mTSBn / 20 h	
		37°C	41.5°C	37°C	41.5°C
0	5	0	0	0	0
1-5	5	4[a]	4[a]	4[a]	5
10	5	5	5	5	5
30-39	4	4	4	4	4
Total number of positive samples	14	13	13	13	14

[a] These three samples were ELISA negative what was in total accordance with the streaking onto CT-SMAC agar: none *E. coli* O157 strain was identified. These negative samples may be explained by the lack of contaminating cells when performing the contamination of the sample.

Based on a numeration done onto TSYE agar, the competing flora in raw minced meat samples evolved from 5×10^5 cells/g to 10^{10} cells/g during the enrichment period with both broths at both temperatures.

With pasteurised soft cheese the competing flora increased from about 6×10^6 cells/g to $3\text{-}7 \times 10^9$ cells/g whatever the enrichment broth (mECn or mTSBn) and the incubation temperature (37°C or 41.5°C). These data demonstrated that the enrichment broths were not able to inhibit the growth of the competing flora, and so their selectivity towards *E. coli* O157 was poor. This competing flora was partly responsible for the difficulties in the confirmation of positive ELISA samples onto CT-SMAC agar, as previously observed.

For both broths, the results obtained at 37°C and 41.5°C were nearly equivalent (Table 4): only three samples turned to be negative, which could be explained by a lack of contaminating cells when performing the contamination of the sample, as it occurred only with the low level of contamination (1-5 cells/ml).

Based on the visual observation of the CT-SMAC plates, the ratio of the characteristic *E. coli* O157 colonies/competition flora looked lower when the samples were incubated at 37°C rather than 41.5°C. In addition, the confirmation of the positive ELISA results obtained after incubation at 37°C were sometimes quite difficult. It required a subculture of the enrichment broth (0.1 ml of enrichment broth incubated at 37°C transferred into 10 ml of fresh enrichment broth and incubated overnight at 41.5°C), in order to isolate *E. coli* O157. For the 26 positive ELISA samples obtained with both broths at 37°C, 9 samples were confirmed positive after subculture at 41.5°C followed by a new plating onto CT-SMAC agar. At the opposite, the positive ELISA samples, previously enriched at 41.5°C, were directly confirmed from the first streaking onto CT-SMAC agar (no subculture needed and lowest number of colonies onto the CT-SMAC plates).

The effectiveness of the incubation temperature at 42°C versus 37°C for recovering *E. coli* O157 from ground beef was previously demonstrated by Blais *et al* [22], with mECn broth. While he showed a shaking during the incubation gave a higher concentration of *E. coli* O157 cells, he recommended a static incubation as the growth of the competing flora was also largely stimulated by shaking. Szabo *et al.* showed also that incubation done at 43°C in mTSB was also providing an optimal recovery of *E. coli* O157 [23].

Using a higher temperature as a selection pressure seemed facilitate *E. coli* O157 isolation onto CT-SMAC, especially because it affected the competing flora. The higher temperature would have stressed the competing flora making them more sensitive to the selective agents of CT-SMAC, or it would have affected the nature of bacteria growing at 41.5°C in mTSBn or mECn, those bacteria being also more sensitive to the selective agents of CT-SMAC agar.

3.5. Study of naturally contaminated food samples

For both incubation times (16 or 20 hours), the ODs and the confirmation onto CT-SMAC agar were equivalent, confirming that a decrease in incubation time did not affect the method performance (Table 5). Two raw meat samples were contaminated by *E. coli* O157, but none of them was producing verotoxins. Nevertheless, both strains showed pink colonies onto CT-SMAC agar, which illustrated fermenting sorbitol bacteria. This sorbitol positive feature was confirmed by a positive test for the sorbitol parameter with the API gallery. So in case of positive ELISA results, the lab operator would have to analyse both colourless and pink colonies onto CT-SMAC, in order to secure the diagnostic.

Table 5
Study of naturally contaminated samples

Food matrix	Number of samples	mECn / 16h		mECn / 20-24h	
		Negative	Positive	Negative	Positive
Raw meat (41.5°C)	30	28	2[b]	28	2[b]
Dairy products (41.5°C)	17	17	0	17	0
Total	47	45	2	45	2

[a] This positive ELISA sample was not confirmed for the presence of *E. coli* O157, whatever the incubation time.
[b] Both positive ELISA samples were confirmed for the presence of *E. coli* O157 by streaking onto CT-SMAC agar. However, these strains did not present characteristic sorbitol negative (colourless) colonies onto CT-SMAC agar and were, at the opposite, sorbitol positive (pink colonies). This biochemical characteristic was confirmed during the identification by API gallery.

Finally, ELISA test for *E. coli* O157 represents an effective alternative method to the cultural one. It shows an objective reading of the results, works with dead cells and is applicable with different enrichment media.

This study has also underlined the difficulties to confirm positive ELISA samples. The use of immunomagnetic beads for extracting *E. coli* O157 cells from the positive ELISA samples, followed by a streaking onto SMAC or CT-SMAC agar plates, might be of help to get less competing flora and improve the identification of the colonies onto the agar plate.

REFERENCES

1. M.A. Karmali, Clin. Microbiol. Rev., 2 (1989) 15.
2. P.M. Griffin and R.V. Tauxe, Epidemiol. Rev., 13 (1991) 60.
3. IFST Position Paper. Int. Food Safety News, 5 (1996) 3.
4. M.P. Doyle and J.L. Schoeni, Appl. Environ. Microbiol., 53 (1987) 2394.
5. A.R. Benneh, S. Mac Phee and R.P. Betts, Lett. Appl. Microbiol., 20 (1995) 375.
6. J.L. Johnson, B.E. Rose, A.K. Sharar, G.M. Ranson, C.P. Lattuada and A.M. McNamara, J. Food Prot., 58 (1995) 597.
7. D.M. Zadick, P.A. Chapman and C.A. Siddons, J. Medical Microbiol., 39 (1993) 153.
8. M. Rocelle, S. Clavero and L.R. Beuchat, Appl. Environ. Microbiol., 61 (1995) 3268.
9. T.S. Hammack, P. Feng, R.M. Amaguana, G.A. June, P.S. Sherrod and W.H. Andrew, JAOAC, 80 (1997) 335.
10. P.T. Feldsine, R.L. Furgey, M.T. Falbo-Nelson and S.L. Brunelle, JAOAC, 80 (1997) 43.
11. A.R Bennett, S. Mac Phee and R.P. Betts, Lett. Appl. Microbiol., 22 (1996) 237.
12. D.J. Whight, P.A. Chapman and C.A. Siddons, Epidemiol. and Infections, 113 (1994) 31.

13. P.M. Fratamico, P.J. Schultz and R.L. Buchanan, Food Microbiol., 9 (1992) 105.
14. A.J.G. Okrend, B.E. Rose and C.P. Lattuada, J. Food Prot., 55 (1992) 214.
15. P.A. Chapman and C.A. Siddons, Food Microbiol., 13 (1996) 175.
16. V.P.J. Gannon, R.K. King, J.K. Kim and E.J.G. Thomas, Appl. Environ. Microbiol., 58 (1992) 3809.
17. M.J. Brian, M. Frosolono, B.E. Murray, A. Miranda, E.L. Lopez, H.F. Gomez and T.G. Cleary, J. Clin. Microbiol., 30 (1992) 1801.
18. L. Easton, Br. J. Biom. Sc., 54 (1997) 57.
19. D.L. Tison, J. Clin. Microbiol., 28 (1990) 612.
20. A.J.G. Okrend, B.E. Rose and C.P. Lattuada, J. Food Prot., 53 (1990) 936.
21. S.D. Weagant, J.L. Bryant and K.G. Jinneman, J. Food. Prot., 58 (1995) 7.
22. B.W. Blais, R.A. Booth, L.M. Philippe and H. Yamazaki, Int. J. of Food Microbiol., 36 (1997) 221.
23. R. Szabo, E. Todd, J. Mac Kenzie, L. Parrington, A. Armstrong, Appl. Environ. Microbiol., 56 (1990), 3546

Food Biotechnology
S. Bielecki, J. Tramper and J. Polak (Editors)
© 2000 Elsevier Science B.V. All rights reserved.

Application of Solid Phase Microextraction (SPME) for the determination of fungal volatile metabolites

H. Jeleń and E. Wąsowicz

Institute of Food Technology, Agricultural University of Poznań,
Wojska Polskiego 31, 60-624 Poznań, Poland

1. INTRODUCTION

Volatile compounds play a significant role amid various fungal metabolites. Not only do they form mushroom aroma so important in food preparation and processing, but they can also impair food flavour or improperly stored agricultural commodities, and as a result serve as an indicator of moulds presence [1]. Volatile metabolites biosynthesised by fungi form a vast array of constituents. The most abundant are compounds originating from fatty acids degradation - aliphatic eight carbon alcohols and ketones, revealed to be responsible for the characteristic flavour of edible mushrooms [2-4]. 1-octen-3-ol has been identified as a predominant compound in edible mushrooms. It is there frequently accompanied by 1-octen-3-one, 3-octanol, 1-octenyl-3-acetate and 3-octanol. These compounds have also been detected as metabolites of filamentous fungi - mainly *Penicillium*, *Aspergillus*, *Alternaria* and *Fusarium* [5-7]. Fungi are capable of producing other alcohols of which fusel alcohols prevail. They are also efficient producers of methyl ketones responsible for the "blue cheese aroma", lactones, esters, hydrocarbons and terpenoic compounds. Monoterpenes influence the flavor of *Basidiomycetes* and have been detected in many wild type mushrooms [8], whereas sesquiterpenes have been detected as typical metabolites of both mushrooms and moulds [9-12].

Biosynthesis of mycelium in liquid cultures can supply a significant amount of biomass, which can be used as a source of mushroom flavour in food industry. The production of *Morchella* in liquid media, patented in USA in the 60-s, is an example of such product. The main obstacle in such processes, despite economical ones, is the correlation of the high yield of biomass with the efficient flavour compounds biosynthesis. Biotechnologists are also interested in microorganisms as a source of pure flavor compounds obtained in either bioconversions or *de novo* syntheses. Fungal volatiles can contribute also to the chemotaxonomical characteristic of species [13].

Therefore, a sensitive, reliable and rapid method for the qualitative and quantitative determination of volatile compounds is demanded.

2. METHODS USED FOR ISOLATION OF FUNGAL VOLATILE COMPOUNDS

There are numerous methods, which have been used for the isolation of volatiles from fungal mycelia for the last 30 years. The techniques utilised for the isolation of volatiles from fungi reflect trends in the development of methods in flavour analysis. Extraction with organic solvents is a method which has been in use for the longest period of time. The choice of solvents which determines to a certain extent the profile of isolated compounds, comprises ethyl ether, pentane, dichloromethane, trimonofluoromethane, to name only few [14, 15, 4]. Extraction processes suffer several drawbacks: a spectrum of extracted compounds is often too rich because of co-extracted extraneous compounds, it is also hard and laborious to perform quantitative extraction, solvent makes the sensory evaluation of extract difficult and it can mask some compounds in the chromatographic analysis. For the analysis of fungal volatiles distillation methods have often been applied [16-18]. The main disadvantage of distillation under atmospheric pressure is a formation of artefacts, compounds degradation and losses. These can be eliminated by carrying out the process under reduced pressure or performing vacuum distillation – then the method becomes more reliable, but it is even more laborious. An interesting method is the simultaneous extraction and distillation (SDE), which usually takes place in Likens-Nickerson apparatus. All described methods require as much as 80% of analysis time to be devoted to the preparation of samples. Therefore, headspace analysis techniques have gained increasing popularity. When sample is placed in a closed system state of thermodynamic equilibrium is reached between the gas phase and the liquid or solid phase. The sampling of vapours of such a system is done with the static headspace technique. In dynamic headspace techniques the state of equilibrium is not reached - neutral gas purges volatiles from liquid or solid matrix (medium). In these techniques gas extracted volatiles are trapped usually on polymers, such as Tenax, Porapak or Chromosorb and subsequently thermally desorbed, or eluted with a solvent and transferred to the gas chromatograph. Cryoconcentration is often required to focus band of compounds entering an analytical column. Both static and dynamic headspace techniques can be automated, which radically improves reproducibility, reduces labour in sample preparation process, but can dramatically increase the cost of chromatographic equipment used for the analysis. A novel and promising technique – solid phase microextraction has been in use for the last few years and its principles and perspectives of applications in analysis of fungal flavour compounds shall be discussed in the following chapters.

2. SPME PRINCIPLES AND METHOD OPTIMISATION

Solid phase microextraction (SPME) has been developed at the beginning of the nineties and since that time its popularity has been constantly increasing. Analysed compounds are adsorbed on the phase which coats an optic fiber mounted to the plunger of a syringe-like holder. In the first stage the fiber is exposed to the sample and the process of extraction goes on until the equilibrium is reached. The extraction can be direct, when fiber is immersed in a liquid phase – or performed in headspace phase. It is usually carried on in capped vials similar to these used in headspace analysis. To protect fiber coating the fiber is then retraced into the needle and the holder removed from the vial. In the next step the holder's needle pierces the septum in the gas chromatograph and by pressing the plunger the fiber is exposed in the injection port. This provides immediate thermal desorption of compounds and their rapid transfer to the column [19]. Small volume of phase coating fiber ensures short time required for the establishment of equilibrium. SPME can be also used as a non-equilibrium method. The extraction times are usually much shorter than those required in static or dynamic headspace techniques, whereas the limits of detection are very low, comparable to purge and trap or even extractions in a CLSA (closed loop stripping analysis) systems. Its big advantage is simplicity – no special desorption equipment is required, the cost of holder is relatively low and the fiber can withstand approximately 50 – 150 analyses. This makes it an interesting tool in flavor analysis. The number of papers where SPME is used in volatiles isolation and quantitation increases each year. This technique has been used mostly by environmental chemists for the analyses of pollutants in water and air but also for the analysis of aroma compounds in fruit, cheese, wine, beer and vodkas. However, a number of applications in the analysis of microbial metabolites is very limited.

Nilsson et al. [20] used headspace SPME for the determination of compounds responsible for a characteristic *Penicillia* odour. Using polydimethylsiloxane (PDMS) fiber for 30 min. and polyacrylate (PA) fiber for 50 min. for extraction they identified among others isopentyl alcohol, 1-octen-3-ol, 3-octanone, 3-octanol, 2-methylisoborneol, geosmin and terpene hydrocarbons. When compared to adsorption on Tenax the authors assumed that SPME was a better method.

The SPME method was also used by Larsen [21] to identify fungal cultures used in a cheese making process. Application of SPME using PDMS fiber for 30 min extraction and mass spectrometry identification using selected ions monitoring allowed species differentiation two days after inoculation. In a mixed culture the differentiation of *P. roquefortii* from *P. commune* (inoculated in ratio 1000:1) was possible after 3 days. The following fungi were tested: *Penicillium commune, P. roquefortii, P. solitum, P. discolor* and *Aspergillus versicolor.*

Pelusio et al. [22] used SPME for the determination of sulphuric compounds present in white and black truffles (*Tuber magnatum* Pico and *Tuber melanosporum* Vitt., respectively). Extractions were performed at room temperature and at 80°C for 30 min. All compounds previously detected in truffles by other researchers were also detected with the SPME method, and in white truffles authors detected three new compounds. In all these applications SPME was used as a qualitative rather than quantitative technique.

In the following part of this paper some remarks concerning steps in method development for volatiles determination of fungi grown on liquid or solid media will be discussed. The evaluation of *Xerocomus badius* volatiles grown in liquid medium, the detection of fungal off-flavour and monitoring volatiles of *Aspergillus ochraceus* in solid medium will be shown. All the examples are based on works carried out in our laboratory.

Biological media form a complex matrix, therefore attention must be paid when parameters of a new SPME method are established. The optimisation of SPME parameters involves several steps:

- Choice of extraction mode
- Choice of fiber coating
- Optimisation of extraction parameters
    - Sample/headspace volume
    - Extraction time and temperature
    - pH changes, derivatization or salt addition
- Optimisation of chromatographic parameters

Extraction of compounds can be performed either in a direct mode or as headspace analysis. Direct extraction can be performed when compounds are dissolved in a relatively clean matrix (for example water). Otherwise, in thermal desorption step many artefacts can be formed - Maillard compounds, thermal degradation of labile compounds other than analyte, present in the medium can occur and as a result fiber lifetime can be dramatically shortened. Biological liquid media are usually relatively rich in organic nutrients and salts, therefore headspace extraction is often the only choice.

There are several types of fiber coatings produced exclusively by Supelco Inc. Different coatings have been developed for various compounds extraction. Phases used as coatings resemble those used in gas chromatography: carbowax, polyacrylate, polydimethylsiloxane, carboxen, divinylbenzene.

Table 1
Extraction of volatile compounds from liquid potato medium by different fibers. The medium was spiked with standards at 1 ppm each and extracted for 60 min. at 20°C using following fiber assemblies: C/PDMS (Carboxen/Polydimethylsiloxane), PDMS (polydimethylsiloxane), CW/D (Carbowax/divinylbenzene) and PA (polyacrylate)

Compound	Fiber type			
	C/PDMS	PDMS	CW/DVB	PA
Ethyl acetate	*123.9	0.8	3.1	2.0
Isoamyl acetate	270.6	34.2	4.5	12.6
2-heptanone	491.3	15.1	3.0	6.9
1-octen-3-ol	88.4	4.7	4.7	9.0
Limonene	5008.5	3854.4	486.2	2061.1
Valencene	3577.6	4927.2	3624.0	5285.3

* - peak areas in relative integrator units

Polarity and volatility of analysed compounds determine the choice of the fiber. Fibers based on Carboxen have recently been developed for the analysis of low molecular weight compounds and flavours. Table 1 illustrates extraction capabilities of different fibers (measured as peak areas) for compounds which have been often detected in fungi.

Certain selectivity can be obtained by choosing different fibers. As can be seen in Table 1. limits of detection will vary dramatically with the type of fiber, physical properties and the character of analysed compounds. The investigated group of compounds is characterized by a big amount of terpene hydrocarbons adsorbed on each fiber. Figure 1 illustrates the influence of matrix on the recovery of volatiles adsorbed on SPME fiber.

SPME is an equilibrium method and time required for reaching this equilibrium shall be established. Due to the small volume of coating on the fiber equilibrium is reached in a relatively short time - usually within 30 minutes. Extraction temperature will naturally influence the migration rate of analyte into the headspace phase and influence the time required to reach equilibrium. Depletion of analyte from the fiber by other compounds sometimes occurs. Figure 2. shows the extraction curves for benzaldehyde run at 20°C and 50°C in a liquid medium.

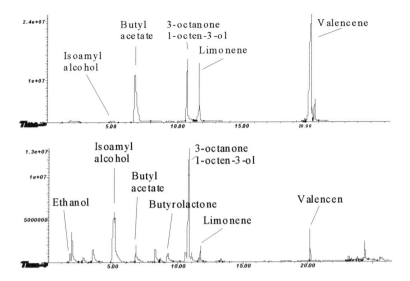

Figure 1. Chromatograms of volatile compounds extracted from liquid and solid matrixes (upper and lower chromatograms respectively). Standards of compounds (3 ppm each) were added to liquid medium (potato extract) and solid medium (autoclaved wheat kernels). Extraction was performed in headspace over 30 g of medium for 20 min. using Carboxen/PDMS fiber.

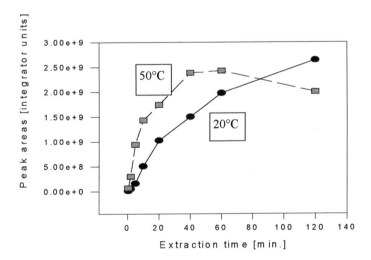

Figure 2. Extraction curves obtained for benzaldehyde using Carboxen/PDMS fiber at 20°C and 50°C. Potato extract (see paragraph 3.1) was used as a liquid medium.

3. APPLICATIONS OF SPME IN DETECTION AND QUANTITATION OF FUNGAL METABOLITES

4.1 Quantitative analysis of volatiles produced by *Xerocomus badius* grown in liquid media

Solid phase microextraction has been used for the quantitative determination of volatile compounds produced by *Xerocomus badius* grown in the potato extract liquid media. Biosynthesis of fungal biomass, which has a strong mushroom flavour can find application as a natural food additive. In order to be meaningful for further scaling-up selected mushroom strains must be characterised with abundant mycelium, as well as its rapid growth and high concentration of produced flavor compounds. Several mushroom strains have been evaluated on a sensory basis: *Pleurotus ostreatus, Pleurotus sajor-caju, Morchella esculenta, Lentinus edodes, Agaricus bisporus, Xerocomus badius, Inonotus obliquus, Phlebia radiata, Phenerochete chrysosporium, Trametes versicolor.*

Table 2.
Concentration of volatile compounds produced by *Xerocomus badius* grown on potato liquid medium

Compound	Concentration [mg/l]		
	5 days	10 days	15 days
1-octene-3-ol	1.278^a	19.562^c	1.72^a
3-octanone	0.954^a	44.499^b	30.678^a
3-octanol	nd	2.546^a	3.075^a
limonene	0.002^a	0.002^a	0.003^a
2-undecanone	nd	0.006^b	nd
α-gurjunene	0.002^b	0.003^c	0.004^c
β-elemene	0.003^c	nd	0.003^b
Mycelium [g/l]	41^a	62^a	76^a

a - RSD<25%, b - RSD 25-50%, c - RSD>50%, n=3

Xerocomus badius has been selected due to its strongest and most pleasant mushroom aroma. It had been grown in a potato extract liquid medium (200g potatoes, 20g glucose, 20g agar, water to 1000ml) and the same medium enriched with yeast extract (0.03%) for 15 days and samples were analysed after 5, 10, 15 days. Volatiles were isolated from the headspace using Carboxen/PDMS fiber (extraction time 20 min., temp. 50°C). The compounds were resolved, identified and quantitated on a HP 5890II/HP 5971 gas chromatograph/mass spectrometer working in the SCAN mode. External standard quantitation was used by adding to potato medium standards of previously identified compounds. Table 2. shows compounds which were identified in *Xerocomus badius* cultures. 3-octanone and 1-octene-3-ol dominate among them and they are responsible for the odour of cultures. Due to excellent "affinity" of terpenes to the fiber phase it was possible to quantitate them in very low concentrations. However, considering odour thresholds their importance in the formation of *Xerocomus badius* aroma is minute.

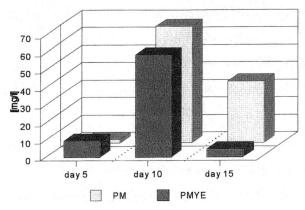

Figure 3. Amount of alcohols and ketones [mg/l] produced by *Xerocomus badius* grown for 15 days on a potato medium [PM] and potato medium enriched with yeast extract [PMYE].

4.2 Analysis of off-odours of stored molded wheat

Fungi can produce compounds which can cause an unpleasant smell. Geosmin (*trans*-1,10-dimethyl-*trans*-9-decalol) is one of the compounds responsible for musty/earthy odour. It has been detected in water, stored grain and vegetables and its low odour threshold (0.015 ppb) makes it an important trace compound in food. It is known as a metabolite of *Actinomycetes* and certain *Penicillia* (*P. expansum*, *P. aethiopicum*). A sample of wheat (50 g), which has an distinct off-odour was placed in a headspace flask, heated to 50°C and volatiles were collected using PDMS fiber for 20 min. To identify compounds mass detector was used (see 3.1). The chromatogram of collected volatile compounds is shown in Fig. 4. Geosmin peak abundance allowed the analysis in the SCAN mode (40-340 a.m.u.) and a clear full spectrum of it could be obtained. However, geosmin concentration in samples is often lower – in this case detection of this metabolite in Selected Ion Monitoring (SIM) is recommended, where m/z 112 can be used as a target ion.

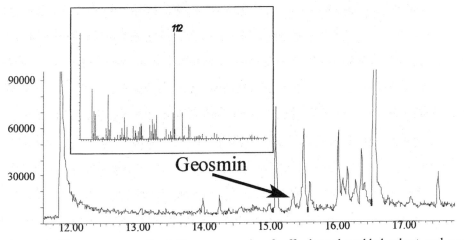

Figure 4. Chromatogram of volatile compounds of off-odoured molded wheat and mass spectrum of detected geosmin. Sample (50g) extracted at 50°C for 20 min. using PDMS fiber.

376

4.3 Monitoring of volatile compounds emitted during the growth of *Aspergillus ochraceus*

Apergillus ochraceus inhabits stored grain together with other "storage fungi" – mainly belonging to *Penicillium* and *Aspergillus*. Under favourable conditions it can produce ochratoxin – a toxic metabolite, the most widespread among storage mycotoxins in a temperate climate. Monitoring fungal volatile metabolites can provide information on fungus presence and activity. Autoclaved wheat was inoculated with a spore suspension (5×10^{6} spores/ml) of *Aspergillus ochraceus* KA 101 (Agric. Univ. of Poznań fungal collection). Volatiles were collected for 30 min. using PDMS fiber. It was possible to detect and identify volatile compounds as soon as 24 hours after the inoculation. Fig. 5. shows the chromatogram of compounds emitted by *A. ochraceus* culture after this time.

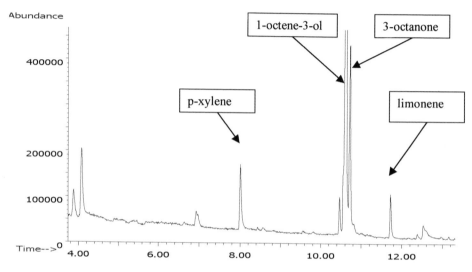

Figure 5. Chromatogram of volatile compounds emitted by *A. ochraceus* strain KA 101 grown on autoclaved wheat. Sample taken 24 hours after medium inoculation (SPME, PDMS fiber).

5. CONCLUSIONS

Solid phase microextraction has been used in numerous applications in environmental chemistry and also in flavour analysis. It has proven to be a rapid, uncomplicated and cheap method offering high sensitivity, good reproducibility and a low cost per sample for most compounds analyses it has been applied to. In monitoring biotechnological processes it can be used on-line, sample do not have to be prepared in a special way, which makes the technique very useful. Recent works in combining SPME with HPLC have enhanced the potential applications of this technique.

REFERENCES

1. H. Jeleń and E. Wąsowicz, Food Rev. Int., 14 (1998) 391.
2. J. Maga, J. Agric. Food Chem., 29 (1981) 1.
3. H. Pyysalo and M. Suihko, Lebensm. Wiss.Technol., 3 (1976) 371.
4. S. Rapior, S. Cavalie, C. Andary, Y. Pellisier and J.M. Bessiere J.Essent. Oil Res., 8 (1996) 199.
5. E. Kamiński, S. Stawicki and E. Wąsowicz, Appl. Microbiol., 27 (1974) 1001.
6. T. Jacobsen and L. Hindrichsen, Food Chem., 60 (1997) 409.
7. T.O. Larsen and J.C. Frisvad, Mycol. Res., 10 (1995) 1153
8. S. Breheret, T. Talou, S. Rapior and J.M. Bessiere, J. Agric. Food Chem., 45 (1997) 831.
9. T. Larsen and J.C. Frisvad, Mycol. Res. 10 (1995) 1167.
10. H.J. Zeringue, D. Bhatnagar, and T. Cleveland, Appl. Environ. Microbiol. 59 (1993) 2264.
11. R.G. Berger, K. Neuhauser and F. drawer, Z. Naturforsch., 41 (1986) 963.
12. H. Jeleń, C.j. Mirocha, E. Wąsowicz, E. Kamiński, Appl. Environ. Microbiol. 61 (1995) 3815.
13. T.O. Larsen, in: J.C. Frisvad, P.D. Bridge and D.K. Arora, Chemical Fungal Taxonomy, Marcel Dekker, Inc., New York (1998) 263.
14. A.F. Halim, J.A. Narciso and R.P. Collins, Mycologia 67 (1975) 1158.
15. T. Kawabe and H. Morita, J. Agric. Food Chem. 42 (1994) 2556.
16. E. Kamiński, L.M. Libbey, S. Stawicki and E. Wąsowicz, Appl. Microbiol. 24 (1972) 721.
17. R. Tressl, D. Bahri and K.H. Engel, J. Agric. Food Chem., 30 (1982) 89.
18. E. Vanhealen, R. Vanhealen-Fastre and J. Geeraerts, Sabouraudia 16 (1978) 141.
19. J. Pawliszyn, Solid Phase Microextraction. Theory and Practice. Wiley-VCH, New York (1997).
20. T. Nillson, T.O. Larsen, L. Montanarella and J. Madsen, J. Microbiol. Methods, 25 (1996) 245.
21. T. Larsen, Lett. Appl. Microbiol., 24 (1997) 463.
22. F. Pelusio, T. Nilson, L. Montanarella, R. Tilio, B. Larsen, S. Facchetti and J. Madsen, J. Agric. Food Chem. 43 (1995) 2138.

Food Biotechnology
S. Bielecki, J. Tramper and J. Polak (Editors)
© 2000 Elsevier Science B.V. All rights reserved.

The HPLC Separation of 6-acyl glucono-1,5-lactones

K. Kolodziejczyk, B. Krol

Institute of Chemical Technology of Food, Technical University of Lodz,
ul. Stefanowskiego 4/10, 91924 Lodz, Poland

The products of enzymatic and chemical esterification of glucono-delta-lactone (GDL) with lauric, myristic, palmitic and stearic acids were analysed with HPLC. The 6-acyl glucono-1,5-lactones were prepared enzymatically using Novozyme 435. The products of reactions were analysed using isocratic HPLC system with an acetonitrile:acetone:water mixture as a mobile phase and Lichrosorb RP18 column. The influence of water content in the mobile phase on chromatographic separation of compounds is discussed. The best results were achieved using a mobile phase containing acetonitrile 50 : acetone 40 : water 10.

1. INTRODUCTION

Gluconic acid and its 1,5-lactone (glucono-delta-lactone, GDL) are cheap and readily available substances. They are prepared by bioconversion from glucose or glucose syrups. This origin makes them renewable.

The GDL (marked E575) is used in some food technologies as an acidulate, a stabiliser, a flavour promoter, a coagulant or a sequestering agent. As a chemical compound, gluconic acid is interesting due to the presence of two kinds of esterifiable functional groups: carboxyl C_1 and hydroxyl, whose primary C_6 is the most reactive. The presence of hydroxyl groups makes gluconic acid and GDL similar to saccharides. Primary esters of sugars and fatty acids have surface acivity due to the hydrophilic character of sugar part (head) and hydrophobic character of fatty chain (tail). Such substances are known and commonly used in food and cosmetic industry. Sugar esters of carboxylic acids were analysed with HPLC using acetonitrile, methanol and water mixture as a mobile phase and UV detector [1,2]. This phase was not suitable for GDL-fatty acid esters because of insufficient solubility of compounds.

This work considers HPLC analysis of GDL esters with lauric, myristic, palmitic and stearic acids using RI detector, acetone:acetonitrile:water mixtures as a mobile phase.

2. MATERIALS AND METHODS

2.1. Synthesis
2.1.1. Enzyme
A commercial preparation of immobilized lipase from Candida antarctica, Novozym 435® (Novo Nordisk) was used in biosynthesis. The esterification of saccharides catalysed with this enzyme was previously described [1].
2.1.2. Chemicals
Crystalline glucono-delta-lactone was USP grade. Lauric, myristic, palmitic acids were over 98% in purity (Merck), stearic acid was 96% in purity. Acetone was an analysis grade. Molecular sieve 0.3 nm (Merck) was used as water absorbing agent. Lauroyl, myristoyl, palmitoyl, stearoyl chlorides and pyridine were synthesis grade.

2.1.3. Methods
Monoesters were prepared by shaking the mixture of 0.02 mol GDL, 0.06 mol of suitable acid, 10 g. molecular sieve and 4g. Novozym 435® in 100 ml of acetone, on orbital shaker at 90 rpm. After the reaction had been completed the enzyme and molecular sieve were separated from the reaction mixture. The monoesters crystallised from the reaction mixture were filtered and re-crystallised from acetone. The total yields was 40-60%.
Diesters were prepared by adding suitable acid chloride to pyridine solution of GDL. Precipitated pyridine hydrochloride was removed, solution was evaporated, the raw product was crystallised twice from acetone.

2.2. HPLC analysis
2.2.1 Eluent components
Acetonitrile HPLC grade (J.T. Baker)
Acetone HPLC grade
Water purified on Millipore water purification system.

2.2.2. HPLC equipment
The reaction products were analysed with HPLC isocratic system: KNAUER WellChrom Mini Star K-500 pump, Rheodyne 7125 injector with 20µl sample loop, Lichrosorb RP18 250x4.6 mm, 5µm. HPLC column in column oven, KNAUER A0298 RI detector, KNAUER WellChrom Interface Box ad PC with KNAUER EuroChrom 2000 Integration Package software.

2.2.3. Analytical procedure
The mixture of acetone, acetonitrile and water was used as a mobile phase in the following proportions:
acetonitrile 59% : acetone 40% : water 1%
acetonitrile 55% : acetone 40% : water 5%
acetonitrile 50% : acetone 40% : water 10%
acetonitrile 45% : acetone 40% : water 15%

The samples of 0.02g of crystalline ester, 0.01g of acid, or 1mL of reaction mixture were dilluted to 10 mL with a mobile phase and analysed with the flow rate of 0.8 ml/min at 40°C.

The following values were used to find the best separation conditions:
R_t - relative retention time: Rt of GDL ester: Rt of 6-lauroyl GDL ratio
R_w - relative peak width: w of GDL ester : w of 6-lauroyl GDL ratio
Rs - resolution calculated:

$$Rs = \frac{2(R_{t1} - R_{t2})}{w_1 + w_2}$$

where Rt - retention time
w - peak width

3. RESULTS AND DISCUSSION

The efficiency of HPLC separation of the compounds strongly depends on the water content in a mobile phase. The more water in a mobile phase, the better separation of ester and acid, the better separation of monoesters of different acids and the longer retention times of all compounds. A mobile phase containing 15% of water is best for the analysis of lauric acid esterification products (Figure 1).

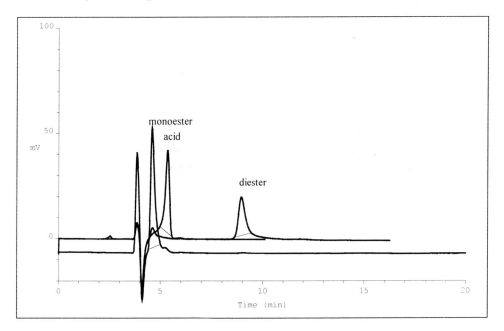

Figure 1. HPLC separations of lauric acid, mono- and di-lauroyl-GDL esters analysed using mobile phases containing 15% of water (optimum HPLC conditions)

For higher esters the solubility problems occurred, the retention time became too long and the peak width became too large (Figure 2 and 3). 5% water content was good enough for stearic acid and its mono and diesters separation, but for lower esters the separation was unsatisfactory (Figure 5). The optimal HPLC conditions can be used to inspect the reaction course in the enzymatic GDL esterification (Figure 4). The phase containing 10% of water was best for the analysis of palmitic acid esters and good enough for other analysed esters.

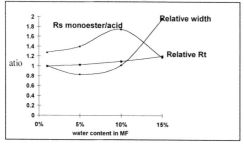

Figure 3. Effect of water content in a mobile phase on relative retention time, peak width and monoester/acid resolution in 6-palmitoyl-GDL HPLC analysis

Figure 4. Effect of water content in a mobile phase on retention time of mono- and diester peaks in HPLC analysis of palmitoyl-GDL

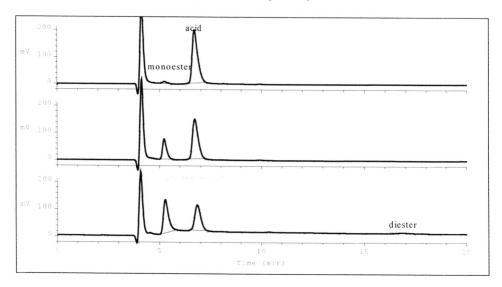

Figure 5. HPLC separations of enzymatic reaction mixture - inspection of ester formation course. Mobile phase with 10% of water

However, for stearic acid esterification, products should be analysed with less water in a mobile phase because of the overincrease of diester peak width (Graph 1) and long, exceeding 20 minutes, retention time of diester peak.

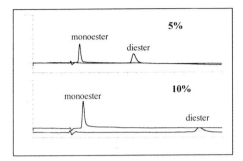

Figure 6. HPLC separations of mono- and di-palmitoyl-GDL esters analysed using mobile phases containing 5 or 10% of water

Figure 7. Effect of water content in a mobile phase on monoester/acid and diester/acid peaks resolution in HPLC analysis of palmitoyl GDL

4. CONCLUSION

The water content in the mobile phase is a key factor in the best HPLC conditions for analysis of GDL esters. The HPLC method with mobile phase acetone 40% : acetonitrile 55, 40, 45% : water 5, 10, 15%, respectively, can be used to analyse GDL mono- and diesters with fatty acids. Retention time depends on the water content in a mobile phase: the higher water content, the longer retention time. However, the optimum peak width appears at a certain water content. The higher the molecular weight of the ester, the lower optimum water content in a mobile phase. 10% water content makes it possible to analyse lauric, myristic and palmitic acids with their mono- and diesters simultaneously. The same phase can be used to analyse stearic acid and 6-stearoyl-GDL. The HPLC analysis at optimum conditions is a useful tool for the inspection of an esterification reaction course.

REFERENCES

1. C. Scheckerman, A.Schlotterbeck, M. Schmidt, V. Wray, S. Lang, Enzyme and Microbial Technology 17 (1995) 157.
2. G. Fregapane, D.B. Sarney, E.N. Vulfson, Enzyme and Microbial Technology 13 (1991) 796.

Food Biotechnology
S. Bielecki, J. Tramper and J. Polak (Editors)
© 2000 Elsevier Science B.V. All rights reserved.

385

Kinetics of activation and destruction of *Bacillus stearothermophilus* spores

Jan Iciek, Agnieszka Papiewska and Lech Nowicki

Institute of Chemical Technology of Food,
Technical University of Łódź, Poland

Kinetics of thermal activation and destruction of *Bacillus stearothermophilus* spores in glass capillaries and flow system was investigated. Additionally, the effect of pH on these processes was analyzed.
It was shown that mathematical models presented in the literature describe inaccurately the real phenomena occurring during thermal inactivation of the spores.
The model proposed by the authors was confirmed to be suitable for the description of these phenomena.

1. INTRODUCTION

It is well-known that the smallest losses of product properties are caused by continuous sterilization accomplished by the UHT method. Such installations permit thermal energy and cooling water to be reduced. Another important issue is the selection of optimum conditions of thermal treatment of the product that will ensure the final sterility accompanied by very good preservation of its valuable components. To solve this problem, the knowledge of both the course of degradation of labile components and the kinetics of inactivation of microorganisms contained in the product sterilized is necessary. When establishing the optimum parameters of the process of thermal sterilization of media, mathematical models describing changes occurring in these products are very helpful [1,2].

In the work, the authors studied the process of thermal inactivation of *Bacillus stearothermophilus* spores as the most heat resistant forms of microorganisms in an attempt to make the mathematical description of this phenomenon more precise.

The models known from the literature do not enable the course of the process of inactivation of spores subjected to high temperature to be described correctly. The non-conformance of the mathematical description of thermal inactivation of spores showing the actual course of biological phenomena, found in the literature, results from the fact that these models do not take into consideration the variety of physiological forms of bacterial spores. In a population of bacterial spores, there can occur individuals in a state of activation, dormancy or deep dormancy, i.e. in the so called super dormant form. These forms are characterised by a varied sensitivity to unfavourable environmental conditions, including the action of high temperature [4,6].

Germination, that is to say the growing of spores in culture media, of individuals which are in dormancy or super dormancy, occurs after they pass into state of the activated spore in a process called bacterial spore activation. Only such a form of the spore is sensitive to the action of high temperature and is easily destroyed to form the inactivated (dead) cell.

A schematic diagram:

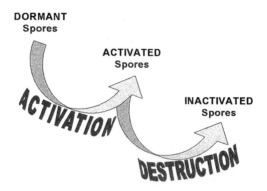

In the literature [4,5], there are models which take into consideration the occurrence of spores only in the activated state and/or in dormancy. However, there are no mathematical descriptions of the process of thermal inactivation of bacterial spores which are in the super dormant state. These spores are characterised by extremely high heat resistance. Their presence in a population can result in the formation, on diagrams of survivor curves, of the so called 'tail' on the final part of their course. The occurrence of this phenomenon causes significant reduction in the efficiency of thermal inactivation of spores at the final stage of the process.

The possibility of predicting the occurrence of this phenomena, and particularly, of eliminating it is essential. Mathematical models found in the literature do not allow the phenomenon of 'tailing' to be analysed.

In this connection, the authors developed their own mathematical model describing the phenomena of activation and destruction of bacterial spores during their thermal inactivation, considering the frequent phenomenon of 'tailing'.

2. AIMS OF THE RESEARCH WORK

⇒ To analyse the course of the process and to verify existing mathematical models describing the inactivation of bacterial spores contained in the media subjected to thermal sterilization.
⇒ To develop the authors' own mathematical model describing the phenomena of activation and destruction of bacterial spores during their thermal inactivation, taking into consideration the frequent phenomenon of 'tailing'.
⇒ To find the relationship of the effect of pH of the environment in which bacterial spores are subjected to thermal treatment on the values of kinetic constants of the process of thermal activation and destruction of these spores.

3. METHODS AND SCOPE OF RESEARCH

The authors' own investigations were carried out in glass capillaries and a flow system operating on a pre-pilot scale. In view of the considerable cost of experiments performed in the system, most of measurement series were made in the capillaries.

Microbiological analyses were made by the plate method using parallel inoculations. To obtain a suspension of bacterial spores, a presporulating TYG and sporulating BPY media were used. The culture was grown following a two-stage procedure of cell multiplication developed by Kim and Naylor [3]. The determination of both the number and kinetics of spore inactivation was made in a regenerative medium of a composition given Kim and Naylor [3]. The following incubation conditions were applied: the temperature 55°C in the time from 10 to 48 hours depending on the culture stage.

Glass capillaries were filled with a water suspension of spores of an initial concentration of about 10^6cell/ml and, following immersion, were heated in an oil bath at a temperature of 110, 115 and 121°C for up to 30 minutes. The level of pH of the water suspension of bacterial spores in the range of 3 to 7 was an additional variable.

At the first stage of handling measurement results, an attempt was made to describe the experimental data obtained using the Shull model. It has been found that the Shull model offers a good description of the course of the spore thermal inactivation process at its initial phase. The model can even be used to predict the phenomenon of activation; however, it does not allow one to predict and describe the retardation of the spore inactivation process called the phenomenon of 'tailing', which actually occurs under certain conditions (Fig 1).

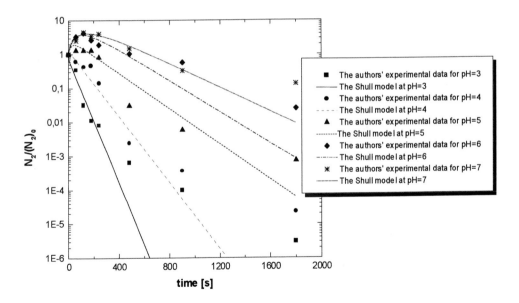

Figure 1. Survival curves plotted according to the Shull model for the authors' own experimental data obtained at a temperature of 115°C.

These observations led the authors to consider it purposeful to build their own mathematical model of the thermal inactivation of spores. A number of variations were analysed and the following pattern was regarded as the most appropriate:

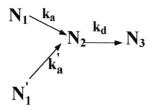

N - spores: $_1$ - dormant, $'_1$ - super dormant, $_2$ - activated, $_3$ - unviable cells
k - the constant reaction rate: $_a$ - of the dormant spore activation,
   $'_a$ - of the super dormant spore activation
   $_d$ - of the activated spore destruction

Spores at the stage of dormancy and super dormancy were identified. For such a pattern of inactivation of spores it has been assumed that activation processes of dormant spores and super dormant spores proceed independent of each other. They can be described by a single reaction of the 1st order. It has been assumed also that the rate of these reactions can be different; on the other hand, the destruction of activated spores can be described by one reaction. With these assumptions, the following equations have been taken for the description of particular phenomena: (1), (2), (3)

$$\frac{dN_1}{dt} = -k_a N_1 \qquad\qquad (N_1)_{t=0} = (N_1)_0 \qquad\qquad (1)$$

$$\frac{dN'_1}{dt} = -k'_a N'_1 \qquad\qquad (N'_1)_{t=0} = (N'_1)_0 \qquad\qquad (2)$$

$$\frac{dN_2}{dt} = k_a N_1 + k'_a N'_1 - k_d N_2 \qquad\qquad (N_2)_{t=0} = (N_2)_0 \qquad\qquad (3)$$

where:

   $(N_1)_0$ – the initial number of dormant spores, $CFU \cdot cm^{-3}$
   $(N'_1)_0$ - the initial number of super dormant spores, $CFU \cdot cm^{-3}$
   $(N_2)_0$ - the initial number of activated spores, $CFU \cdot cm^{-3}$
   k_a - the constant of the dormant spore activation reaction rate, s^{-1}
   k'_a - the constant of the super dormant spore activation reaction rate, s^{-1}
   k_d - the constant of the activated spore destruction reaction rate, s^{-1}
   t - the time of the thermal treatment of the medium, s

After solving the above system of equations, the following equation was obtained: (4)

$$\frac{N_2}{(N_2)_0} = \exp(-k_d t) + \frac{(N_1)_0}{(N_2)_0} \cdot \frac{k_a}{k_a - k_d}[\exp(-k_d t) - \exp(-k_a t)] + \frac{(N_1')_0}{(N_2)_0} \cdot \frac{k_a'}{k_a' - k_d}[\exp(-k_d t) - \exp(-k_a' t)] \quad (4)$$

Basing on this equation and the experimental data obtained in the capillaries, the values were determined of kinetic constants describing particular processes, i.e. the activation of dormant spores k_a, activation of super dormant spores k_a' and the destruction of activated spores k_d. The results have been shown on the diagrams (Fig.2 and 3), in which the experimental results are described by the authors' own model and, for comparison, by the Shull model.

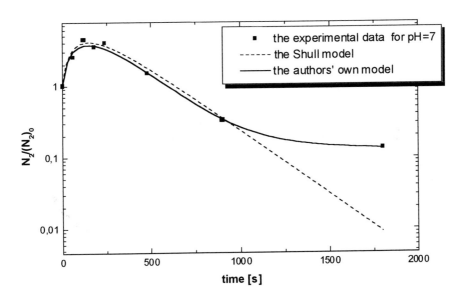

Figure 2. Survival curves plotted according to the authors' own model and the Shull model for the experimental data obtained during a thermal treatment at a temperature of 110°C.

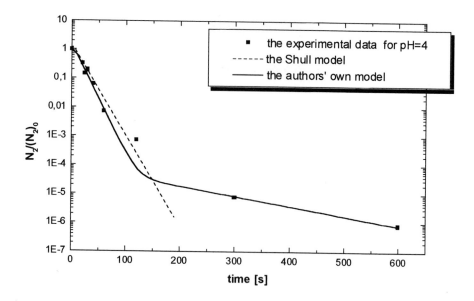

Figure 3. Survival curves plotted according to the authors' own model and the Shull model for the experimental data obtained during a thermal treatment at a temperature of 121°C.

3. CONCLUSIONS

The following can be considered the most important results obtained in this research:
- ♦ Demonstrating that in a population of spores of *Bacillus stearothermophilus* bacteria there can occur, at a very small concentration, particularly heat-resistant cells that are referred to as super dormant spores;
- ♦ Proving that mathematical models quoted in the literature of the subject are not capable of describing all the phenomena occurring during the thermal inactivation of bacterial spores;
- ♦ Developing the author's own mathematical model and demonstrating that this model allows one to predict phenomena of activation and 'tailing', which frequently occur during the process of thermal inactivation of bacterial spores.

REFERENCES

1. J. Iciek: Major issues, Ch. 5, pp. 139-164. In: "Thermal processing of bio-materials", in a series "Topics in Chemical Engineering", vol. 10, T. Kudra and Cz. Strumiłło (eds.), Gordon and Breach Sci. Publishers, Amsterdam, 1998.
2. J. Iciek: Sterilization and pasteurization of media, Ch. 6, pp. 165-192. In: „Thermal processing of bio-materials", in a series "Topics in Chemical Engineering", vol. 10, T. Kudra and Cz. Strumiłło (eds.), Gordon and Breach Sci. Publishers, Amsterdam, 1998.

3. Kim J., Naylor H.B., Appl. Microbiol., 14 (1966) 690.
4. V. Sapru, G.H. Smerage, A.A. Teixeira and J.A. Lindsay, J.Food Sci., 58 (1993) 223.
5. J.J.Shull, G.T.Cargo and R.R Ernst, Appl. Microbiol., 11 (1963) 485.
6. L.M. Prescott, J.P. Harley and D.A. Klein; E. M. Sievers (ed.), Microbiology, 3rd edition, Wm. C. Brown Publishers, London, 1996.

Food Biotechnology
S. Bielecki, J. Tramper and J. Polak (Editors)

Dielectric permittivity as a method for the real time monitoring of fungal growth during a solid substrate food fermentation of Quinoa grains

P. Kaminski [A], J Hedger [A], J. Williams [B], C. Bucke [A], I. Swadling [A]

[A] Fungal Biotechnology Group, Department of Biotechnology, School of Biosciences, University of Westminster, 115 New Cavendish Street, London, W1M 8JS, UK

[B] Aber Instruments Ltd, Science Park, Aberystwyth, LL60 6HR, UK

This paper describes a method for the direct, real time measurement of fungal biomass during a novel solid substrate food fermentation of Quinoa grains (*Chenopodium quinoa* Willd). The technique is based on the dielectric behaviour of living cells and uses a Biomass Monitor (Aber Instruments Ltd.) to measure biomass-dependent capacitance. The design of new probes dedicated for use in solid substrate systems is discussed and first results are presented.

1. INTRODUCTION

Solid substrate fermentation (SSF) has advantages over submerged processes. Cheaper and simpler, it is being increasingly used for various applications, including food production and preservation processes. Tempe is an example of a solid substrate food fermentation and is the staple diet in Indonesia, but is also becoming very popular in USA and Europe. In the fermentation the fungus *Rhizopus oligosporus* overgrows the substrate (typically soya beans) to form a highly nutritious product. We have used *R. oligosporus* to develop a novel fermentation of the S.American grain Quinoa (*Chenopodium quinoa*) at a laboratory and pilot plant level. There are two distinctive stages in the process. The Stage 1 process prepares the quinoa grain for the subsequent Stage 2 solid substrate fermentation. It consists of submerged fermentation of pasteurised Quinoa grain using Lactic Acid Bacteria (LAB) inocula, followed by repasteurisation and preparation with additives and inocula. The inoculated acidified grain is then added to fermentation containers and in Stage 2 is converted to a solid mass by growth of the fungus. The product can then be sold fresh, chill stored or processed into a variety of foods.

In earlier studies of a similar solid substrate fermentation W. Penalosa *et al* [1] showed that real time monitoring of capacitance of the fermentation medium at a frequency range of 0.3 MHz, using a Biomass Monitor (Aber Instruments Ltd., Wales) was a suitable technique for measuring biomass. They correlated dielectric spectroscopy measurement of biomass with direct measurement of hyphal length by microscopy. However the probe used was designed for cell suspensions in liquid.

The purpose of this paper is to demonstrate that measurement of biomass-dependent capacitance using a Biomass Monitor can be used to control the SSF of Quinoa. We will discuss the need for specific probes designed for SSF. Prototype sensors will be demonstrated and the test results discussed.

2. PRINCIPLES OF DIELECTRIC SPECTROSCOPY

If the ion solution containing biological cells is placed into an electric field, the field pushes along the ions both inside and outside the cells, until they come in contact with the cell membrane. As a consequence, the poles of cells become coated with the layer of oppositely charged ions (Figure 1) and the cells act as capacitors. The measure of this charge separation induced by the electric field is capacitance (expressed in picofarads). If the frequency of applied electric field is within the range of 0.1–10 MHz, the cell-membrane capacitance dominates the dielectric properties of the medium. Therefore measured capacitance changes occurring in the system reflect changes in cell concentration in the medium. Thus, if the cell concentration increases, so does the extent of the charge separation induced in the system, i.e. capacitance. As only living cells can have the charge separation induced, the changes in capacitance reflect only changes in viable biomass.

This simplified summary describes the basis for the biomass measurement method used by the Biomass Monitor. Many groups working with the Biomass Monitor reported a linear relationship between capacitance and biomass concentration measured off-line [2].

Although the biomass monitor was designed for submerged fermentation, the concept is applicable for solid substrate systems as well. Here we discuss the application of capacitance measurements in monitoring fungal SSF of Quinoa.

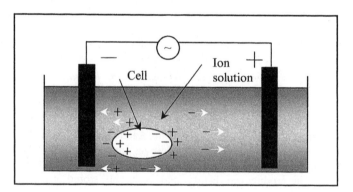

Figure 1. Representation of charge separation occurring at the poles of the cell when placed in electric field.

3. MATERIALS AND METHODS

The Biomass Monitor was used for SSF process control of both laboratory and pilot plant scale Quinoa fermentation. In both cases, prior to the Stage 2 solid state process, the grain was thoroughly washed, pasteurised in water at 80°C for 60 seconds and inoculated with freeze dried culture of LAB. Subjected to a 21 hours fermentation, the pH of the fermentation liquid drops to 3.6 within 4 hours, whereas the internal pH of the grain only reaches 4.2-4.3 at 21 hours. Grain in the fermentation liquid is then subjected to a second pasteurisation at 80°C for 5 minutes. The grain is then processed into an optimal consistency, whereby grains remain separate, yet still moist on the surface, and have a water content of 60%. The prepared substrate is then inoculated with freeze dried inoculum of *Rhizopus oligosporus* and packed into the fermentation containers for a 24 hour fermentation at 35°C. A qualitative scale of 1 (Poor)–10 (Excellent) was used to judge the final product, based partly on the density of the mycelial growth.

In the lab scale operation petri plates of various diameters were used. Optimal conditions for the process were determined and applied at pilot scale for various types of fermentation containers. Finally, adapted sausage casings were chosen for the final scaled-up fermentation system.

For both types of containers i.e. petri plates and sausage casings the dielectric properties of the fermentation medium were monitored, using a standard, four pin sensor (Sensor E, Figure 2) and sensors specially designed for SSF (A-D, Figure 2)

A   B   C   D   E

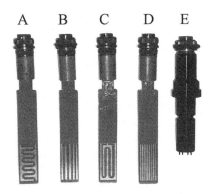

Figure 2. Four probes (A–D) designed and built by Aber Instruments Ltd. for monitoring fungal growth in a solid substrate compared with the standard probe (E), which is used for submerged fermentation.

During the fermentation the product was sampled and assessed off-line for hyphal length using the membrane filter technique [3].

Individual samples (2 g fresh weight) were added to a solution of 19 ml of distilled water (filtered through Whatman No 1 filter paper) and 1 ml 1% Tween 20 solution, acting as a dispersant. The solution was then homogenised in an Omni Mixer Homogeniser at speed 2 for 15 seconds. The homogenate was then diluted to 100 ml in filtered distilled water. Further dilutions were prepared in distilled water of which 10 ml was transferred to the filter holder (Fisher Scientific, 47 mm diameter) containing a membrane filter, 1.2 μm pore size, white, plain (Millipore RAWP04700). Twenty ml of filtered water and two drops of Methylene Blue were added and after 30 sec the stained suspension was filtered using a filter pump. After filtration the filter was blotted dry and transferred to a microscope slide, followed by a drop of immersion oil and a No. 1 cover slip. Hyphal lengths were estimated using an Image Analysis System (Leica Ltd) attached to the microscope using 125× magnification. Total hyphal lengths in 20 randomly selected fields of view was determined for each slide. Three replicates for each sample were prepared and analysed in this way. The measured hyphal length was averaged and used to calculate fungal biomass, expressed as m. Hyphae/g. Dry Wt. of product.

4. RESULTS

For both fermentation containers; petri plates and modified sausage casings, the capacitance, conductance and temperature profiles recorded using standard probe showed the same pattern (Figure 3). The data obtained showed an initial one hour adaptation period followed by 6 to 7 hours without a change in capacitance, suggesting that no mycelial growth occurred during this period. In fact, studies of fungal spores adhering to slides incorporated in the substrate, and removed at 30 minute intervals for microscopy, showed that they germinated after 3 hours of incubation and produced very dense mycelium within 4-5 hours. After 11-12 hours of incubation the capacitance data indicated very extensive microbial growth, plateauing after 22 to 23 hours of incubation. This data was calibrated against the product quality assessment. The product showed improving internal and external mycelial density from 16 hours onwards. After 22 hours the quality of the product was excellent (Scale 10) and showed no further improvements in mycelial density after additional incubation.

The data on capacitance correlates very well with the temperature profile of the fermentation (Figure 3). Typically for SSF, where no or very limited mixing is applied, heat build-up occurs due to metabolic activity inside the medium. In the case of Quinoa fermentation, the heat build–up is recorded at the same time as the extensive growth, reflected by changes in capacitance. Also, when the capacitance profile tends to plateau, at the same time the heat build-up ceases and temperature starts to drop.

A very similar profile was recorded for conductance (Figure 3), which could reflect the changes in ionic composition of the medium as a function of microbial growth.

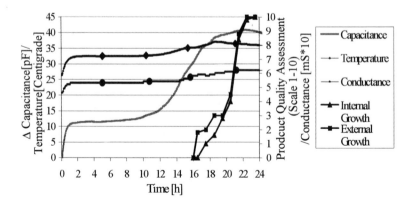

Figure 3. Fungal growth, capacitance, conductance and temperature profiles of the quinoa solid substrate fermentation in sausage casings.

The data on hyphal length of the product sampled during the growth phase, assessed off-line using filter membrane technique, shows very similar pattern to the profile of changes in capacitance (Figure 4). Until 21 hours we observed good correlation between the changes in hyphal length and capacitance. After 21 hours, although the hyphal length was still increasing, we observed drop in capacitance.

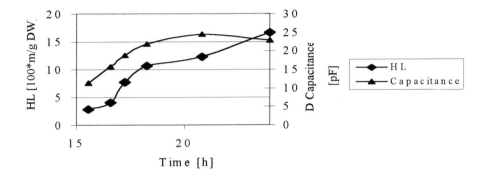

Figure 4 Hyphal length/capacitance correlation for standard probe.

All experimental probes were tested using an identical setup and the standardised protocol for substrate preparation. Data obtained indicates that the probe providing the most reproducible results is the sensor C (Figure 2). For all fermentations monitored the profile of changes in capacitance was remarkably similar (Figure 5).

Unlike the capacitance profiles obtained using a standard probe, following the initial one hour-long adaptation period, we observed a continuos increase in capacitance. These changes could reflect the initial activity of the fungus, not detected by the standard probe. After that initial, 10-11 hours long period, the changes in capacitance indicated a phase of intensive mycelial growth. However, after 16-18 hours of incubation the capacitance dropped.

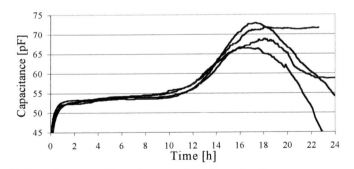

Figure 5. Capacitance profiles using experimental probe C.

When mycelial length was measured off-line during this period biomass appeared to be still increasing as capacitance fell (Figure 6).

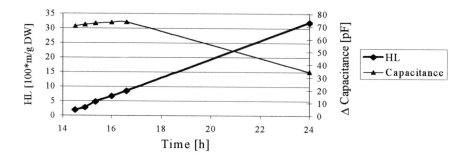

Figure 6. Hyphal length/capacitance correlation for experimental probe C.

5. DISCUSSION

The results presented clearly show that capacitance measurements can give useful, qualitative information on microbial growth in SSF. The use of a standard Biomass Monitor Probe designed for submerged fermentation proved useful for preliminary studies. However the sensitivity seemed to be limited, given there was no capacitance response in the first 6-7 hours of the fermentation, although direct observation by microscopy showed mycelial growth occurred in this period. The small size of the electrode pins in the standard probe also made them highly sensitive to air voids created during the substrate settling. All this lead to the design of improved, more sensitive probes, dedicated for use in solid state systems.

The preliminary tests performed on the new probes proved that the new design ensures not only higher reproducibility of the capacitance profiles, but also provides the information on the initial stage of fermentation, although clearly more development work is needed. The information on the later stages of the fermentation is also interesting. The apparent continued increase in mycelial biomass as measured directly seems to conflict with the decline in capacitance. A possible explanation may be that although growth is occurring during this period, an increasing proportion of the older hyphae die and would therefore not register capacitance, showing the technique is a true measure of biomass rather then biomass + necromass, which are not distinguished by the off-line technique of hyphal extraction.

REFERENCES

1. W. Penaloza, C.L. Davey, J.N. Hedger and D.B. Kell, World Journal of Microbiology and Biotechnology, 7 (1990) 248.
2. G.D. Austin, R.W.J. Watson and T. D'Amore, Biotechnol. Bioeng., 43 (1994) 337.
3. J.F. Hanssen, T.F. Thingstad and J. Goksoyr, Oikos, 25 (1974) 102.

Food Biotechnology
S. Bielecki, J. Tramper and J. Polak (Editors)
© 2000 Elsevier Science B.V. All rights reserved.

Method of *Lactobacillus acidophilus* viable cell enumeration in the presence of thermophilic lactic acid bacteria and bifidobacteria

M. Bielecka, E. Biedrzycka, A. Majkowska, El. Biedrzycka

Department of Food Microbiology, Division of Food Science,
Institute of Animal Reproduction and Food Research of the Polish Academy of Sciences,
ul. Tuwima 10, 10-747 Olsztyn, Poland

Six media were evaluated taking into consideration their selectivity for enumeration of *Lactobacillus acidophilus* in the presence of traditional yoghurt cultures (*Lactobacillus delbrueckii* subsp. *bulgaricus, Streptococcus thermophilus*) and bifidobacteria strains commercially used as well as freshly isolated from human and animal gut. The minimal nutrient agars BCP and MNA were used as base media. The BCP medium was supplemented with penicillin and sodium pyruvate as selective agents or with 0.25-1.0% salicin instead of 0.1% glucose. The MNA medium contained 1% salicin. Nutritive media: MRS, M17 or the Garche's were used as controls. The BCP agar with glucose replaced with 0.5 or 1.0% salicin was chosen as the most suitable for *L. acidophilus* selective enumeration in bio-yoghurts due to high recovery of *L. acidophilus* strains (89-100%), along with good visibility of their colonies surrounded by yellow aureole, and enough prevention from forming colonies by concomitant bacteria.

1. INTRODUCTION

L. acidophilus is a natural inhabitant of the human gastrointestinal tract and traditional fermented food and it takes part in maintaining healthy microecology. The results of several investigations show a potential benefit of consuming milk fermented with *L. acidophilus* or supplemented with these bacteria [1,2,3]. In order to ensure beneficial effects, the recommended minimum level of probiotic bacteria in yoghurts is 10^6-10^7 viable cells per gram or millilitre of the product [4,5,6]. Therefore, several scientists undertook elaboration of a selective medium for enumeration of *L. acidophilus* in the presence of concomitant microflora, particularly the thermophilic lactic acid bacteria and bifidobacteria. Gilliland and Speck [7], and Collins [8] used media containing bile salts to enumerate *L. acidophilus* in yoghurt or sweet acidophilic milk. However, the use of bile salts as a selective inhibitory agent was found to reduce the recovery of viable *L. acidophilus* cells. Lankaputhra and Shah [9] and Lankaputhra et al. [10] found that bile salts allow growing of several *Bifidobacterium* spp. used mainly in AB yoghurts, thus limiting the selectivity of that medium. Hull and Roberts [11] modified MRS agar replacing glucose with maltose for selective enumeration of *L. acidophilus* in yoghurt. This medium however, was not sufficiently selective to *Bifidobacterium* spp., which also formed colonies in it. Hunger [12] used esculin cellobiose agar for *L. acidophilus* counting and noted that *L. acidophilus*, *L. casei* subsp. *rhamnosus* and *L. helveticus* could not be distinctly differentiated. Subsequently, Kneifel and Pacher [13] used X-glu agar (Rogosa medium supplemented with 5-bromo-4-chloro-3-indolyl-β-D-glucopyranoside). The authors concluded that the X-glu agar was more selective than other commonly used media (MRS, TGV, Rogosa agar). The two latter methods based on chromogenicity of esculin or X-glu are expensive and their application for routine analyses could be limited. Lankaputhra and Shah [14] found a comparable number, shape and size of *L. acidophilus* colonies formed on MRS and minimal nutrient agar (MNA) containing salicin. The MNA-salicin medium ensured not only high recovery rate of *L. acidophilus* but was also selective enough to *S. thermophilus*, *L. delbrueckii* subsp. *bulgaricus* and some species of *Bifidobacterium*.

Taking advantage of the experience gained by other authors, we have attempted to find selective factors inhibiting the growth of bacteria closely related to *L. acidophilus*, as well as to exploit differences in fermentation abilities of these bacteria. The aim of the undertaken study was to develop a simple method for selective enumeration of *L. acidophilus* in yoghurt containing *S. thermophilus*, *L. delbrueckii* subsp. *bulgaricus* and *Bifidobacterium* spp.

2. MATERIAL AND METHODS

2.1. Bacteria strains

Industrial strains of *L. acidophilus*, *L. delbrueckii* subsp. *bulgaricus* and *S. thermophilus*, six strains of each species, and nine strains of *Bifidobacterium* spp. were used in this study. Additionally, seven fermenting salicin bifidobacteria strains belonging to the species: *B. angulatum*, *B. adolescentis*, *B. pseudocatenulatum* and *B. animalis* were included. The above mentioned strains were freshly isolated from human and animal gastrointestinal tract.

2.2. Selective and nutrient media

Bromocresol purple agar BCP [15] with minimised nutritive composition (yeast extract 2.5 g, peptone 5.0 g, glucose 1 g, Tween 80 1.0 g, cysteine hydrochloride 0.3 g, bromocresol purple 0.04 g, pyruvic acid 1.0 ml, dipotassium hydrogenortophosphate 3.0 g, magnesium sulphate heptahydrate 0.575 g, ferrum sulfate heptahydrate 0.034 g, manganese sulphate tetrahydrate 0.12 g, agar 15 g, distilled water to 1000ml, pH 6.8-7.2) was used as a base medium. The medium was supplemented with: sodium pyruvate – 5 g, and penicillin sodium salt G (Sigma) 30 or 40 U/L (GP30P or GP40P). The BCP medium with 0.25, 0.5 or 1.0% salicin instead of 0.1% glucose as the only source of carbon and energy was also examined. A minimal nutrient agar (MNA) (composition: tryptone 20 g, yeast extract 2,0 g, Tween 80 1.0 g, dipotassium hydrogenortophosphate 2.0 g, sodium acetate trihydrate 5.0 g, trisodium citrate 2.0 g, magnesium sulphate heptahydrate 0.2 g, manganese sulphate tetrahydrate 0.05 g, agar 11.0 g, distilled water to 1000 ml, pH 6.6-6.8) with 1% salicin [14] was tested comparatively. Nutrient media M17 [16] for *S. thermophilus*, MRS [16] for *L. delbrueckii* subsp. *bulgaricus* and *L. acidophilus* and Garche's [15] for bifidobacteria were used as controls.

2.3. Enumeration of bacteria on selective and nutrient media

All strains were multiplied in sterile reconstituted 10% non fat dry milk with a 0.3% yeast extract (*S. thermophilus* strains - in milk unsupplemented) at 37°C until a stationary growth phase occurred. The cultures multiplied in milk were diluted in 0.1% sterile peptone water. Enumeration was carried out using the pour plate technique. Duplicate plates were incubated anaerobically (Gas Pak Anaerobic System CO_2+H_2, Oxoid) at 37°C for 72 h. The colonies were counted on selective and nutrient media and the numbers were expressed as the colony forming units (cfu) per ml. The recovery rate of bacteria strains was expressed as the per cent of their number on control media.

3. RESULTS AND DISCUSSION

The obtained results were evaluated according to the recovery rate of *L. acidophilus* as well as selectivity to *S. thermophilus*, *L. delbrueckii* subsp. *bulgaricus* and bifidobacteria. The results presented in Table 1 show a high recovery rate of each *L. acidophilus* strain in all tested media. The recovery rate ranged from 81% to 100% on MNA with 1% of salicin, from 95% to 100% on the BCP medium with selective supplements (GP30P or GP40P), and from 89% to 100% on the BCP medium with salicin. The salicin concentration ranging from 0.25% to 1.0% did not significantly affect the colony numbers or their size and shape. The presence of bromocresol purple in the BCP medium made counting of acidifying bacteria colonies easier than in MNA agar developed by Lankaputhra & Shah [14] because the colonies were surrounded by a yellow aureole. The yellow colour was less visible on BCP medium containing low concentration (0.1%) of glucose not enough to produce acid changing the indicator colour. Dave & Shah evaluating 12 different media for enumeration of *L. acidophilus*, found the best recovery on MRS with 1% salicin or sorbitol. The medium was however not selective to bifidobacteria. Therefore, the authors offered it for the estimation of total *L. acidophilus* and bifidobacteria count or for the *L. acidophilus* enumeration provided that the product did not contain bifidobacteria [13]. Lankaputhra & Shah [14] comparing MRS and MNA containing salicin, found colonies of *L. acidophilus* similar in shape and size on both media. However, *Bifidobacterium*, *S. thermophilus* and *L. delbrueckii* subsp. *bulgaricus* formed colonies on MRS, but not on MNA medium with salicin.

The tested MNA and BCP media with salicin inhibited colony forming by commercial strains of bifidobacteria as well as those freshly isolated from gastrointestinal tract (Table 2). However, BCP agar supplemented with GP30P or GP40P was insufficiently selective to the majority of bifidobacteria strains which formed small or pinpoint (pp) colonies on this medium. The strains which were able to ferment salicin formed small or pp colonies also on MNA medium with 1% salicin. Sufficient suppression of the growth of bifidobacteria strains was found on BCP medium with salicin in concentration ranged from 0.25 to 1.0%, except of one strain *B. pseudocatenulatum* KD15 which formed pp or small colonies on this medium which were visible only with loop. These colonies were not surrounded with yellow aureole and could be easily differentiated from *L. acidophilus*. *B. pseudocatenulatum* strain formed small colonies on all tested media. This strain was isolated from human gut and was able to ferment salicin. All tested media with salicin were inhibitory to *S. thermophilus* and *L. delbrueckii* subsp. *bulgaricus* (Table 3 & 4), whereas BCP agar media supplemented with penicillin were rejected as insufficiently inhibitory.

Table 1

Recovery of *Lactobacillus acidophilus* on the selective media expressed as cfu number x10[7] and as %

L. acidophilus strain	MRS (control)	MNA salicin 1%		BCP					
				GP30P		GP40P		salicin 1%	salicin 0.5%	salicin 0.25%
8/4	160[1]	152	95[2]	160	100	157	98	145 91	146 91	142 89
V	76	62	81	76	100	75	99	76 100	76 100	73 96
H	52	52	100	50	96	49	94	52 100	51 98	49 94
A	63	63	100	62	98	62	99	63 100	63 100	63 100
B	105	104	100	106	101	105	100	102 97	101 96	100 95
K$_2$	200	196	98	199	100	195	98	201 100	200 100	194 97

[1] cfu number x10[7];
[2] recovery rate of bacteria on the selective medium in comparison to the number of bacteria counted on control MRS medium (100%) expressed as %.

Table 2

Selectivity of MNA and BCP modified media to bifidobacteria

Strain	MEDIA						
	G control	MNA salicin 1%	GP30P	GP40P	BCP salicin 1%	salicin 0.5%	salicin 0.25%
B. angulatum KD1[1][3]	110[4]	pp	nd	nd	nd	nd	nd
B. adolescentis KD2[1][3]	270	pp	pp	pp	nd	nd	nd
B. animalis KD10[1][3]	170	nd	pp	pp	nd	nd	nd
B. pseudocatenuatum. KD15[1][3]	130	100s	15s	15s	102s	pp	pp
B. animalis J38[1][2]	230	nd	230s	210s	nd	nd	nd
B. animalis KS1b2[1][3]	280	0.07s	pp	pp	nd	nd	nd
B. animalis 30[1][2]	130	0.013	pp	pp	nd	nd	nd
B. species 61[2]	120	nd	22s	22s	nd	nd	nd
B. species 62	130	nd	nd	nd	nd	nd	nd
B. species 64	340	nd	120s	120s	nd	nd	nd
B. species 65	130	nd	8s	pp	nd	nd	nd
B. species 66	79	nd	20s	17s	nd	nd	nd
B. species 67	120	nd	15s	15s	nd	nd	nd
B. species 68	90	nd	30s	30s	nd	nd	nd
B. species 69	110	nd	110s	110s	nd	nd	nd
B. species 70	350	nd	260s	260s	nd	nd	nd

G - Garche's medium; pp – pinpoint colonies; s - small colonies; nd - not detected in the dilutions: 10^{-4} - 10^{-7};
[1] strains fermenting salicin; [2] *B. sp.* isolated from commercial bio-yoghurts;
[3] bifidobacteria isolated from the gastrointestinal tract; [4] cfu x 10^7.

Table 3
Selectivity of MNA and BCP modified media to *Streptococcus thermophilus*

Strain	M17 (control)	MNA salicin 1%	BCP				
			GP30P	GP40P	salicin 1%	salicin 0.5%	salicin 0.25%
D	47[4]	nd	nd	nd	nd	nd	nd
2K	67	nd	nd	nd	nd	nd	nd
MK-10	89	nd	nd	nd	nd	nd	nd
E	60	nd	0.06s	nd	nd	nd	nd
B	50	nd	nd	nd	nd	nd	nd
BS	62	nd	3s	0.13s	nd	nd	nd

Abbreviations like in Table 2.

Table 4
Selectivity of MNA and BCP modified media to *Lactobacillus delbrueckii* subsp. *bulgaricus*

Strain	MRS (control)	MNA salicin 1%	BCP				
			GP30P	GP40P	salicin 1%	salicin 0.5%	salicin 0.25%
DL	83	nd	0.68[4]	0.6	nd	nd	nd
E	81	nd	0.1	0.05	nd	nd	nd
151	86	nd	nd	nd	nd	nd	nd
Db$_3$	67	nd	5s	3s	nd	nd	nd
b$_9$	45	nd	45s	42s	nd	nd	nd
9lb$_1$	40	nd	nd	nd	nd	nd	nd

Abbreviations like in Table 2.

4. CONCLUSIONS

BCP agar containing salicin in the concentrations of 0.5% and 1.0% as the only source of carbon and energy was useful for the selective enumeration of *L. acidophilus* from yoghurts containing *L. delbrueckii* subsp. *bulgaricus*, *S. thermophilus* and *Bifidobacterium* spp. Although some bifidobacteria strains were able to ferment salicin, minimal nutritional composition of BCP medium unabled them to form well visible colonies. Additionally, the formation of fast fermenting salicin *L. acidophilus* surrounded with yellow aureole can be easily differentiated from others.

ACKNOWLEDGEMENT
The authors thank Mrs. Maria Śmieszek for technical assistance.

REFERENCES

1. M.F. Bernet, D. Brassart, J.R. Nesser, A.L. Servin, Gut, 35 (1994) 483.

2. A. Lidbeck, E. Overvick, J. Rafter, C.E. Nord, J.A. Gustafsson, Microbial Ecol. Health Dis., 5 (1992) 59.
3. B.K. Mital, S.K. Garg, Crit. Rev. Microbiol., 21 (1995) 175.
4. J.A. Kurmann, J.L. Rasic, *In* Therapeutic Properties of Fermented Milks (ed. R.K. Robinson), Elsevier Applied Food Sciences Series, London (1991) 117.
5. N. Ishibashi, S. Shimamura, Food Technol., 47 (1993) 126.
6. R.K. Robinson, South African J. Dairy Sci., 19/1 (1987) 25.
7. S.E. Gilliland, M.L. Speck, J. Dairy Sci., 60 (1977) 1394.
8. E.B. Collins, J. Food Prot., 41 (1978) 439.
9. W.E.V. Lankaputhra, N.P. Shah, Cult. Dairy Prod. J., 30 (1995) 2.
10. W.E.V. Lankaputhra, N.P. Shah, M.L. Britz, Milchwissenschaft, 51/2 (1996) 65.
11. R.R. Hull, A.V. Roberts, Austr. J. Dairy Technol., 34/1 (1984) 160.
12. W. von Hunger, Milchwissenschaft, 41/5 (1986) 283.
13. W. Kneifel, B. Pacher, Int. Dairy J., 3/3 (1993) 277.
14. W.E.V. Lankaputhra, N.P. Shah, Milchwissenschaft, 51/8 (1996) 446.
15. Bulletin IDF, 252 (1990) 24.
16. FIL/IDF Intern. Standard (provisional), 146 (1991) 1.
17. R.I. Dave, N.P. Shah, J. Dairy Sci., 79 (1996) 1529.

LEGAL AND SOCIAL ASPECTS OF FOOD BIOTECHNOLOGY

Food Biotechnology
S. Bielecki, J. Tramper and J. Polak (Editors)
© 2000 Elsevier Science B.V. All rights reserved.

Some aspects of plant and food biotechnology

Stefan Malepszy

Warsaw Agricultural University, Nowoursynowska 166, 02-787 Warszawa, Poland

1. INTRODUCTION

Biotechnology may provide very different situations in the field of plant and food production (Figure 1). Some of them are noncontroversial, others, however, are hard to accept or even are not approved of, at least in some societies. The greatest doubts concern genetically modified organisms (GMOs), that is transgenic plants and products obtained from them. There are fewer objections concerning products obtained by means of genetically modified microoorganisms. Therefore, only in few countries, mainly in the USA and Canada, the producers may cultivate transgenic plants and consumers find them in shops. The European Union countries treat GMOs with various degrees of acceptance even though none of the member countries has accepted the introduction of transgenic varieties into cultivation.

Food Biotechnology
- use of plant material from transgenic varieties**
  - fresh**
  - processing*
- use of transgenic strain of microorganisms for food processing
- both – transgenic plant material and transgenic microorganisms for processing**
- bioreactor produced substances from
  - transgenic*
  - nontransgenic

Plant Biotechnology
- artificial seeds (clonal propagation)
- transgenic varieties**
- protoplast wide hybridization derived varieties*

Figure 1. Acceptation of biotechnology in various cases of food and plant production
** - strong controversy; * - low controversy, star-free – no controversy.

The varying acceptability of transgenic plants depends on a number of factors and sometimes on several factors acting simultaneously. These are: the fight for markets, cultural factors, the general level of contemporary biological and technological knowledge in a society, the scientific and economic potential, the limited confidence in science and in economic cartels, a strong increase of concern about the state of the natural environment and different risk perception.

What are GMOs, and what distinguishes them from the organisms used so far? There is one fundamental difference, i.e. the way they are made. GMOs are a result of introduction of additional genetic information (called a transgene) into a given organism by means of methods described as genetic engineering. These methods are a natural development of the methods hitherto used to obtain new varieties, strains and races and are free from their limitations, i.e. barriers to crossing and the long time required to obtain practical effects. GMOs make it possible to obtain better production results without the necessity to use special, additional treatments (e.g. spraying of plants with chemical substances in order to protect from disease and pests, better nutrition etc.). These treatments are considered to be the main cause of the ecological harmfulness of plant production. Therefore GMOs are an expression of the aim to obtain production based on the realization of natural predispositions due to the composition of genetic material and do not require burdening the environment with the effects of agrotechnical procedures, in particular the use of chemicals.

The aim of the current presentation is to show the state of the introduction of transgenic plants into cultivation and to present a short discussion of the main problems.

2. TRANSGENIC PLANTS - CURRENT SITUATION

Transgenic varieties have been introduced into cultivation in 16 species, in some species several kinds (Table 1). They represent 15 new properties in which herbicide and insect resistance predominate.

Only in two cases properties are directly linked to the presented food quality. It is worth stressing that in two species varieties with two improved properties have been introduced. The list of species in which field experiments have been performed is much longer and encompasses at least 60 (Table 2).

The number of these experiments during the last 10 years has been evaluated to be 25 000 and they have been performed in 25 countries. The participation of particular categories of new characters is as follows: herbicide resistance 29.2%, insect resistance 23.8%, product quality 21.4%, virus resistance 9.5%, fungal resistance 9.5% and other 11%.

3. WHO USES THE NEW TRAITS IN TRANSGENIC VARIETIES

The main task of contemporary agriculture is the production of highly valuable food while maintaining the principles of environmental protection. Therefore the benefits from this type of production can be divided into two groups. The first is addressed at specific recipients and the second is rather general. The latter includes benefits derived from environmental protection, which are also difficult to calculate precisely. In Table 3 the benefits hitherto derived from the GMO culture are shown which can be ascribed to five recipients, above all the owner of the variety and the producer. There is no difference here in comparison with non-transgenic varieties. However, in the case of transgenic varieties there is an important difference - the environment is a frequent beneficiary.

Table 1
Plant species with novel traits in transgenic varieties approved for commercial release by at least one regulatory agency (OECD Product Database)

Beta vulgaris	herbicide [a,b,c] resistance
Brassica napus	herbicide [a,b,c] resistance, male sterility and herbicide [b] tolerance, modified fatty acid profile
Brassica rapa	herbicide [a] resistance
Carica papaya	virus resistance
Cichorium intybus	male sterility and herbicide [b] resistance
Cucurbita pepo	virus resistance
Dianthus caryophyllus	flower colour, extended vase life
Glycine max	herbicide [a,b] resistance
Gossypium hirsutum	lepidopteran insect resistance, herbicide[a,c,d] resistance; lepidopteran insect resistance and herbicide [c] resistance
Linum usitatissimum	novel herbicide [d] soil residue resistance
Lycopersicon esculentum	modified ripening, lepidopteran insect resistance
Nicotiana tabaccum	herbicide [b,c] resistance
Solanum tuberosum	coleopteran insect resistance, virus resistance, coleopteran insect resistance and virus resistance
Zea mays	herbicide [a,b,c,d] resistance, lepidopteran insect resistance; lepidopteran insect resistance and herbicide [a,b] resistance male sterility and herbicide [b] tolerance

[a] Glyphosate herbicide resistance
[b] Phosphinothricin (glufosinate ammonium) herbicide resistance
[c] Oxynil herbicide resistance
[d] Sulfonyl urea herbicide resistance

In order to prepare of the Table 3, the benefits to the environment have been evaluated very cautiously taking into consideration the reservations of some social groups, and therefore they are underestimated. The discussion about the dangers of transgenic varieties led by associations acting for plant protection does not consider the commonly known general conditions such as that: (1) transgenic plants do not change the hybridization barriers that normally exist; (2) cultivated plants do not survive under conditions other than cultivated fields; (3) genetic information introduced into transgenic plants is derived from organisms living on earth (it is not new for our planet); (4) the risk of failure to survive is included into each existence; (5) selection and breeding of plants have accompanied man's development since the beginning.

Table 2
Plant species by which transgenic lines/varieties has been experimentally tested in confined field trials

Actinidia deliciosa	Kiwi fruit	*Ipomea batatas*	Sweet potato
Agrostis stolonifera	Creeping bentgrass	*Juglans sp.*	Walnut
Allum cepa	Onion	*Lactuca sativa*	Lettuce
Arabidopsis thaliana	Thale Cress	*Linum usitatissimum*	Flax
Arachis hypogea	Peanut	*Liquidambar sp.*	Sweetgum
Asparagus officinalis	Asparagus	*Lupinus angustifolius*	Lupin
Atropa belladona	Belladonna	*Lycopersicon esculentum*	Tomato
Beta vulgaris	Sugar beet	*Malus domestica*	Apple
Betula pendula	Silver Brich	*Medicago sativa*	Alfalfa
Brassica carinata	Ethiopian mustard	*Nicotiana benthamiana*	Tobacco
Brassica junacea	Mustard	*Nicotiana tabaccum*	Tobacco
Brassica napus	Canola/Oilseed rape	*Oryza sativa*	Rice
Brassica oleraceae	Cabbage /Broccoli	*Pelargonium sp.*	Pelargonium
Brassica rapa	etc.	*Picea abies*	Norway spruce
Capsicum annuum	Canola/Oilseed rape	*Picea sp.*	Spruce
Carica papaya	Pepper	*Pinus sylvestris*	Scots Pine
Castanea sp.	Papaya	*Pisum sativum*	Pea
Cichorium intybus	Chestnut	*Populus sp.*	Poplar
Citrullus lanatus	Chicory	*Prunus domestica*	Plum
Cucumis melo	Watermellon	*Rosa sp.*	Rose
Cucumis sativus	Melon	*Saccharum officinarum*	Sugar cane
Cucurbita pepo	Cucumber	*Sinapsis alba*	White mustard
Daucus carota	Squash	*Solanum melongera*	Eggplant
Dianthus	Carrot	*Solanum tuberosum*	Potato
caryophyllatus	Carnation	*Tagetes sp.*	Marigold
Eucalyptus sp.	Eucalyptus	*Trifolium subterraneum*	Clover
Fragaria sp.	Strawberry	*Triticum aestivum*	Wheat
Gladiolus sp.	Gladiolus	*Vaccinium oxycoccus*	Cranberry
Glycine max	Soybean	*Vitis vinifera*	Grape
Gossypium hirsutum	Cotton	*Zea mays*	Corn/Maize
Helianthus annuus	Sunflower		
Hordeum vulgare	Barley		

One of the most frequent ecological objections concerning transgenic varieties states that new characters present in the transgene may be transferred to other groups of organisms and that the plants themselves might spread in an uncontrolled way. The latter is not realistic, as cultivated plants do not have the chance to survive outside the conditions created in a cultivated field. If a field is left without human intervention, the cultivated plants will disappear within a year to a few years depending on the species. As uncontrollable recipients of the transgene, weeds or other groups of plants forming the surroundings of cultivated fields, microorganisms in the human digestive tract and animals feeding on transgenic plants are mentioned.

Table 3
Direct beneficiars of new characters present in till now released transgenic varieties

New character	Beneficiars					
	Producer	Consumer	Trader	Variety Owner	Environment	Other
Herbicide resistance	+	0	0	+	(+)	+
Insect resistance	+	0	0	+	(+)	0
Insect resistance	+	0	0	+	(+)	0
Male sterility	+	0	0	+	0	0
Flower extended vase life	0	+	(+)	+	+	0
Flower colour	+	+	+	+	0	0
Modified ripening	+	(+)	(+)	+	+	0

+ - clear benefit, (+) – not clear benefit, 0 – absence of benefit

The transfer of the transgene to digestive tract micoorganisms does not create a new danger as (a) an appropriate gene has already existed in the environment; (b) the property of resistance to the antibiotic kanamycin which has been introduced into most varieties obtained so far will only become manifest if two phenomena take place - the bacteria integrate this gene and this anitbiotic is used for treatment (in fact its use has become rarer). In presently created transgenic plants markers causing resistance to chemical substances are no longer used and therefore many of these objections will no longer be valid.

4. THE FUTURE OF TRANSGENIC PLANTS

Futurology is a very inexact science, however I would like to skill some near tendencies. On the basis of Table 3 it can be foreseen that in the coming years transgenic varieties will occur in many other species and that they will represent many new properties and not only the resistance to disease, pests and herbicides. The number of varieties having two or more improved properties will continue to increase. It may also be assumed that the number of countries in which cultivation of transgenic varieties will be allowed will increase considerably. New genes will be captured for particular categories of varieties. For example in the case of insect resistance the following novel insecticidal genes will be used:

1. cholesterol oxidase = CO (*Streptomyces* culture filltrate, lysis of midgut epithelial cells,
   - highly toxic to boll weevil (*Anthomus grandi*) [1],
   - also activity against – southern corn rostworm (*Diadroytica undecimpuncata*), tobacco budworm (*Heliothis virescens*), yellow mealworm (*Tenebrio molitor*) [2],
2. vegetative insecticidal proteins = Vips (not sporulating Bt endotoxins), midgut cell lysis
   - Vip 3A, toxic to black cutworm (*Agrotis ipsilon*), fall armyworm (*Spodoptera frugiptera*), beet armyworm (*S. exigua*), tobaco budworm (*H. virescens*), corn earworm (*Helicoverpa zea*) [3],
3. toxins from bacterium *Photorhabdus luminescens*, toxic for several orders of insects, damaged midgut epithelial cells [4],

4. chitinases – digest chitin, constituent of insect exoskeleton and gut linings. Broad range of insects. Seems to be synergistic with Bt Cry toxins [5].

Most probably the cumulation of more various resistant genes and combination of quite different transgenic characters will happen in one variety.

Our knowledge about nutrition appears to be one-sided. We know a great deal about the chemical composition of cultivated plants and the nutritional action of individual components. We know much less about the relations between the chemical composition of food and the genetic differentiation of the human population. Bacterial genes from microorganisms on which eukaryotic cells fed at the beginning of their evolution are known to occur in the human genome [6]. We also know that particular human blood groups appeared in periods of drastic changes in the mode of nutrition and today they decide about individual dietetic predisposition [7]. Some experimental data show that foreign DNA ingested by mice is not completely degraded in their gastrointestinal tracts [8]. These data on the one hand suggest that our knowledge should be broadened, and on the other hand this should make us to analyze individual cases of introducing new characters into plants very carefully. Certainly the traits, which are represented in contemporary transgenic varieties cannot radically change the nutritive value of particular cultivated plants and thus start new evolutionary changes. We should, however, try to explain what large changes in properties would have to take place and what specific properties of food could cause different food reactions in people and harm or favor the creation of new traits. However, the question arises whether the velocity of changes of the properties of transgenic varieties will not be larger than the minimal time required for the fixation of appropriate genetic combinations. The explanation of these possibilities would allow an appropriate evaluation of the effects of several new traits in GMOs, and simultaneously could be an indication as to what nutritive properties of cultured plants should be improved and what the direction of future activtities should be.

REFERENCES

1. D. Bowen, T.A. Rocheleau, M. Blackburn, O. Andreev, E. Golubeva, R. Bhartia, R.H. Fferench-Constant, Science, 280 (1998) 2129.
2. P.J.D. D'Adamo, C. Whitney, 4 Blood Types, 4 Dicts. Cop. By Hoop.A.Joop, LLV (1997).
3. W. Doerfer, R. Schubbert, H. Heller, C. Kammer, K. Hilger-Ewersheim, M. Knoblauch, R. Remus, Trends in Biotechnol., 15 (1997) 297.
4. W.F. Doolittle, Trends in Genetics, 14 (1998) 307.
5. J.J. Estruch, G.W. Warren, M.A. Mullins, G.J. Nye, J.A. Craig, M.G. Koziel, Proc. Natl. Acad. Sci. USA, 93 (1996) 5389.
6. K.J. Kramer, S. Muthukrishnan, Insect Biochemistry and Molecular Biology 27 (1997) 887.
7. J.P. Purcell, J.T. Greenplate, M.G. Jennings, J.S. Ryerse, J.C. perhing, S.R. Sims, M.J. Prinsen, Biochem. Biophys. Res. Commun., 196 (1993) 1406.
8. Z. Shen, D.R. Corbin, J.T. Greenplate, R.J. Grebenok, D.W. Galbraith, J.P. Purcell, Archives of Insect Biochemistry and Physiology, 34 (1997) 429.

Food Biotechnology
S. Bielecki, J. Tramper and J. Polak (Editors)
© 2000 Elsevier Science B.V. All rights reserved.

413

Animal biotechnology – methods, practical application and potential risks

Z. Smorąg, J. Jura

National Research Institute of Animal Production
Department of Animal Reproduction, 32-083 Balice/Kraków, Poland.

Animal biotechnology is primarily based on reproductive methods, with advances in gene technology being increasingly used. It is also concerned with methods of using the effects and possibilities of molecular biology for stimulating growth, lactation and DNA recombination of rumen microorganisms.

This paper presents the most important methods in animal biotechnology, their practical applications and potential dangers. As already mentioned, reproductive methods play a very important role in animal biotechnology. They make greater use of both male and female reproductive material possible, while enabling manipulation of gametes and embryos that also includes genetic manipulation leading to the creation of new genotypes. The creation of new genotypes is highly correlated with production of new food related products.

1. ARTIFICIAL INSEMINATION AND SPERM SEXING

Artificial insemination, which is most widely used in cattle breeding, makes it possible to use the reproductive potential of sires to a larger extent. An estimated 3/4 of the world population of cattle is inseminated using this method. One factor, which helps to use this method in practice, is the possibility of freezing and storing sperm taken from the majority of mammalian species in liquid nitrogen.

Among the gametes of more than 50 species, which have so far been successfully frozen, the bull's semen proved to be the easiest to freeze.

In the other species of farm animals, i.e. sheep, pigs, horses and rabbits, artificial insemination has been used on an incomparably smaller scale than in cattle. This is for two reasons:
- the reproductive capacity of males representing these species is low (the number of insemination doses obtained from one ejaculate is a dozen to several dozen smaller than that produced from bull's ejaculate);
- the conservation, and particularly the freezing of semen, poses some problems.

Artificial insemination, especially when used in cattle or pigs has played and still plays a major role in the genetic improvement of animals. It is estimated that the milk yield of cattle has doubled over the past 30 years thanks to the use of artificial insemination. The insemination of animals has never provoked any fears or protests against its possible effects, presumably because it did not change the essence of the previously used breeding methods while making their implementation quicker. On a global scale, in addition to the genetic improvement of animal populations, it led to many inefficient local breeds being eliminated, which meant a decrease in animal biodiversity. This may be regarded as a negative consequence of this technology. Breeders soon realised this danger and launched special genetic reserve programmes aimed at preserving the diversity of breeds among farmed animals. This action partly offset the negative consequences of using this technology in animals.

Another risk of using artificial insemination is the possibility of spreading pathogenic microorganisms on a global scale. However, strict sanitary and veterinary regulations concerning trade in isolated genetic material applied in particular countries have reduced such risks to the minimum. This goal is achieved in two ways:

1. Semen donors must not carry pathogenic microorganisms (they should be Specific Pathogen Free)
2. Antibiotics are added into conserved semen in order to neutralise specific and environmental pathogens.

Sperm and embryo sexing [1-3] has recently received a great deal of attention from many research centres. Sex control in the progeny may be of practical application for livestock breeding. This is for at least two reasons. First, the increased proportion of females in the herd has a positive influence on the breeding effects. Second, the possibility of increasing the male ratio when slaughter animals are needed makes it possible to increase the meat content by more than ten percent.

The best solution would be to sex sperm by segregating semen into the "male" and "female" fractions. Research launched in the mid-1980s has concentrated on using flow cytometry methods for this purpose.

The idea of using a flow cytometer for the sex segregation of semen is based on the assumption that the majority of mammals show only 3-5% differences in the amount of DNA in the chromosome X and chromosome Y carrying sperm. Differences in the DNA content of sperm among the three principal farm animals, i.e. bulls, boars and rams are 3.9, 3.7 and 4.2%, respectively [4].

Despite some encouraging results with segregating semen into "male" and "female" fractions using the flow cytometry method in rabbits, boars and bulls, the technique is not completely problem free. It is necessary to better define the conditions of sperm staining and the technique for the DNA analysis of sperm, and to improve the sorting capacity of cytometers. The results reported by the X, Y company from Colorado are evidence that considerable progress has been made in this respect [5].

Sperm staining for flow cytometry analysis gave rise to some objections. Our own studies have shown that the staining of semen to be analysed in a flow cytometer with fluorochrome Hoechst 33342 did not affect its survival rate [6]. However, the fertilizing capacity, estimated on the basis of attempts at in vitro fertilization, may in some cases be reduced. Cytogenetic analysis of embryos produced, as a result of stained semen fertilization in vitro did not show any karyotype abnormalities [7].

2. SUPEROVULATION AND EMBRYO TRANSFER

Like insemination in males, much the same role in females is played by superovulation and embryo transfer. By using the reproductive potential of females to the full, these methods resulted in more rigorous elimination of less valuable females for the breeder. The possible negative effects of these methods are similar to those in the A.I. method, although the scale of the embryo transfer method differs considerably from that of artificial insemination.

The usefulness of the embryo transfer method in farm animals stems from the possibility of increasing the reproductive efficiency of the female, leading, in consequence, to the advancement of the breeding progress. This method can also be used for purposes other than genetic improvement, such as production of twins in cattle or the overcoming of sanitary and veterinary problems.

Whatever the species, embryo transfer is an expensive method. Different stages of embryo transfer, i.e. superovulation, embryo collection, manipulation, embryo conservation and synchronization of recipients involve high financial inputs. This method has found the widest application in cattle, and although much less common than the A.I. method in extent, it is an important means of exerting practical influence on cattle breeding. The embryo transfer method is much less common in other species of farm animals such as sheep, pigs or mares which have little potential to increase their reproductive performance, and because of the technological obstacles to embryo collection and transfer.

When combined with non-surgical collection and transfer of embryos, superovulation makes it possible to take greater advantage of the reproductive potential of cows. However, the gametogenic potential of bovine ovaries is greater than that involved when superovulation and embryo transfer are used. Recent progress in controlling the in vitro embryo collection methods in cattle is making them a popular choice in practical breeding programme.

This comprehensive method involves in vitro maturation of oocytes, fertilization in vitro and several-day embryo culture in vitro up to the blastocyst stage. As a result of intensive research conducted for the last decade in many research centres around the world, including Poland, this method is now brought under firm control [8-14].

Today it is possible to obtain more than 90% mature oocytes in vitro, of which over 80% are fertilized, over 70% undergo embryonic development, and about 40% embryos develop to the blastocyst stage. In terms of the number of oocytes used for fertilization, this makes it possible to produce 30 to 40% of blastocysts. Before in vitro production of bovine embryos can be used in breeding programmes, oocytes have to be repeatedly retrieved from the same live female. This can be achieved using the ultrasound-guided follicular aspiration method, developed in recent years [15-18].

In our own studies in stimulated and non-stimulated heifers and cows, the number of retrieved oocytes was similar in all the variants, ranging from 35 to 40%. Tests with the maturation and fertilization of oocytes have shown a decreased developmental ability of the zygotes [19].

The possibility of conservation is essential to the practical application of embryos. Cryoconservation by freezing or vitrification is the most interesting method of embryo conservation. In addition to making the organizational procedures of embryo transfer easier, cryoconservation also enables the creation of genetic reserves, and this is of key importance for the breeding programme aimed at preserving endangered or threatened breeds.

3. EMBRYONIC AND SOMATIC CLONING

Cloning is a field of biotechnology enabling non-sexual methods of mammalian reproduction and production of genetically identical individuals. Before Wilmut and his team [20] produced a sheep born after transfer of an embryo obtained by combining the nucleus of an adult sheep with the cytoplasm of a denucleated sheep oocyte, the cloning of mammals in practice meant embryo cloning. There are several methods for producing a clone through embryo cloning. These methods vary according to the number of clones obtained and efficiency.

Cloning by blastomere isolation

This technique of cloning embryos of farm animals was developed in the late 1970s. Twins and triplets in cattle, and quadruplets and quintuplets in sheep were produced using this method [21].

Despite the possibility of producing genetically identical triplets, quadruplets or quintuplets, this method was not applied in practice and has never gone beyond the experimental stage.

Cloning by embryo bisection

To date this is the only method, which has found practical application, as it is used in some bovine embryo transplantation centres. However, there is a serious limit to the embryo bisection method: a clone of only two identical individuals can be obtained with this method [22].

Cloning by nucleus transplantation

This method makes it possible to obtain clones of many more individuals in mammals than with the methods described above. It involves transplanting embryo cell nuclei (at least this procedure was used before a British experiment involving somatic cell nuclei) into a mature oocyte or, less frequently, a zygote, from which their own genetic material was previously removed [23,24].

Another variant of cloning by nucleus transplantation is to clone using embryonic stem cells [25]. At least in theory, this method enables mass production of identical embryos in mammals. To date, it was used with success only in mice, where inner mass cells able to make clones were produced in vitro. They are grown in an in vitro culture in numbers exceeding millions. At the present moment, the potential use of embryonic stem cells in farm animals is not clear, although much progress has been made in this field [26].

Somatic cloning

All the cloning techniques mentioned above depend on reproductive cells for their genetic material. Wilmut and his team were the first to show that clones in mammals can be obtained using genetic material from the somatic cells taken from an adult animal.

A number of reports published over the last two years have confirmed the possibility of somatic cloning in farm animals [27-31].

The importance of cloning for animal breeding

In the first place, different individuals, which form a clone, are not completely identical because of the differences resulting from the cytoplasmatic effect. From the viewpoint of breeding methods, cloning is not attractive in the sense that it reduces diversity. Therefore its application is limited to obtaining large semen amounts from best bulls and to increasing the reliability of the results of breeding.

Female clones may be more important from a purely commercial point of view, for the individual breeder may quickly improve the genetic value of his herd. With a stock of identical dairy cows, the breeder can produce milk of uniform i.e. standardised quality. What is more, the sex of embryos obtained as a result of cloning is known.

There are very good prospects for the use of cloning in the production of transgenic animals.

4. Transgenic animals – potential risks

Potential risk to a species

Transgenic animals are animals produced from embryos into which a foreign gene has been transferred. If the foreign gene is introduced into a one-cell embryo and if integration occurs, the transgene becomes a dominant Mendelian genetic characteristic that is inherited by the progeny of the founder animal. The ability to manipulate farm animals genetically has enormous potential, with almost unlimited applications in basic and applied research [32]. There are many available techniques to introduce foreign genes into an embryo. The most widely used technique, and still the most effective one in the production of transgenic farm animals is the microinjection of DNA into one of the pronuclei of a zygote [33]. As a result of introduced gene being integrated into one-cell embryo, the exogenous information is contained in every cell of a newborn animal. The creation of giant mice in 1982 [32] made researchers realise that transgenic technology could be applied to the production of transgenic farm animals. The giant mice harbouring the metallothionein rat growth hormone fusion gene served as a fundamental model for the production of transgenic farm animals with various growth hormone genes integrated. As a result of the introduction of bovine or human growth hormone genes driven by mouse metallotionein liver specific promoter, several transgenic farm animals were produced. The experiments carried out on pigs showed that the introduction of extra copies of exogenous growth hormone genes led to accelerated growth of transgenic animals. Some of these animals indeed grew faster, converted feed to body weight more efficiently and even reduced backfat thickness [34,35]. Despite some desirable effects there were many adverse ones. Transgenic animals lowered their reproductive capacity, had gastric ulcers and were diabetic [34]. Also transgenic sheep carrying metallothionein growth hormone gene have been produced [36,37]. Little or no growth benefits in these animals were accompanied by serious health problems, such as diabetes and premature death [37]. It thus appeared that instead of modifying the normal physiology of the whole animal, a targeted, tissue specific expression of the transgene seems a much more attractive approach [32]. Recent developments in molecular biology techniques opened new possibilities to use many tissue specific promoters driving foreign genes into desired tissue. This applies especially to the mammary gland. The genetic modification of milk composition made it possible to use farm animals as bioreactors to produce many of the very important proteins in the human therapy. There have been many transgenic animals reported secreting with the milk numerous important therapeutic compounds which have improved or will improve our quality of life: cattle producing human lactoferrin and erythropoietin; sheep secreting human α_1-antitripsin, human coagulation factor X and VII; goats producing human tissue-type plasminogen (tPA), growth hormone (GH), pigs secreting with milk human protein C and rabbits producing human insulin like growth factor 1 [38]. The most significant, however, is the fact that the tissue specific promoters used for driving foreign genes into the mammary glands cause no health problems to transgenic animals. The milk-derived pharmaceuticals are easy to collect and of high quality. The major benefits of biopharming over cell culture-based bioreactors are the lower cost of production and the relative ease to scale up [39]. Milk is the best natural nutritive product. Modified milk containing new milk components, e.g. proteins better digested by human organism, may make milk even more attractive as a nutritive product. Another important task increasing the practical role of milk in human nutrition is to eliminate or significantly reduce the milk component lactose, which is responsible for about 10% cases of intolerance in humans.

Potential risk to the environment

Transgenic farm animals do not pose any potential risks to humans and their environment. The whole process of the production of transgenic animals is strictly controlled in highly specialised laboratories. Transgenic animals represent great value for basic or applied research and also for commercial purposes. Farm animals and especially the transgenic ones would not survive if released to the natural environment, for they are too heavily dependent on human care. Products obtained from transgenic farm animals are natural and so biodegradable. Moreover, their sources, milk, blood or urine are also easy to utilize. Finally, many countries where biotechnology is widely applied have adopted legal regulations preventing potential dangers. Moreover, the problems involved in increased utilisation of farm animals are still limited, as many laboratory results are not automatically used in practical or commercial situations. Special attention should be paid to the production of transgenic fish and insects. Some species are known to produce toxins. It is feared that in some cases a transgene could integrate toxic genes, which normally could never be expressed. To eliminate that potential risk, it is necessary to prevent them from penetrating into the natural environment or to produce transgenic fish and insects that are not able to transmit the transgene into next generations.

Human risk.

In the last decade, biotechnology has fully emerged and has been applied to biomedical, diagnostic, and food-related products. In terms of human medicine, recombinant DNAs have been introduced into a variety of expression systems, including farm animals, which have yielded important therapeutic proteins. Many recombinant proteins such as human growth hormone (hGH), tissue plasminogen activator (tPA), coagulation factors VIII and IX, purified, have completed clinical trials and been approved for commercial use. A plethora of other recombinant proteins are in various stages of clinical trials. In addition, genetic engineering strategies have been used in the development of products used for the diagnosis of human and animal diseases [38]. One of the major applications of transgenic animals is to model human diseases. When it is not ethical or practical to do studies on humans, having an animal model of the disease is essential. However, a fairly large number of human diseases are not mimicked accurately by any animal model. Among the models used are the following: humanized mice for AIDS research, models for diabetes, cancer and immune function studies [40]. The last application has many implications for many other studies of diseases such as autoimmune reaction, arthritis, allergies, multiple sclerosis, and finally for the xenotransplantation projects. In view of the above, there is no doubt that transgenic animals could be a source of danger to humans, but transgenic technology should be seen in its benefit aspects rather than as a source of danger.

To sum up, transgenic technology poses limited danger to individual transgenic animals rather than the species as a whole. There is no evidence that transgenic farm animals could be a danger to the environment and especially to humans. Also, at the present stage we have no evidence that food products obtained from genetically modified animals will cause side effects such as allergies, in humans.

REFERENCES

1. L.A. Johnson, J.P. Flook, H.W. Hawk, Biol. Reprod., 41 (1989) 199.
2. L.A. Johnson, Reprod. Dom. Anim., 26 (1992), 308.
3. L.A. G.R. Welch, D.L. Garner, Cytometry, 7 (1994) 83.
4. L.A. Johnson, R.N. Clark, Gamete Res., 21 (1988) 335.
5. G.E. Jr Seidel, Theriogenology, 51 (1999) 5.
6. Z. Smorąg, B. Ryńska, L. Kątska, E. Slota, Anim. Sci., Pap & Rep., 11 (1993) 117.
7. Z. Smorąg, L. Kątska, E. Słota, B. Ryńska, Ann. Anim. Sci., 26 (1999) 67.
8. J.J. Parish , J.L. Susko-Parish, M.A. Winer, N.L. First, Biol. Reprod., 38 (1988) 1171.
9. W.H. Eyestone and N.L. First, J. Reprod. Fertil., 85 (1989) 715.
10. Y. Aoyagi, Y. Fukui, Y. Iwazumi, M.Urajawa, H.Ono, Theriogenology, 34 (1990) 749.
11. H.W. Hawk and R.J. Wall, Theriogenology, 41 (1994) 1571.
12. L. Kątska, B. Ryńska, Z. Smorąg, Theriogenology, 43 (1995) 859.
13. L. Kątska, B. Ryńska, Z. Smorąg, Anim. Reprod. Sci., 44 (1996) 23.
14. C. Galli and G. Lazzari, Anim. Reprod.Sci., 42 (1996), 371.
15. M.C. Pieterse, P.L.A.M. Vos, T.A.M. Kruip, Y.A. Wurth, Th.H. Van Beneden, A.H. Williemse, M.A.M. Taverne, Theriogenology, 35 (1991) 857.
16. T.A.M. Kruip, M.C. Pieterse, Th.H. Van Beneden, P.L.A.M. Vos, Y.A. Wurth, M.A.M. Taverne, Vet. Rec., 128 (1991) 208.
17. M.C. Pieterse, P.L.A.M. Vos, T.A.M. Kruip, A.H. Willemse, M.A.M. Verne, Theriogenology, 37 (1992) 273.
18. R.B. Stubbings and J.S. Walton, Theriogenology, 43 (1995) 705.
19. Z. Smorąg, P. Gogol, L. Kątska, J. Jażdżewski, Medycyna Wet., 55 (1999) 317.
20. I. Wilmut, A.E. Schnieke, J. McWihr, A.J. Kind, K.H.S. Campbell, Nature, 395 (1997) 810.
21. S.M. Willadsen, J. Reprod. Fert., 59 (1980) 357.
22. M. Skrzyszowska and Z. Smorąg, Biotechnologia 2 (1998) 137.
23. J.A. Modliński and Z. Smorąg, Mol. Reprod. Dev., 28 (1991) 361.
24. I. Heyman and J.P. Renard, Anim. Reprod. Sci., 42 (1996) 427.
25. J.A. Modliński, M.A. Reed, T.E. Wagner, J. Karasiewicz, Anim. Reprod. Sci., 42 (1996) 437.
26. J.A. Modliński and J. Karasiewicz, Anim. Sci. Pap. & Rep., 16 (1988) 11.
27. X. Vingnon, P. Chesne, D. LeBourhis, J.E. Flechon, Y. Heyman, J.P. Renard, Compt. Rend. Acad. Sci., 321 (1998) 375.
28. J.B. Cibelli, S.L. Stice, P.J. Golueke, J.J. Kane, J. Jerry, C. Blackwell, F.A. Ponce de Leon, J.M. Robl, Science, 280 (1998) 1256.
29. Y. Kato, T. Tani, Y. Sotomaru, K. Kurokawa, J. Kato, H. Doguchi , H. Yasue, Y. Tsunoda, Science, 282 (1998) 2095.
30. K. Shiga, T. Fujita, K. Hirose, Y. Sasae, T. Nagai, Theriogenology, (1999) (in press).
31. A. Baguisi, E. Behboodi, D.T. Melican, J.S. Pollock, M.M. Destrempes, Ch. Cammuso, J.L. Williams, S.D. Nims, C.A. Porter, P. Midura, M.J. Palacios, S.L. Ayres, R.S. Denniston, M.L. Hayes, C.A. Ziomek, H.M. Meade, R.A. Godke, W.G. Gavin, E.W. Overstromöm, Y. Echelard, Nature Biot., 17 (1999) 456.
32. J. Janne, J-M. Huttinen, T. Peura, M. Tolvanen, L. Alhonen and M. Halmekyto, Annals of Medicine 24 (1992) 273-280.
33. J. Jura, Z. Smorąg, L. Kątska, M. Skrzyszowska, B. Gajda, B. Ryńska, Biotechnologia 2(41), (1998), 101-109.

34. V.G. Pursel, R.E. Mammer, D.J. Bolt, R.D. Palmiter, R.L. Brinster. J. Reprod. Fert. Suppl 41 (1990) 77-87.
35. Z. Smorąg, J. Jura, J.J. Kopchick, B. Gajda, M. Skrzyszowska, M. Różycki, J. Pasieka, Biotechnologia 2 (41), (1998), 145-152.
36. J.D. Murray, C.D. Nancarrow, J.T. Marshall, I.G. Hazelton, K.A. Ward, Reprod. Fertil. Dev. 1 (1989) 147-55.
37. C.E Rexroad, R.E. Hammer, R.R. Behringer, R.D. Palmiter, R.L. Brinster, J. Reprod. Fert. Suppl. 41 (1990) 119-24.
38. J.J. Kopchick, J. Jura, P. Mukerji, B. Kelder, Biotechnologia 2 (33) (1996) 31-51.
39. D.E. Kerr, F. Liang, K.R. Bondioli, H. Zhao, G. Kreibich, R.J. Wall, and T-T. Sun, Nature Biotechnology 16 (1998) 75-79.
40. W. Bains, Biotechnology from A to Z, Oxford University Press, 1994.

Food Biotechnology
S. Bielecki, J. Tramper and J. Polak (Editors)
© 2000 Elsevier Science B.V. All rights reserved.

Public perception and legislation of biotechnology in Poland

Tomasz Twardowski

Institute of Bioorganic Chemistry, Polish Academy of Sciences
ul. Noskowskiego 12/14, 61-704 Poznań, Poland
Institute of Technical Biochemistry, Technical University of Łódź
ul. Stefanowskiego 4/10, 90-924 Łódź, Poland.

1. INTRODUCTION

In the case of biotechnology - in any country - we should talk about a specific "value added chain" [VAC]; which is a sum of the following three factors:

VAC = [science and technology] + [legislation and IPR] + [public perception]

We have to take into account the fact that in the case of food biotechnology science and technology are much more advanced than the development of legislation and public acceptance of novel food.

Surveys of public perception of biotechnology are quite common today [see 1,2 and papers cited herein]. However, in Poland these are very first tests related to our earlier data [3] and relevant to other publications [4,5]. Legal aspects of biotechnology in Poland are also a subject of very limited number of publications [see 6, 7 and articles cited] and most of them are in Polish.

2. PUBLIC PERCEPTION SURVEYS

One of the most interesting themes of biotechnology nowadays is the public perception of novel food ("GMO food"). Public perception is very important for commercial progress (development) of biotechnology. Because of that we have decided to survey on a representative sample of the Polish society and Polish biotechnologists.

The surveys were conducted by OBOP (Centre for Public Perception Research). Our tests were surveyed between 27 - 30 June 1998 and in March 1999. The sounding method was used on people over 15, who were selected aleatory, laminary proportional, from the whole country. There were 982 and 1080 interviews carried out, respectively. The data obtained were elaborated statistically (using the SPSS DOS method) with the measurement error ± 3% and appreciation reliability 0,95.

The questions were addressed to a representative group of Polish society and were also shown to Polish biotechnologists (150 experts answered our surveys; in this case the survey was conducted by mail).

Of course, it is a big risk to give the same questions to a group of specialists and to lay people who do not face these problems on a daily basis. Yet this was the only way to compare standpoints of experts and common people.

2.1. Biotechnology use in food production

We asked about the use of biotechnological methods in the production of food and beverages. 14% of the respondents said that they were interested in the problem, 48% had only heard about it but they were not interested in it and 38% had never heard about it. The next question was:

Have you ever heard about the use of biotechnology in the production of food and beverages production?

Predominantly, people who were interested in biotechnology were with secondary and higher education. It is usually bound up with size of abode and employment - the bigger searching place the problem is more interesting. Young people, especially 20 year old people, had often heard about it but they were not interested.

2.2. Opinions about using biotechnology in food and drinks (beverages) production.

In order to learn about the opinion on using biotechnology in food and drinks production, we have presented contrary statements concerning this problem. We asked respondents to judge these statements according to five-note scale:

- **it is useful,**
- **it should be developed,**
- **it is acceptable because we could change nature to get better food.**

It appears that respondents more often favour the use of biotechnology in food production (Figure 1). In each opinion about half or more than half of the respondents voiced their applause for biotechnology use. The data show that 41% of the respondents voiced their positive opinion in all three cases, 13% - in two, 12% - in one and 35% - in any of them.

The highest degree of appreciation earned biotechnology as useful (38.9%). The opinion that this domain should be developed gained similar results (38.6%). The degree of acceptation for biotechnology methods is not much lower (34.9%).

Those who were better educated, bigger abode, and less religion expressed slightly more positive opinion on biotechnology use in food and beverage production more often. Furthermore, they believed that biotechnology use in food production was of great interest and that traditional production methods of agro-food industry are worse in comparison with genetic engineering methods.

Among experts more than 90% accept biotechnology and think that it is useful.

The question: **Should we permit or ban production and sale of transgenic food?** is a critical one. More than half of the respondents - 57% think that production and sale of transgenic food should be permitted. 23% is opposed and 20% do not have any opinion. However, two third of biotechnogists is accept „GMO food" production, and one fourth is opposed (Figure 2).

The higher the educational level of the respondents, the better their financial situation and the weaker their religious affiliation the more positive their opinion that production and sale of transgenic food should be permitted.

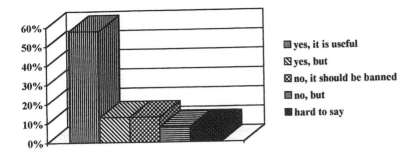

Figure 1. Is the use of biotechnology in food production positive or negative?

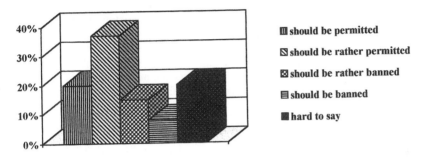

Figure 2. Permit or ban production and sale of transgenic food?

2.3. Traditional and genetic methods of food production

The Poles have different opinions concerning the traditional methods for agro-food production industry in comparison with methods using genetic engineering. 36% of respondents opt for genetic engineering. The similar percent of respondents affirm that none of these methods is either better or worse - 34%. 9% of the surveyed prefer traditional methods, and 21% do not have any opinion about the problem. Similarly two thirds of experts have positive opinions and one fifth consider biotechnology as worse or (also one fifth) do not have any opinion about it.

The opinions concerning traditional methods of food production in comparison with genetic engineering methods do not vary according to the social-demographic variables.

The next question refers to the problem of special labelling of transgenic food. There were five kinds of these food presented. In each case about four fifth of the respondents opted for its special labelling. It is understandable that in experts' opinion the necessity of marking such products which do not contain GMO (for example sugar) is judged superfluity much more often. As an example the following question is presented and the results are given in Figure 3.

To learn what kind of features of transgenic food may encourage consumers to purchase it, in the next question five basic parameters were presented to the respondents. The most treasured features, as was shown, are: good quality, which was indicated by 68% of the respondents and 90% of the experts. The next were: longer durability (63%), in opinion of experts - 68%; flavour 62% and 79%; such as appearance (50%), in biotechnologists' opinion - 54%. Comparatively (46% and 43%) indicated lower price.

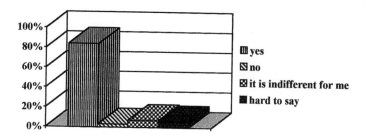

Figure 3. In your opinion, should food obtained with genetic engineering methods (i.e. transgenic food) be specifically marked?

The discussion concerning the future and the perspectives of biotechnology is related to general acceptance by the society and legislation. To find the answer we asked next 4 questions in our survey (Figures 4-7).

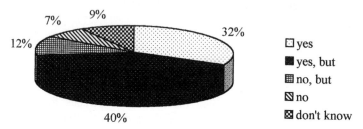

Figure 4. Do you agree with the statement "We should continue research on food using genetic engineering"?

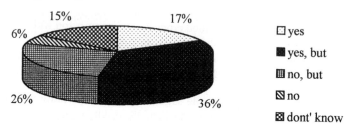

Figure 5. Do you agree with the statement: "Such research could be danger for human health and for environment."

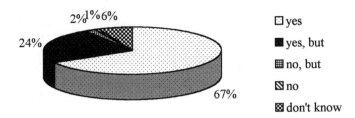

Figure 6. Do you agree with the statement: "The research should be carried out under the government supervision and regulated by law."

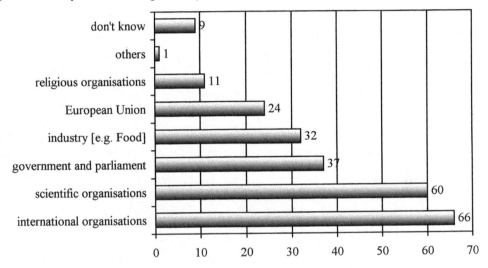

Figure 7. What is your opinion: Please indicate the organisation/s/ of choice for cooperation in legislation of modern biotechnology?

3. LEGAL ASPECTS OF BIOTECHNOLOGY IN POLAND (A/ INTELLECTUAL PROPERTY RIGHTS, B/ PROTECTION OF NATURAL GENOMIC RESOURCES, C/ BIOSAFETY AND BIOHAZARD REGULATIONS)

In our country many legislation aspects have been successfully solved. However, many problems have to be advanced as soon as possible and this is an obligation of the state authorities.

The following milestones are strictly connected with legal aspects of biotechnology and are of special interest in correlation with the Agenda 21:

- Protection of biodiversity of species and genetic resources (Polish official journal announcing current legislation: Dziennik Ustaw, October 16,1991, No 114).

- Protection by patents: drugs, chemical compounds, food and food additives, techniques of isolation and identification of natural compounds, gene technology: modification and transfer; new biological systems (cf. microorganisms). (Polish official journal announcing current legislation: Dziennik Ustaw, October 30, 1993, No 4).
- Signing of the "Budapest Treaty" concerning the deposition of microorganisms (September 22, 1993).
- Membership of the "Australian Group" and act of December 2,1993 on the control of goods and technology in foreign trade resulting from international agreements and obligations as well as the Order of the Minister of Foreign Economic Cooperation of 23 March 1994 on products and technologies subject to particular control in foreign trade concern, inter alia, such products as chemical compounds, microorganisms, viruses, bacteria and toxins, and such apparatus and technologies which could be used to develop chemical and biological weapons.
- Protection of authorship's rights (February 4,1994).
- New law on environment protection [in power from January 1, 1998, and the article 37a concerning GMO on January 1, 1999 in the following aspects: releasing to the environment, biohazard and food labelling]
- Poland's membership in the OECD and association with the European Union.

The experts Committee for Genetically Modified Organisms was established by the Ministry of Agriculture in cooperation with the Ministry of Environment Protection, State Committee for Scientific Research and Ministry of Health. Poland is a member of the OECD and an associate member of the European Union. Besides, Poland has signed the „Biodiversity Convention". Therefore, the Polish Government is obliged to establish appropriate regulations and setting up the experts' board is the first step to fulfil this obligation.

The aims of the Committee are:
- to draft legal regulations;
- to evaluate the applications concerning the release of GMO (registration, permit, technical guidelines).

The members of the Committee are the delegates from the scientific community and representatives of the state administration.

The following issues have to be dealt with:
A) harmonization of national legislations with EU directives;
B) priority of international regulations over national ones,
C) obligatory licensing for any activities involving GMO,
D) free access to information,
E) public safety as the top priority.

4. CONCLUSIONS

Unequivocal we may frame the conclusion of our public perceptions surveys: „GMO food" - yes, but under the law „supervision". Poland will "copy" the legislation system of the

European Union and will take into account international conventions, like "Biodiversity convention". The ministry of environment in cooperation with ministries of: health, agriculture and science will play the leading role, most probably. The agenda for legal actions is strictly related with these unification procedures. In my opinion, today, we should stress the positive aspects of modern biotechnology to accelerate its further progress, particularly in the so-called "transition" states.

The recent development of biotechnology in Poland has been connected with political, economic and sociological changes, which have occurred in the country over the last 10 years. We have to take into account the transformation to the market economy, the government program of privatization and the goal: joining United Europe. Legal aspects are particularly important for the cooperation and integration with the European Union. In the last years we have observed a significant modifications in the Polish law towards the West European standards and norms.

ACKNOWLEDGEMENTS

The surveys were supported by KBN within SPUB project and GF 1200-98-84 UNEP-IHAR. This is a part of the project sponsored by EU within the project "Biotechnology and the European Public (concerted action)"
I appreciate the help of Ms. Maria Brochwicz and OBOP staff.
Part of these of data were published in journal „Biotechnologia" in Polish, (Dec. 1998).

REFERENCES

1. W. Wagner, H. Torgerson, E. Einsiedel, E. Jelsoe, H. Fredrickson, J. Lassen, T. Rusanen, D. Boy, S. De Cheveigne, J. Hampel, A. Stathopoulou, A. Allansdottir, C. Midden, T. Nielsen, A. Przestalski, T. Twardowski, B. Fjaestad, S. Olsson, A. Olofsson, G. Gaskell, J. Durant, M. Bauer and M. Liakopoulos, Nature, 387 (1997) 845.
2. A. Przestalski, B. Suchocki and T. Twardowski in: Biotechnology in the Public Sphere, Eds. J. Durant, M. W. Bauer, G. Gaskell, Science Museum, (1998), 118.
3. A. Twardowska-Pozorska and T. Twardowski, Biotechnologia, 43 (1998) 20.
4. T. Rusanen, B. Suchocki, T. Twardowski and A. Von Wright, Biotechnologia, 1 (1996) 106.
5. A. Przestalski, B. Suchocki, T. Twardowski, Ruch Prawniczy, Ekonomiczny i Socjologiczny, 1(1998) 167.
6. A. Zabża and S. Ułaszewski, Biotechnologia, 31 (1995) 13.
7. „Analiza porównawcza z przepisami międzynarodowymi stanu uregulowań prawnych w obszarze zastosowań genetycznie modyfikowanych organizmów (GMO), ocena zagrożeń wynikających z rozwoju biotechnologii oraz dostosowanie polskiego prawa do nałożonych na RP zobowiązań" [„Biotechnology; proposal for the Polish gene law", in Polish with extended English summary], ed. T. Twardowski, 1997, Poznań.

Index of authors